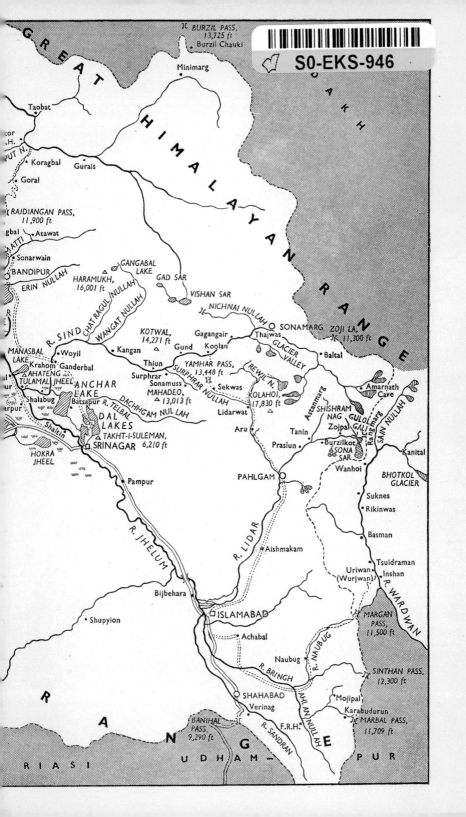

BREEDING BIRDS OF KASHMIR

PLATE I

1. Indian Moorhen. 2. Himalayan Snow-Cock. 3. Impeyan Pheasant (adult male).
4. Indian Purple Coot. 5. Kashmir Koklas (adult male). 6. Black-eared Kite.
7. Long-legged Buzzard. 8. Booted Eagle

BREEDING BIRDS OF KASHMIR

R. S. P. BATES
Lieutenant-Colonel, Indian Army (Retired)

AND

E. H. N. LOWTHER
Members of the British Ornithologists' Union

*Illustrated with
151 photographs by the Authors
and 5 coloured plates by*

D. V. COWEN

GEOFFREY CUMBERLEGE
OXFORD UNIVERSITY PRESS

Oxford University Press, Amen House, London, E.C.4

GLASGOW NEW YORK TORONTO MELBOURNE WELLINGTON
BOMBAY CALCUTTA MADRAS CAPE TOWN

Geoffrey Cumberlege, Publisher to the University

First published 1952

PRINTED IN GREAT BRITAIN BY HEADLEY BROTHERS LTD
109 KINGSWAY LONDON WC2 AND ASHFORD KENT

To
Molly and Shuna

CONTENTS

	PAGE
INTRODUCTION	xvii
SIZE KEY AND BOOK REFERENCES	1

ORDER PASSERES

FAMILY CORVIDÆ

Corvus corax tibetanus Hodgson. Tibetan Raven	343
,, *corone orientalis* Eversmann. Eastern Carrion Crow	3
,, *macrorhynchos intermedius* Adams. Himalayan Jungle Crow	4
,, *splendens zugmayeri* Laubmann. Sind House Crow	6
,, *monedula soemmeringii* Fischer. Eastern Jackdaw	8
Urocissa flavirostris cucullata Gould. Western Yellow-billed Blue Magpie	11
Dendrocitta formosae occidentalis Ticehurst. Western Himalayan Tree-Pie	343
Garrulus lanceolatus Vigors. Black-throated Jay	13
Nucifraga multipunctata Gould. Larger Spotted Nutcracker	15
Pyrrhocorax pyrrhocorax himalayanus (Gould). Eastern Red-billed Chough	17
,, *graculus forsythi* Stoliczka. Yellow-billed Chough	19

FAMILY PARIDÆ

Parus major caschmirensis Hartert. Kashmir Grey Tit	21
,, *monticolus monticolus* Vigors. Simla Green-backed Tit	23
Lophophanes melanolophus (Vigors). Crested Black Tit	25
,, *rufonuchalis rufonuchalis* (Blyth). Simla Black Tit	27
Ægithaliscus concinnus iredalei Stuart Baker. Red-headed Tit	344
,, *niveogularis* (Gould). White-throated Tit	29

FAMILY SITTIDÆ

Sitta caesia cashmirensis Brooks. Brooks' Nuthatch	31
,, *leucopsis leucopsis* Gould. White-cheeked Nuthatch	32

FAMILY TIMALIIDÆ

Garrulax albogularis whistleri Stuart Baker. Western White-throated Laughing-Thrush	344
Trochalopteron variegatum simile Hume. Western Variegated Laughing-Thrush	34
,, *lineatum griseicentior* Hartert. Simla Streaked Laughing-Thrush	36
Ianthocincla rufogularis occidentalis (Hartert). Kumaon Rufous-chinned Laughing-Thrush	344
Turdoides terricolor sindianus Ticehurst. Sind Jungle Babbler	344

FAMILY PYCNONOTIDÆ

Microscelis psaroides psaroides (Vigors). Himalayan Black Bulbul	39
Molpastes leucogenys leucogenys (Gray). White-cheeked Bulbul	41

FAMILY CERTHIIDÆ

Certhia himalayana limes Meinertzhagen. Himalayan Tree-Creeper	43
,, *familiaris hodgsoni* Brooks. Hodgson's Tree-Creeper	345
Tichodroma muraria (Linnæus). Wall-Creeper	46

FAMILY TROGLODYTIDÆ
Troglodytes troglodytes neglectus Brooks. Kashmir Wren — 48

FAMILY CINCLIDÆ
Cinclus cinclus cashmeriensis Gould. Kashmir White-breasted Dipper — 50
,, *pallasii tenuirostris* Bonaparte. Indian Brown Dipper — 52

FAMILY TURDIDÆ : SUBFAMILY BRACHYPTERYGINÆ
Luscinia brunnea brunnea (Hodgson). Indian Bluechat — 54
Hodgsonius phoenicuroides (Gray). Hodgson's Shortwing — 56

SUBFAMILY SAXICOLINÆ
Saxicola caprata bicolor Sykes. Northern Indian Pied Bushchat — 58
,, *torquata indica* (Blyth). Indian Stonechat — 60
Rhodophila ferrea ferrea (Gray). Western Dark-grey Bushchat — 63

SUBFAMILY ENICURINÆ
Enicurus maculatus maculatus Vigors. Western Spotted Forktail — 64
Microcichla scouleri scouleri (Vigors). Little Forktail — 67

SUBFAMILY PHŒNICURINÆ
Phoenicurus frontalis Vigors. Blue-fronted Redstart — 68
,, *ochruros phoenicuroides* (Horsfield & Moore). Kashmir Redstart — 70
Chaimarrhornis leucocephalus (Vigors). White-capped Redstart — 71
Rhyacornis fuliginosa (Vigors). Plumbeous Redstart — 73
Calliope pectoralis pectoralis Gould. Western Himalayan Rubythroat — 76
Ianthia cyanura pallidiora Stuart Baker. Kashmir Red-flanked Bush-Robin — 77
Adelura coeruleocephala (Vigors). Blue-headed Robin — 79
Saxicoloides fulicata cambaiensis (Latham). Brown-backed Indian Robin — 345
Copsychus saularis saularis (Linnæus). Indian Magpie-Robin — 81

SUBFAMILY TURDINÆ
Turdus merula maximus (Seebohm). Central Asian Blackbird — 83
,, *rubrocanus rubrocanus* Gray. Grey-headed Thrush — 85
,, *unicolor* Tickell. Tickell's Thrush — 87
Arceuthornis viscivorus bonapartei Cabanis. Himalayan Mistle-Thrush — 89
Monticola cinclorhyncha (Vigors). Blue-headed Rock-Thrush — 91
,, *solitaria pandoo* (Sykes). Indian Blue Rock-Thrush — 92
Myiophoneus coeruleus temminckii Vigors. Himalayan Whistling-Thrush — 94

SUBFAMILY PRUNELLINÆ
Laiscopus collaris whymperi Stuart Baker. Garhwal Accentor — 96
Prunella strophiata jerdoni (Brooks). Jerdon's Accentor — 97

FAMILY MUSCICAPIDÆ
Hemichelidon sibirica gulmergi Stuart Baker. Kashmir Sooty Flycatcher — 99
Siphia hyperythra Cabanis. Kashmir Red-breasted Flycatcher — 101
Muscicapula tricolor tricolor (Hodgson). Western Slaty-blue Flycatcher — 103
,, *superciliaris superciliaris* Jerdon. White-browed Blue Flycatcher — 105
Eumyias thalassina thalassina (Swainson). Common Verditer Flycatcher — 345
Alseonax ruficaudus (Swainson). Rufous-tailed Flycatcher — 107
Culicicapa ceylonensis pallidior (Ticehurst). Simla Grey-headed Flycatcher — 346
Terpsiphone paradisi leucogaster (Swainson). Himalayan Paradise Flycatcher — 109

FAMILY LANIIDÆ
Lanius schach erythronotus (Vigors). Rufous-backed Shrike — 112

FAMILY PERICROCOTIDÆ
Pericrocotus brevirostris brevirostris (Vigors). Indian Short-billed Minivet 114

FAMILY DICRURIDÆ
Dicrurus macrocercus albirictus (Hodgson). Himalayan Black Drongo 346
,, *leucophaeus longicaudatus* Jerdon. Indian Grey Drongo 116

FAMILY SYLVIIDÆ
Acrocephalus stentoreus brunnescens (Jerdon). Indian Great Reed-Warbler 118
,, *concinens haringtoni* Witherby. Witherby's Paddy-field Warbler 120
Tribura major (Brooks). Large-billed Bush-Warbler 123
Sylvia althoea Hume. Hume's Lesser Whitethroat 124
Phylloscopus affinis (Tickell). Tickell's Willow-Warbler 127
,, *tytleri* Brooks. Tytler's Willow-Warbler 129
,, *collybita sindianus* Brooks. Sind Chiffchaff 346
,, *proregulus simlaensis* Ticehurst. Ticehurst's Willow-Warbler 131
,, *inornatus humei* (Brooks). Hume's Yellow-browed Willow-Warbler 133
,, *magnirostris* Blyth. Large-billed Willow-Warbler 134
,, *occipitalis occipitalis* (Blyth). Large Crowned Willow-Warbler 136
Seicercus xanthoschistos albosuperciliaris (Jerdon). Kashmir Grey-headed Flycatcher-Warbler 347
Homochlamys pallidus pallidus (Brooks). Pale Bush-Warbler 138
Suya criniger criniger Hodgson. Nepal Brown Hill-Warbler 347

FAMILY REGULIDÆ
Regulus regulus himalayensis Jerdon. Himalayan Goldcrest 139
Cephalopyrus flammiceps flammiceps (Burton). Fire-capped Tit-Warbler 141

FAMILY ORIOLIDÆ
Oriolus oriolus kundoo Sykes. Indian Oriole 143

FAMILY STURNIDÆ
Sturnus vulgaris humii Brooks. Himalayan Starling 145
Temenuchus pagodarum (Gmelin). Black-headed Myna 148
Acridotheres tristis tristis (Linnæus). Common Myna 149

FAMILY FRINGILLIDÆ : SUBFAMILY COCCOTHRAUSTINÆ
Perissospiza icteroides (Vigors). Black and Yellow Grosbeak 152
Pyrrhula aurantiaca Gould. Orange Bullfinch 154
Pyrrhospiza punicea humii Sharpe. Western Red-breasted Rosefinch 347
Propasser thura blythi Biddulph. Kashmir White-browed Rosefinch 156
,, *rhodochrous* Vigors. Pink-browed Rosefinch 157
Carpodacus erythrinus roseatus (Blyth). Hodgson's Rosefinch 159
Carduelis caniceps caniceps Vigors. Himalayan Goldfinch 161
Callacanthis burtoni (Gould). Red-browed Finch 163
Metoponia pusilla (Pallas). Gold-fronted Finch 347
Hypacanthis spinoides spinoides (Vigors). Himalayan Greenfinch 166
Passer domesticus griseigularis Sharpe. Kashmir House Sparrow 167
,, *rutilans debilis* Hartert. Kashmir Cinnamon Sparrow 169
Fringilauda nemoricola altaica (Eversmann). Stoliczka's Mountain-Finch 171

SUBFAMILY EMBERIZINÆ
Emberiza fucata arcuata Sharpe. Indian Grey-headed Bunting 173
,, *stewarti* Blyth. White-capped Bunting 175
,, *cia stracheyi* Moore. Eastern Meadow-Bunting 177
Melophus lathami subcristatus (Sykes). Crested Bunting 179

FAMILY HIRUNDINIDÆ

Delichon urbica cashmeriensis (Gould). Kashmir House Martin	181
Riparia rupestris (Scopoli). Crag Martin	348
Hirundo rustica rustica Linnæus. Common Swallow	184
,, *smithii filifera* Stephens. Indian Wire-tailed Swallow	348
,, *rufula scullii* Seebohm. Scully's Red-rumped Swallow	186

FAMILY MOTACILLIDÆ

Motacilla alba alboides Hodgson. Hodgson's Pied Wagtail	188
,, *cinerea melanope* Pallas. Eastern Grey Wagtail	190
,, *citreola calcarata* Hodgson. Hodgson's Yellow-headed Wagtail	192
Anthus trivialis haringtoni Witherby. Witherby's Tree-Pipit	194
,, *similis jerdoni* Finsch. Brown Rock-Pipit	196
,, *roseatus* Blyth. Hodgson's Pipit	197

FAMILY ALAUDIDÆ

Eremophila alpestris longirostris (Moore). Long-billed Horned Lark	199
Alauda gulgula lhamarum Meinertzhagen. Kashmir Skylark	201

FAMILY ZOSTEROPIDÆ

Zosterops palpebrosa occidentis Ticehurst. Western White-Eye	203

FAMILY NECTARINIIDÆ : SUBFAMILY NECTARINIINÆ

Cinnyris asiatica (Latham). Purple Sunbird	348

ORDER CORACIIFORMES
SUB-ORDER PICI
FAMILY PICIDÆ

Picus squamatus squamatus Gould. Scaly-bellied Green Woodpecker	205
Dryobates himalayensis albescens Stuart Baker. Kashmir Pied Woodpecker	207
,, *brunifrons* (Gould). Brown-fronted Pied Woodpecker	209

SUBFAMILY JYNGINÆ

Jynx torquilla japonica Bonaparte. Japanese Wryneck	210

SUB-ORDER CUCULI
FAMILY CUCULIDÆ : SUBFAMILY CUCULINÆ

Cuculus canorus telephonus Heine. Asiatic Cuckoo	212
,, *optatus optatus* Gould. Himalayan Cuckoo	217
,, *poliocephalus poliocephalus* Latham. Small Cuckoo	218
Clamator jacobinus pica (Hemprich & Ehrenberg). Northern Pied Crested Cuckoo	221

SUB-ORDER PSITTACI
FAMILY PSITTACIDÆ

Psittacula himalayana himalayana (Lesson). Himalayan Slaty-headed Parakeet	223

SUB-ORDER CORACII
FAMILY CORACIIDÆ

Coracias garrula semenowi (Loudon & Tschusi). Kashmir Roller	226
,, *benghalensis benghalensis* (Linnæus). Indian Roller	349

FAMILY MEROPIDÆ

Merops apiaster Linnæus. European Bee-Eater	229

FAMILY ALCEDINIDÆ

Ceryle rudis leucomelanura Reichenbach. Indian Pied Kingfisher	231
,, *lugubris guttulata* Stejneger. Himalayan Pied Kingfisher	233
Alcedo atthis pallasii Reichenbach. Central Asian Kingfisher	234
Halcyon smyrnensis smyrnensis (Linnæus). White-breasted Kingfisher	237

FAMILY UPUPIDÆ

Upupa epops epops Linnæus. European Hoopoe	239

SUB-ORDER APODES
FAMILY APODIDÆ

Apus melba melba Linnæus. Alpine Swift	241
Micropus apus pekinensis (Swinhoe). Eastern Swift	243
,, *pacificus leuconyx* (Blyth). Blyth's White-rumped Swift	349
,, *affinis affinis* (Gray). Common Indian House Swift	349
Chaetura caudacuta nudipes Hodgson. Indian White-throated Spinetail	349

SUB-ORDER STRIGES
FAMILY STRIGIDÆ

Tyto alba stertens Hartert. Indian Barn Owl	350
Asio otus otus (Linnæus). Long-eared Owl	245
Strix aluco biddulphi Scully. Scully's Wood Owl	247
Bubo bubo turcomanus (Eversmann). Turkestan Great Horned Owl	248
Glaucidium cuculoides cuculoides (Vigors). Western Himalayan Barred Owlet	350
,, *brodiei brodiei* (Burton). Western Collared Pigmy Owlet	250

ORDER ACCIPITRES
FAMILY PANDIONIDÆ

Pandion haliaëtus haliaëtus (Linnæus). Osprey	350

FAMILY ÆGYPIIDÆ

Ægypius monachus (Linnæus). Cinereous Vulture	350
Gyps himalayensis Hume. Himalayan Griffon Vulture	252
Pseudogyps bengalensis (Gmelin). White-backed Vulture	351
Neophron percnopterus percnopterus (Linnæus). Neophron, or Large White Scavenger Vulture	254

FAMILY FALCONIDÆ : SUBFAMILY GYPAËTINÆ

Gypaëtus barbatus hemachalanus Hutton. Himalayan Bearded Vulture, or Lämmergeier	255

SUBFAMILY FALCONINÆ

Falco subbuteo centralasiae Buturlin. Central Asian Hobby	258
,, *tinnunculus* Linnæus. Kestrel	260
Aquila chrysaëtos daphanea (Gray). Himalayan Golden Eagle	351
Hieraëtus pennatus Gmelin. Booted Eagle	262
Haliaëtus leucoryphus (Pallas). Pallas' Fishing-Eagle	263
Milvus migrans govinda Sykes. Common Pariah Kite	351
,, ,, *lineatus* (Gray). Black-eared Kite	265
Circus aeruginosus aeruginosus (Linnæus). Marsh Harrier	351
Buteo rufinus rufinus (Cretzschmar). Long-legged Buzzard	267
Accipiter nisus melanoschistos Hume. Indian Sparrow-Hawk	269
,, *virgatus affinis* Hodgson. Northern Besra Sparrow-Hawk	352

ORDER COLUMBÆ
FAMILY COLUMBIDÆ : SUBFAMILY COLUMBINÆ

Columba livia neglecta Hume. Hume's Blue Rock-Pigeon	271
,, *leuconota leuconota* Vigors. Nepalese Snow-Pigeon	273
Dendrotreron hodgsonii (Vigors). Speckled Wood-Pigeon	352
Streptopelia orientalis meena Sykes. Himalayan Rufous Turtle-Dove	275
,, *chinensis suratensis* (Gmelin). Indian Spotted Dove	352
,, *senegalensis cambayensis* (Gmelin). Indian Little Brown Dove	352
,, *decaocto decaocto* (Frivalszky). Indian Ring-Dove	277

ORDER GALLINÆ
SUB-ORDER ALECTOROPODES
FAMILY PHASIANIDÆ : SUBFAMILY PHASIANINÆ

Catreus wallichii (Hardwicke). Cheer Pheasant	353
Ceriornis macrolophus biddulphi (Marshall). Kashmir Koklas	278
Gennaeus leucomelanos hamiltonii (Griffith & Pidgeon). White-crested Kaleej	353
Lophophorus impejanus (Latham). Impeyan Pheasant, or Monal	280

SUBFAMILY PERDICINÆ

Tragopan melanocephalus (Griffith & Pidgeon). Western Tragopan	353
Coturnix coturnix coturnix (Linnæus). Common or Grey Quail	353
,, *coromandelicus* (Gmelin). Black-breasted or Rain Quail	353
Alectoris graeca chukar (Griffith & Pidgeon). Chukor	282
Francolinus francolinus asiae Bonaparte. Indian Black Partridge	284
Tetraogallus himalayensis himalayensis Gray. Himalayan Snow-Cock	286

ORDER GRALLÆ
SUB-ORDER FULICARIÆ
FAMILY RALLIDÆ

Rallus aquaticus korejewi (Zarudny). Turkestan Water-Rail	288
Porzana pusilla pusilla (Pallas). Eastern Baillon's Crake	290
Amaurornis fuscus bakeri (Hartert). Northern Ruddy Crake	292
Gallinula chloropus indicus Blyth. Indian Moorhen	294
Porphyrio poliocephalus poliocephalus (Latham). Indian Purple Coot	296
Fulica atra atra Linnæus. Coot	298

SUB-ORDER JAÇANÆ
FAMILY JAÇANIDÆ

Hydrophasianus chirurgus (Scopoli). Pheasant-tailed Jaçana	300

SUB-ORDER ROSTRATULÆ
FAMILY ROSTRATULIDÆ

Rostratula benghalensis benghalensis (Linnæus). Painted Snipe	302

ORDER CHARADRIIFORMES
SUB-ORDER LARI
FAMILY LARIDÆ

Chlidonias hybrida indica (Stephens). Indian Whiskered Tern	304

Sub-Order Limicolæ
Family Charadriidæ : subfamily charadriinæ

Charadrius dubius curonicus Gmelin. European Little Ringed Plover	306

Subfamily Vanellinæ

Lobivanellus indicus aigneri Laubmann. Mekran Red-wattled Lapwing	309
Himantopus himantopus himantopus (Linnæus). Black-winged Stilt	311
Ibidorhyncha struthersii Gould. Ibis-bill	314

Family Scolopacidæ : subfamily tringinæ

Actitis hypoleucos (Linnæus). Common Sandpiper	317

Subfamily Scolopacinæ

Scolopax rusticola Linnæus. Woodcock	319
Capella gallinago gallinago (Linnæus). Fantail Snipe	323

Order Steganopodes
Family Phalacrocoracidæ

Phalacrocorax carbo sinensis (Shaw & Nodder). Indian Large Cormorant	354

Order Herodiones
Sub-Order Ardeæ
Family Ardeidæ

Ardea cinerea rectirostris Gould. Eastern Grey Heron	325
Egretta garzetta garzetta (Linnæus). Little Egret	354
Bubulcus ibis coromandus (Boddaert). Indian Cattle Egret	354
Ardeola grayii (Sykes). Indian Pond Heron	327
Nycticorax nycticorax nycticorax (Linnæus). Night Heron	329
Ixobrychus minutus minutus (Linnæus). Little Bittern	332
Botaurus stellaris stellaris (Linnæus). Bittern	355

Order Anseres
Family Anatidæ : subfamily anatinæ

Anas platyrhyncha platyrhyncha Linnæus. Mallard, or Wild Duck	335
,, *crecca crecca* Linnæus. Common Teal	355

Subfamily Nyrocinæ

Aythya rufa rufa (Linnæus). White-eyed Pochard	338

Order Pygopodes
Family Podicepidæ

Podiceps ruficollis capensis Salvadori. Indian Little Grebe, or Dabchick	339

INDEX	357

ILLUSTRATIONS

COLOURED PLATES

I. Indian Moorhen. Himalayan Snow-Cock. Impeyan Pheasant (adult male). Indian Purple Coot. Kashmir Koklas (adult male). Black-eared Kite. Long-legged Buzzard. Booted Eagle *Frontispiece*

II. Ticehurst's Willow-Warbler. Long-billed Horned Lark. White-cheeked Nuthatch. Blue-headed Robin (adult male). Scully's Red-rumped Swallow. White-throated Tit. Hodgson's Shortwing (adult male). Indian Short-billed Minivet (adult male). Little Forktail. Himalayan Goldcrest. Wall-Creeper. Hodgson's Yellow-headed Wagtail (adult male) *facing page* 40

III. Northern Indian Pied Bushchat (adult male). Crested Bunting (adult male). Western Collared Pigmy Owlet. Orange Bullfinch (adult male). Himalayan Goldfinch. Indian Grey-headed Bunting. Eastern Swift. Western Slaty-blue Flycatcher (adult male). Japanese Wryneck. Pink-browed Rosefinch (adult male). Himalayan Greenfinch (adult male). Stoliczka's Mountain-Finch *facing page* 136

IV. Black and Yellow Grosbeak (adult male). Brown-fronted Pied Woodpecker (adult male). Northern Pied Crested Cuckoo. Black-throated Jay. Himalayan Pied Kingfisher. Indian Grey Drongo. Himalayan Black Bulbul. Indian Blue Rock-Thrush (adult male). Himalayan Cuckoo. Larger Spotted Nutcracker *facing page* 232

V. Indian Black Partridge (adult male). Kestrel (adult male). Eastern Red-billed Chough. Hume's Blue Rock-Pigeon. Indian Sparrow-Hawk. Nepalese Snow-Pigeon. Long-eared Owl. Ibis-bill. Central Asian Hobby *facing page* 328

PHOTOGRAPHS

facing page

1. Himalayan Jungle Crow. Sind House Crow — 4
2. Western Yellow-billed Blue Magpie — 5
3. Eastern Jackdaws at nesting hole. Indian Brown Dipper — 12
4. Kashmir Grey Tit leaving nesting hole. Simla Green-backed Tit approaching nest in roof of forest hut — 13
5. Simla Black Tit at nesting hole in grassy slope. White-cheeked Bulbul on nest — 20
6. Brooks' Nuthatch at nesting hole. Himalayan Tree-Creeper at nesting site — 21
7. Simla Streaked Laughing-Thrush at nest. Western Variegated Laughing-Thrush at nest — 28
8. Rufous-backed Shrike. Indian Magpie-Robin, female, leaving nest — 29
9. Indian Bluechat at nest — 36
10. Western Dark-grey Bushchat, male. Indian Stonechat, male, at nest — 37
11. Western Dark-grey Bushchat — 44
12. Western Spotted Forktail on nest. Western Spotted Forktail, female with worn tail-feathers, approaching nest — 45
13. Blue-fronted Redstart — 52
14. Plumbeous Redstart at nest — 53
15. White-capped Redstart taking food to young. Blue-headed Robin, female, on nest — 60
16. Nest and eggs of Indian Blue Rock-Thrush in quarry face. Nest and eggs of Blue-headed Robin — 61
17. Western Himalayan Rubythroat at nest — 68

xiv

18.	Kashmir Red-flanked Bush-Robin, female, leaving nest. Central Asian Blackbird approaching nest	69
19.	Grey-headed Thrush. Tickell's Thrush at nest	76
20.	Himalayan Mistle-Thrush. Nest and eggs of Himalayan Jungle Crow, built on a clump of mistletoe in a walnut tree	77
21.	Blue-headed Rock-Thrush	84
22.	Nest and eggs of Himalayan Whistling-Thrush. Himalayan Whistling-Thrush taking food to young	85
23.	Jerdon's Accentor on lichen-covered nest in tree. White-browed Blue Flycatcher, male, at nesting cavity	92
24.	White-browed Blue Flycatcher	93
25.	Western Slaty-blue Flycatcher, female, on nest. Kashmir Sooty Flycatcher at nest	100
26.	Kashmir Red-breasted Flycatcher	101
27.	Rufous-tailed Flycatcher on nest. Himalayan Paradise Flycatcher, female, on nest	108
28.	Himalayan Paradise Flycatcher, adult male. Himalayan Paradise Flycatcher, immature male in red phase	109
29.	Witherby's Paddy-field Warbler : nest among weeds on sloping ground. Indian Great Reed-Warbler	116
30.	Hume's Yellow-browed Willow-Warbler near nest. Hume's Lesser Whitethroat	117
31.	Tickell's Willow-Warbler building in juniper. Large Crowned Willow-Warbler with food for young	124
32.	Indian Oriole at nest	125
33.	Himalayan Starling at nesting hole. Kashmir House Sparrow, male, leaving nest	132
34.	Common Myna with food for young. Black-headed Myna at nesting hole	133
35.	Hodgson's Rosefinch, female, on nest. Nest and eggs of Hodgson's Rosefinch	140
36.	Kashmir Cinnamon Sparrow at nesting hole	141
37.	White-capped Bunting at nesting site	148
38.	Eastern Meadow-Bunting. Nest and eggs of Eastern Meadow-Bunting	149
39.	Nest and eggs of Himalayan Goldfinch. Nest and eggs of Hodgson's Yellow-headed Wagtail	156
40.	Common Swallow. Young Kashmir House Martins ready to receive food	157
41.	Nest of Scully's Red-rumped Swallow. Eastern Grey Wagtail, female, leaving nest	164
42.	Hodgson's Pied Wagtail. Hodgson's Yellow-headed Wagtail, female, leaving nest	165
43.	Brown Rock-Pipit removing fæces from nest. Hodgson's Pipit leaving nest	172
44.	Kashmir Skylark. Young Asiatic Cuckoo in nest of Blue-fronted Redstart	173
45.	Scaly-bellied Green Woodpecker at nesting hole. Kashmir Pied Woodpecker at nesting hole	180
46.	Himalayan Slaty-headed Parakeets ; immature bird on perch	181
47.	Kashmir Roller with frog for young. European Bee-Eater outside its tunnel	188
48.	Indian Pied Kingfisher	189
49.	Central Asian Kingfisher. Indian Pied Kingfisher	196
50.	White-breasted Kingfisher with insect food for young. European Hoopoe	197
51.	Scully's Wood Owl. Turkestan Great Horned Owl on nest	204
52.	Himalayan Griffon Vulture. Himalayan Bearded Vulture at nest	205
53.	Himalayan Bearded Vulture in flight. Large White Scavenger Vulture	212

54.	Pallas' Fishing-Eagle	213
55.	Himalayan Rufous Turtle-Dove on nest. Nest and eggs of Impeyan Pheasant	220
56.	Chukor on nest. Nest and eggs of Chukor	221
57.	Turkestan Water-Rail on nest. Northern Ruddy Crake on nest	228
58.	Eastern Baillon's Crake repairing nest canopy	229
59.	Moorhen on nest	236
60.	Painted Snipe, male, on nest	237
61.	Pheasant-tailed Jaçana at nest	244
62.	European Little Ringed Plover with two eggs displaced by bird Indian Whiskered Tern at nest	245
63.	Mekran Red-wattled Lapwing on nest. Nest and eggs of Mekran Red-wattled Lapwing	252
64.	Black-winged Stilt approaching nest	253
65.	Common Sandpiper on nest. Nest and eggs of Common Sandpiper	260
66.	Woodcock on nest. Newly-hatched Woodcock chicks in nest	261
67.	Fantail Snipe on nest. Ibis-bill chick in down	268
68.	Eastern Grey Heron at nest	269
69.	Night Heron at nest. Indian Pond Heron sheltering young	276
70.	Little Bittern on nest	277
71.	Protective camouflage : Little Bittern with neck fully extended Little Bittern's chick seizing parent's bill to induce regurgitation	284
72.	Mallard on nest	285
73.	White-eyed Pochard on nest. Nest and eggs of White-eyed Pochard	292
74.	Indian Little Grebe on nest. Coot on nest	293
75.	Brown Hill-Warbler. Nest of Pale Bush-Warbler	300
76.	Among the chir pines in the lower Kishenganga Valley, elevation 3,000 ft ; a transitional zone where certain Plains' species are in evidence The distribution in Kashmir of the Slaty-headed Parakeet is correlated with the deodar, a cedar growing (between 4,000 and 8,000 ft) only in districts west of the Wular Lake	301
77.	Taobat, lower Gurais, in the Blue Pine and fir zone which stretches up to nearly 11,000 ft A corner of the Dal lakes, typical of the marshes and village areas of the Vale	308
78.	The Naubug Valley, characteristic of the lower reaches of the side valleys Above the tree line except for scattered birches, the haunt of Rubythroats, Redstarts, and many upland species	309
79.	Ibis-bill islands near Inshan in the Wardwan Valley Crow's nest, occupied by Kestrel, also holds fresh egg of Sparrow-Hawk	316
80.	Delta at Basman, Wardwan Valley, which supports a great concentration of Hodgson's Rosefinches Weed-covered areas near Suknes, Wardwan Valley, elevation 9,500 ft, in which the Paddy-field Warbler nests in some numbers	317

MAP

Sketch Map of area dealt with *Endpapers*

INTRODUCTION

In many countries of Europe, and in the United States of America, a large section of the population may be termed nature-minded, a fact borne out by the great demand for books dealing with every branch of natural history, each district, each county, each state, moreover, demanding its own works. Considering its huge population, India as a whole has so far acquired little of this nature-mindedness. Nevertheless, scattered throughout India and Pakistan, there are a large number of Indians and Europeans who would become so were they provided with the necessary incentives and encouragement. Admittedly there are now a few popular works on India's flora and fauna, as, for instance, the late Hugh Whistler's unsurpassable *Popular Handbook of Indian Birds*, and those most useful books compiled by Mr Sálim Ali, namely *The Book of Indian Birds* and *Indian Hill Birds*. It is only now, however, that works dealing with the birds or flowers of any particular Province or State have begun to make a belated appearance.

The Himalayas provide a vast area whose riches, botanical and zoological, are immense, and Kashmir in particular is a part of that rich area which is both pleasant and easy to work and is visited annually by large numbers of people in search of rest and change from their work in the Plains. Although Blatter's *Beautiful Flowers of Kashmir*, and Coventry's series of coloured plates of that State's flowers and flowering shrubs, made their appearance some years ago now, there is still no book on its trees, while the seeker after knowledge of its diverse bird-life has perforce to wade through innumerable works dealing with the whole of India, Burma and Ceylon, or hunt out a few articles scattered here and there throughout the fifty odd volumes of the Bombay Natural History Society's Journal. It is most surprising to us that no separate volume dealing with Kashmir's really striking bird-life has been published. Our intention in this small work is, therefore, to do what we can to rectify this omission.

The territories at present comprising the State of Kashmir and Jammu cover more than 85,000 square miles. In such an expanse of country the number of breeding birds is bound to be great, but in Kashmir this number is augmented by a climate ranging from tropical to arctic, and the types of country from dry, hot plains lying at only a few hundred feet above sea level, through scrub-covered foothills and monsoon-drenched outer ramparts culminating in the 15,000 ft peaks of the Pir Panjal mountains. Within this outer barrier there lies at 5,000 ft the Vale of Kashmir, a plain encircled by magnificent forests, with its temperate climate and an intricate system of canals, marshes, and lakes. Still farther in, beyond the mighty wall of the main Himalayan range which rises in parts to eternally snow-capped peaks of well over 20,000 ft, there lie the arid uplands of Ladakh, Baltistan, and like Provinces. These have a scant flora owing to the lack of rainfall and a consequent desert fauna. Small wonder is it then that ornithologically the State of Kashmir is wealthy in the extreme.

To deal exhaustively with this great store would require a work of much greater proportions than the authors consider expedient, both on account of the higher cost of such a work and also of its bulk. We have considered it essential, therefore, in order to produce a book of sufficiently handy proportions, to restrict our scope to the breeding birds of only certain areas. The reasons for the selection of these areas are set out below.

The State of Kashmir may conveniently be divided as follows :

(*a*) Jammu and Poonch, which lie without the Pir Panjal range and vary in altitude from Plains level to 14,000 ft.

(*b*) Kishtwar and Bhadarwar, likewise outside this range, not easy of access, and ranging from 3,000 ft to about 16,000 ft.

(*c*) The Vale of Kashmir and the Jhelum Valley gorge leading to it, easily reached by the Murree, Abbottabad, and Banihal motor roads ; elevation from 5,000 ft in the Vale to surrounding peaks of approximately 17,000 ft.

(*d*) The Kishenganga Valley, including Gurais, lying chiefly between 2,000 ft and 8,000 ft with enclosing ridges of 15,000 ft, and accessible either from Domel or from Bandipur in the Vale.

(*e*) The Districts and Provinces beyond the Great Snowy range: Gilgit, Astor, Baltistan, Ladakh, and Rupshu, consisting of high plateaux and bare ridges whose climate is dry and wind-swept in summer and bitterly cold in winter.

The birds of Jammu and Poonch include a great many species common to the Plains of India. These are fully dealt with in Hugh Whistler's and Sálim Ali's excellent books already mentioned. Kishtwar and Bhadarwar also contain many birds of the foothills and are little visited as they are rather out of the way. There is, however, within the boundaries of Kishtwar the Wardwan Valley. This valley contains a number of birds of exceptional interest, and it is now easy of access from the head of the Vale via a fair-weather motor road running from Achabal to the top of the charming little Naubug Valley whence a good bridle-path carries the traveller over the Margan Pass to Inshan.

Astor, Ladakh, Rupshu, etc., would not add more than thirty or forty birds to the list, but these Provinces lie far afield, and are penetrated by but a small proportion of the visitors who throng Kashmir every summer. Besides, a considerable number of these wanderers are bent on what is, to them, of greater moment than the study of birds, namely the pursuit of big game.

To cater for the majority ; to avoid confusion between Plains' birds and those of Kashmir proper ; and to keep the book within reasonable proportions, we have, therefore, confined ourselves to dealing with the nesting species only of the following areas, ignoring passage migrants which have chiefly passed through before the main influx of human visitors commences :

(*a*) The Vale of Kashmir and the Jhelum Valley from Kohala to Baramullah.

(*b*) The slopes and side valleys around the Vale of Kashmir up to the Passes over the Great Snowy range.

(*c*) The Kishenganga Valley inclusive of Gurais.

(*d*) The Upper Wardwan Valley of Kishtwar.

We feel this work will have achieved its purpose if it helps the bird-lover to recognize the nesting birds which he is likely to meet in spring and summer in any of these areas, and to realize what is known, and what remains to be learnt, of their distribution and general habits.

We do not claim that our work is exhaustive ; it is not. For one thing there are remarkably few published records referring to the birds of the Lolab, the Pir Panjal mountains—except in the neighbourhood of Gulmarg—and of the Kishenganga Valley. Also, there are numbers of birds not included, which we know to occur in certain of the areas in summer but of which there is as yet little or no reliable record of nidification. These are by no means few in number, so we have included a list, together with short notes, of those birds concerning which we feel further data is required before admitting them to the role of fully-described breeding species.

We have kept before us another object in the preparation of this book. While the five coloured plates contain the majority of the most gaily dressed birds of Kashmir, as well as portraits of a number of species whose photographs we have as yet been unable to obtain, we have aimed at including such photographic illustrations as will help in the identification of each bird and its nest, thereby rendering long descriptions in the text unnecessary. We are only too well aware that our efforts in that respect are by no means complete, but we hope that this book will stimulate others to take up the photography and study of Kashmir birds to fill in the gaps to their own and everyone else's satisfaction and benefit.

A series of photographs obtained without damage or cruelty to the birds, and the knowledge of their habits which one gains in the process, are surely of greater use to the ordinary bird-lover than a few skins and a cabinet full of eggs. Large numbers of eggs are taken annually without sufficient data to be of any real use in scientific research and which, in fact, are never put to any scientific purpose whatsoever. Sufficient eggs of many species have now found their way into the museums, so to collect any more of these, except as replacements, is not only a waste of effort but of detriment to the species. We cannot bring ourselves to agree with the dictum that in India there is never any chance of the activities of the egg-collector endangering the existence of any species. Particularly is there danger in Kashmir from the shikari egg-collector from whom people buy eggs without inquiring too closely into their origin and method of collection. Not only are such collectors generally unscrupulous, but their productions, often of ' made-up ' clutches with equally ' made-up ' data, are quite valueless.

Lately we received a letter giving two blatant examples of the harm certain local bird-shikaris are causing in the Vale. These men offer to show novices the bird-life of the Valley and reap rich rewards not only from filching all the eggs they can lay hands upon, of common as well as of rare birds, but also by staging special finds. These ' finds ' are often made up from the contents of two or more nests, and even a nest itself may be moved to a more convenient site! On revealing the discovery a wealth of completely fictitious detail is poured forth to delude the victim into parting with a high fee. One of the above cases involved a previous year's clutch of Streaked Laughing-Thrush's eggs placed on a pad of straw and shown as the eggs of the extremely rare (*sic*) Blue Thrush; the other discloses the reason for the probable disappearance of the Ibis-bill from its last haunts in Kashmir short of Kishtwar. An even more shameful fact is that considerable sums have been offered to local shikaris by oversea collectors for Ibis-bills' eggs from this area. We are aware that it is virtually impossible for any one man to make a completely representative collection of skins and eggs from a wide area, but if the oologist does employ paid or even casual collectors, let him do so with the very greatest circumspection and with a scientific and not a purely acquisitive object in view. Surely it is downright criminal under the circumstances disclosed above for oologists to offer money for eggs whose collection they are not in a position to supervise.

Thoughtless egg-collectors are not, of course, the sole offenders ; bird-photographers may at times not be guiltless of causing havoc amongst nesting birds. We cannot condemn too strongly the photographer whose carelessness and lack of thorough preparation cause his subjects to desert their nests. Carried out on proper lines there can be no finer and healthier hobby than bird-photography ; especially in the Western Himalaya where it is a delight, and of the greatest benefit to health, to trek the many hundreds of miles at all heights up to the perpetual snows in one's efforts to run to earth species which breed under such widely differing conditions. Bird-photography can, in fact, be as arduous as big-game shooting and may demand just as much nerve and physical endurance as that or any other pursuit sought after by the young and energetic ; while for those who feel as the years creep by that they can no longer get a kick out of that 2,000-ft climb before a chilly dawn, the hobby can be equally fascinating, and, even taken in a leisurely manner, productive of great benefit to all workers in the sphere of ornithology.

In Kashmir, and indeed in India as a whole, there remains a wide field still to be covered. Stuart Baker, in his *Nidification of Birds of the Indian Empire*, informs us that our knowledge of the nesting habits of nearly 12 per cent of the birds of India is still a complete blank—an astounding proportion ; while there are many more birds about which our knowledge is incomplete, even as to the simplest facts. Much, he says, still remains to be learnt by our field naturalists regarding incubation and other details. Bird-photographers are essentially field naturalists and many of these required details are to be gleaned through observation from their hiding-tents. Unfortunately there are numbers of naturalists who never venture into print. Would they only impart their knowledge through some such medium as that mine of information, the *Journal of the Bombay Natural History Society*, the percentage quoted above would be greatly reduced if not altogether eliminated.

A few words about bird song will not be out of place. Many Kashmir birds can be recognized from their songs and call-notes, but, to quote from Turnbull's *Bird Music*, ' Bird songs cannot be written down. Most attempts to transcribe them into words or music notation are sheer gibberish. . . . One cannot commit to paper a complex of sibilant or contralto sounds such as, for instance, the songs of skylark and blackbird, because one cannot relate them to any alphabetical pattern. Even if you can, or think you can, point some vowels, the " consonants " bewilder us. The score has got to be kept in one's brain.' After all, *tzeet-tze-tze-tzeet* may convey to us the alarm-notes of a Whistling-Thrush simply because we are so familiar with that bird, but such a jumble of letters may present a totally different meaning to anyone else. There is, however, one means by which the reader can be helped towards recognizing some of the birds from their notes. B. B. Osmaston states that the call of the Pale Bush-Warbler consists of two loud phrases, both commencing with a long-drawn-out whistle. These two phrases he expresses in words :

> You . . . mixed-it-so-quick
> He'll . . . beat-you.

This simple expedient of fitting words to music is most certainly an aid towards recognition in this case, although, after all, the words of a song really tell us little if anything about the composition of the music to which they are fitted, and nothing of its timbre.

The song of the Large-billed Willow-Warbler is a remarkably far-reaching melodious whistle, a phrase of five clear notes round which, were we competent to do so, nothing would give us greater pleasure than to compose a symphony. It is, alas, one of the few bird songs for which an accurate musical score can be written. After due consideration, therefore, we have come to the conclusion that it is advisable to attempt the transcription of the songs of but few of our Kashmir birds into either words or music, although enthusiasm has, we fear, at times got the better of our judgement.

Although we have been photographing Kashmir's birds since the year 1920 with this book in view, the text commenced to take its final shape only in 1943 when one of us had some enforced leisure in hospital. The writing up was continued throughout the war whenever circumstances permitted, but war conditions, when of necessity reference libraries and one's own books and notes cannot be consulted, are not conducive to exhaustive inquiry. We therefore crave our readers' indulgence for any omissions. Since leaving India we have, of course, checked up the text with those works on Indian ornithology now available to us.

It would be invidious to pick out any names from that great band of past and present ornithologists to whom we are indebted for all they have written bearing on Kashmir's birds, but we must express our special gratitude to Mr Sálim Ali for the many ways in which he has rendered assistance, particularly with nomenclature and also with the selection of suitable skins from which to prepare the coloured plates. Our thanks are also due to Mrs D. V. Cowen for the great trouble she took to attain accuracy in her beautiful paintings for the five coloured plates which depict so many of our most colourful birds. The work of one ornithologist stands out in regard to Kashmir. Without Mr B. B. Osmaston's numerous articles and notes on Kashmir bird-life, this book would lack much of its meat, particularly about the winter population of the Vale. We must also express our indebtedness to Mr Samsar Chand Koul for many of the Kashmiri bird names. A number of these are admittedly of his own invention, but we are sure they have every chance of universal adoption throughout Kashmir through the medium of the many boys who passed through his enthusiastic hands at the C.M.S. School in Srinagar.

Our own contributions to the habits and distribution of the birds in the area dealt with are the result of sixteen visits to Kashmir, each of from one to three months' duration between the months of April and September. The area has been well covered on our treks with the exception of the Pir Panjal slopes and karewas from some five miles east of the Ferozpur nullah to the Banihal Pass. Incidentally, notes from this stretch of country, with the exception of a few mentions of Tosha Maidan, are conspicuously lacking. We would also point out that we have had no chance of studying Kashmir's bird population in winter, and wish to take this opportunity to urge upon field naturalists who visit the Vale at that season the necessity of publishing their observations. Finally, our acknowledgements are due to the staff of the Bombay Natural History Society for assistance so freely given both at the Society's office and at the Prince of Wales' Museum, and to the Society for the provision of a number of half-tone blocks used at different times to illustrate articles in its Journal.

SIZE KEY

THE number in brackets immediately preceding the FIELD IDENTIFICATION paragraph of each species denotes its approximate size in relation to the following universally known birds. The plus or minus sign denotes respectively slightly larger or slightly smaller than the key species:

    (1) Very small birds, much smaller than a House Sparrow.
    (2) Birds the size of the House Sparrow.
    (3) Birds the size of Tickell's Thrush.
    (4) Birds the size of the House Crow.
    (5) Birds the size of the Kite.
    (6) Large birds much bigger than a Kite.

BOOK REFERENCES

To enable the reader to refer quickly to the following standard works on Indian birds, abbreviated references have been included under each bird thus:

Fauna of British India Series, second edition: *Birds,* Volumes I to VI, by E. C. Stuart Baker.
    Example: *F.B.I.,* IV, 46.

Nidification of Birds of the Indian Empire, Volumes I to IV, by E. C. Stuart Baker.
    Example: *Nidification,* III, 260.

Popular Handbook of Indian Birds, fourth edition, by the late Hugh Whistler, revised and enlarged by Norman B. Kinnear.
    Example: *Handbook,* 25.

J.B.N.H.S., or simply the *Journal,* used at times in the text, stands for the *Journal of the Bombay Natural History Society.*

EASTERN CARRION CROW

Corvus corone orientalis Eversmann

KASHMIRI NAMES. *Diva Kāv, Pantsol Kāv.*
BOOK REFERENCES. *F.B.I.*, I, 24; *Nidification*, I, 5; *Handbook*, 3.
FIELD IDENTIFICATION. (4+) The entire plumage is black but in the Carrion Crow the feathers of the hind neck are firm and glossy with glistening shafts, whereas in the Jungle Crow they are soft and almost glossless. Even with the finest of binoculars such a characteristic is quite indistinguishable with any certainty in the field, where under the best of circumstances the light plays many tricks upon the eyesight.
DISTRIBUTION. The Eastern Carrion Crow's chief habitat is north of the main Himalayan range, this race being distributed throughout Eastern Asia from the Yenisei River to the Pacific and southwards to Persia, Baluchistan, Afghanistan, Baltistan, Ladakh and Tibet. Failing a careful survey entailing the collection of many skins, it is most difficult to say to what extent it penetrates southwards into our area. That it does occur and has bred is beyond doubt. We do not know, however, whether it nests, or even occurs, in the Pir Panjal range as well as in the mountains on the northern side of the Valley. So far as we know it is a resident bird, though it is said to drop to lower levels in the winter. Nothing, however, is on record as to its winter distribution in Kashmir territory.
HABITS AND NESTING. Unfortunately the habits, nest, and eggs of this bird are quite similar to those of the Jungle Crow, so, except where the parent has been collected with the eggs, it is impossible to give authoritatively instances of its breeding in Kashmir proper. At the moment, therefore, the only good record still remains that of Brooks (Hume's *Nests and Eggs*, second edition, Vol. I, p. 4), who took eggs, and secured the parent bird, from a nest obtained on 30 May at Sonamarg. It is, of course, possible that a certain number of nests and eggs taken and recorded as those of the much more common Jungle Crow were those of this bird, but it is safest, for the present at any rate, to look upon it as a straggler into our area whose status still requires elucidation. Stuart Baker gives the average size of 60 eggs, collected from the entire range of this form, as 43·6 by 30·9 mm.

HIMALAYAN JUNGLE CROW

Corvus macrorhynchos intermedius Adams
(See Plate 1 facing p. 4 and Plate 20 facing p. 77.)

KASHMIRI NAMES. *Diva Kāv, Pantsol Kāv.*
BOOK REFERENCES. *F.B.I.*, I, 28; *Nidification*, I, 10; *Handbook*, 3.
FIELD IDENTIFICATION. (4+) Entirely black. Could be confused only with the Eastern Carrion Crow and the Tibetan Raven. The latter bird is much larger and not likely to be seen within our area in the breeding season, although occasional stragglers may appear in the neighbourhood of Baltal at the head of the Sind Valley.
DISTRIBUTION. Is a widely distributed resident except in the middle of the main Vale where it is scarce, although in the winter it becomes more common, even venturing into gardens in Srinagar. It occurs in all the better wooded parts, in the thick forests, and above them on the margs and uplands to about 13,000 ft or even higher, whither it follows the Gujars and their flocks to the limits of the grazing, descending with them from the higher altitudes in the autumn. In the winter it shows some vertical movement, but does not seem to reach the Punjab Plains.

This Himalayan race occurs in the hills of Eastern Afghanistan and as far west along the North-West Frontier as the Kurram Valley where Whitehead found it breeding freely in the Sufed Koh. To the east it reaches Bhutan where it links up with the race found in Assam and Burma.

HABITS AND NESTING. The Jungle Crow is as black as he is painted. He is indeed the 'bold, bad baron' of the Kashmir forests and uplands, for his depredations amongst the eggs and young of weaker vessels are little short of appalling. It is a fact that we never approach a newly-spotted nest without first looking to see whether our movements are being shadowed by one of these black-hearted rascals. We have witnessed the eggs of even a Monal Pheasant being destroyed by this crow, and on another occasion saw one force an immature turtle-dove to the ground and behead it ruthlessly before we could go to its rescue. They are bold yet full of wariness and out in camp take the place of the house crows in hanging around the tents to see what they can pick up in the way of scraps.

They are quite omnivorous and do not eschew a bit of bread or scraps of kitchen refuse, as well as frogs, eggs, young of other

PLATE I

Sind House Crow

Himalayan Jungle Crow

PLATE 2

Western Yellow-billed Blue Magpie

birds, and the pickings left by vultures on the bones of a rotting carcase. They have, too, developed a thorough drill for hunting out *Cicadae*. Several together will work a group of trees, hopping down and round the trunks from branch to branch and stump to stump, sometimes picking off a sluggish insect, but more often flushing the quarry which is seized by another of the gang ere it can alight elsewhere. Though found even in the deepest forests, they are equally at home in mixed forests and well-wooded country, and on the many margs whence they follow the Gujars to the bare uplands when they take up their flocks to graze. If not particularly gregarious, a dozen or so may collect to feed or to fly about together high above the trees, carrying out graceful evolutions at which they are adept. In the winter, according to a reliable big game shikari, they have been noticed about Surphrar in the Sind Valley congregating for roosting purposes in large numbers in the walnut and other trees. Their notes are a fairly loud *caw* with various other harsh noises when occasion demands, and they are also prone to a habit of sitting alone on a shaded perch talking nonsense to themselves in subdued quarks and gurglings.

They nest early, and in consequence not, we believe, at elevations much above 8,000 ft. Building commences in April—many birds will be seen carrying material halfway through that month—but eggs may be taken towards the end of April and throughout May. At the extremes we have found fresh eggs on 18 April (4) and 3 June (2). In the latter case the birds were still laying. Ward says they begin in March, but his notes include Jammu and the outer ranges. The number of eggs laid varies from 3 to 6; we took a nest containing the latter number at Thiun in the Sind Valley on 30 April. They average 44·8 by 31·3 mm. in size, being slightly larger than eggs from farther east along the Himalayas. They are often rather elongated and have a blue-green ground splashed closely with streaks and blotches of umber and blackish brown, often somewhat longitudinally. There may also be inconspicuous secondary markings of a neutral tint.

The nest is a large affair placed more often than not, in Kashmir at any rate, in a pine or fir tree, close to the trunk at a considerable height from the ground, seldom as low as 30 ft, often higher near the crest of the tree where the closer-growing foliage affords good concealment; in fact, in spite of their size they are generally not easy to locate and can often be overlooked. Sticks are employed, rather untidily put together, but the inner nest and lining are compact and thick, and comprise a varied assortment of materials, such as wool, hair or pine-needles. We have indeed seen a Jungle Crow pulling wool and hair off the backs of a sheep and a yearling calf. The cavity is deep and in the lining fine twigs or rootlets are often intermingled, and these last we have watched a Jungle Crow extracting from freshly ploughed ground.

SIND HOUSE CROW

Corvus splendens zugmayeri Laubmann

(See Plate 1 facing p. 4.)

KASHMIRI NAME. *Kāv*.

BOOK REFERENCES. *F.B.I.*, I, 34 ; *Nidification*, I, 18 ; *Handbook*, 5.

FIELD IDENTIFICATION. (4) The Sind House Crow belongs to a species with which everyone in India is thoroughly conversant. In this race the grey of the neck is very pale, in some lights appearing almost white, but Kashmir's birds are often not quite so pale as typical birds from Sind. The House Crow is smaller than the all-black Jungle Crow, but a little larger than the perky Jackdaw from which it is easily distinguished by the heavier bill and the lack of the white iris.

DISTRIBUTION. The distribution of the Sind House Crow in Kashmir is restricted and peculiar. It is not particularly common even in and around Srinagar where we are certain that for many years after 1920 its numbers showed a steady fall. It seems probable, however, that it is once more on the increase as it is again to be seen in Srinagar city and around the Dal lakes. It appears now to have shifted its headquarters to the Anchar and its surrounding villages where it nests commonly in the extensive willow plantations. It occurs at Baramullah and to a lesser extent at intermediate villages along the Jhelum, but elsewhere its occurrence is restricted to a small colony at Garhi. We have never seen this bird above Srinagar, i.e. at Pampur, Bijbehara, Islamabad or Achabal to which it might well have penetrated, nor are there any at Domel and Muzaffarabad or in the Kishenganga. It is resident throughout the year. This is the race of *splendens* which is found from South Baluchistan through Karachi, Sind, the North-West Frontier Province, and the northern Punjab.

HABITS AND NESTING. It is hard to say why House Crows have failed to gain as prominent a place in the Vale of Kashmir as Eastern Jackdaws, which are so very numerous, especially as they do not seem to mind the cold and remain in their haunts throughout the winter, surviving its rigours with apparent ease. Whatever the reason, they now form an isolated body which, as Osmaston says, may well develop in time into a geographical race of their own. We are not at all certain that the Garhi colony has not ceased to exist ; we have seen none there for some years, but when merely passing through in a car their presence might easily have been overlooked. Colonel Unwin, in the list which he compiled for Sir Walter Lawrence's wonderful book *The Valley of Kashmir*, wrote as follows : ' This well known pest, with his thoroughly inappropriate Latin title, is unfortunately much more common at Srinagar and the towns on the Jhelum now than formerly, and has thus obtained a right to a place among the birds of Kashmir.' Now that was written approximately fifty years ago. Perhaps, therefore, it is a bird whose population is prone to considerable fluctuation, brought about perhaps by epidemics, perhaps by other causes, due possibly to its being an interloper in the Valley which has never really obtained a sufficiently strong footing. As, however, it remains there throughout the winter, and stands the severe frosts of December and the snows and grey days of January and February with apparent nonchalance, it is hard to understand why it is not more common.

Its habits differ in no way from those of its Plains' brethren, that is, it is just such a cheeky ruffian out to gain its livelihood by attaching itself to the haunts of man whence it can pick up its omnivorous diet with the least trouble. Some spilt grains of cooked rice, a piece of dry bread, a nestful of young birds whose whereabouts have been given away by a careless bulbul parent, alike attract its watchful eye, and a plateful of scraps thrown from a houseboat on to the bank will soon attract a squad of jackdaws and those crows which happen to be in the neighbourhood. Unfortunately it does not confine its cannibalistic depredations to the prolific bulbuls alone, but is destructive of both the eggs and young of the Little Bittern, Pond and Night Herons, and in fact of the contents of any unguarded nests it may come across of the large and small birds of the Vale.

It does not form rookeries, and nests are placed where convenient to the owner in most varied positions, but where House Crows occur in sufficient numbers they may be seen flying about and feeding in small parties. Willows and poplars around the villages or along the canal banks are the most popular sites, particularly the former trees in which the nests are often quite low down ; we have both of us, in fact, reached nests with our hands when merely standing up in a shikara. Along the main river we have also noticed one or two nests tucked away in the angles on the wooden struts holding up the projecting upper stories of the houses. The nest is the usual rather untidy collection of sticks with not much depth or shape about it, and is generally lined with fine twigs, roots and fibres, and at times a few miscellaneous bits of refuse such as pieces of rag, or wads of hair or wool.

The eggs number 4 to 6, seldom the latter, and are rather elongated ovals. They vary considerably in both size and coloration, but the general description given in the *Fauna* under the Common Indian House Crow fits them as well as any : ' they are typical Crow's eggs and run through the same range of variations as do those of all the Corvidæ. The ground is any shade of blue-green and the markings are of dull reddish and brown with secondary markings of grey and neutral tint, usually they are small and irregular in shape and are scattered profusely over the whole egg.' Nesting commences in May and continues throughout the month of June. Osmaston gives the following measurements of 17 eggs taken by him which exemplify the great variation in size. Longest egg 43·2 by 27·4 mm. ; shortest 37·0 by 26·1 ; broadest 40·5 by 27·7 ; narrowest 40·5 by 26·0 ; average 39·0 by 26·6 mm.

EASTERN JACKDAW

Corvus monedula sœmmeringii Fischer

(See Plate 3 facing p. 12.)

KASHMIRI NAME. *Kavīn.*
BOOK REFERENCES. *F.B.I.*, I, 36 ; *Nidification*, I, 20 ; *Handbook*, 8.
FIELD IDENTIFICATION. (4—) Slightly smaller than the House Crow. Perhaps we should say more dumpy rather than smaller, as it is less slender and has a perky upright carriage. The bill is short and nearly straight, but the most prominent characteristic of the Jackdaw is that almost white iris which gives it such a penetrating mischievous expression.
DISTRIBUTION. In its breeding season the Jackdaw is exceedingly numerous throughout the Vale, chiefly around the towns and villages, and less so, but still common, along the subsidiary valleys up to about 6,500 ft. For instance, it does not breed as far up the Lidar Valley as Pahlgam, nor beyond Gund or perhaps Koolan

in the Sind. It is exceedingly common along the lower course of the Pohru River and in the Lolab, also for some miles up all those numerous little valleys draining the Kazinag mountains. It is common at Tangmarg, but does not breed in Gulmarg. The Bringh, Naubug, etc., all have their quota. In the Jhelum Valley it is first met with a few miles below Uri whence it is common, but it is entirely absent from the Kishenganga Valley. As soon as nesting is finished, it at once commences to invade higher levels. We have seen a few birds in Pahlgam at the latter end of July, but it does not appear in any numbers at such places as Sonamarg or Gulmarg until the first week in August. It then rapidly extends its range to the higher margs on both sides of the Vale and may even be come upon at elevations of 12,000 ft or more. The on-coming winter drives it back again to the main Valley and some even descend to the Punjab. The majority, however, collect in and around the larger towns and villages of the Vale.

The Jackdaw is of course widespread as a species in Europe, Western Asia, and North Africa. The breeding range of this particular race is South-Eastern Europe and Western Asia, south to Persia, Afghanistan and Kashmir. In the cold season migrants appear sparingly amongst the rooks around Quetta and in larger numbers in the North-West Frontier Province and in the North Punjab along the base of the Himalayas.

HABITS AND NESTING. There are always certain points which stand out in first impressions of a place, and these in connexion with our introduction to the Vale of Kashmir can be summarized thus. A jumble of colourful scenes, sights, and smells. Canals and kingfishers; reed-beds, bitterns, noisy warblers, and vast stretches of vivid green rice; carpets of irises; orchards harbouring Paradise Flycatchers, Rufous-backed Shrikes and Orioles; willow groves, tall spindly poplars and massive chenar trees, and everywhere amongst them, on the banks of the channels, in the willows, amongst the rickety houses, quartering the fields and peopling every chenar tree, the ubiquitous Jackdaws, not in ones and twos but populating the land almost more thickly than us human beings. A single chenar assuredly provides shelter and nesting sites for as many pairs as there are holes in its massive old branches, but the Jackdaws do not confine their occupation to the great chenars, for holes in willows, walnut and mulberry trees, sometimes in cavities in steep sandy banks and even in the walls of gardens and houses, are likewise appropriated and made use of year after year to swell the numbers of that most successful colonist. For the Jackdaw is a cheeky, bold, yet cunning and wary crow, which exploits its contact with mankind to the full, both to augment its natural food-supply from whatever scraps and morsels it can pick up, and to obtain nest material in the form of old rags and other village rubbish, for nesting sites in holes in old walls and buildings. One of the pleasantest features in the groves and gardens of the Mogul Emperors is the cheerful *jack* of the Jackdaws, for scores nest in the many cavities in the colossal chenar trees.

In describing their nidification we cannot do better than start with a verbatim quotation from B. B. Osmaston's *Bird Notes in Kashmir in Winter*. ' Throughout the winter they leave Srinagar daily in large flocks of hundreds or even thousands of individuals. They leave in a vast stream every morning about half an hour before sunrise, flying roughly south along the Jhelum River and return in the evening about quarter of an hour after sunset. Presumably they go to feed in the extensive areas of cultivation found upstream along the Jhelum River. Early in February they pair off and select nesting holes in trees or buildings but they do not lay until the first week in May.' Although pairing off takes place in February and nesting sites are at once taken up, the serious job of nest-building does not appear to receive much attention before April. In any case this entails merely a few repairs, for the same sites are used year after year. As can well be imagined when one thinks of the girth of the huge trunks and branches of some of the chenars, the nest is often many feet in from the entrance, and where these ancient trees are well riddled with cavities, the Jackdaw population of a single tree may be considerable. Sometimes the edge of the nest may be clearly in view, as was the case of a nest we investigated two years running in the Bandipur Nullah, which was in a slight hollow of a sandy bank about 15 ft high on the outskirts of a village near the river bed. Another quaint nest had its entrance in the wall of a house in the Erin Nullah ; quite a small hole, but it happened to widen out into the gloomy room beyond from which the sitting bird could be seen plainly hardly a foot distant. Incidentally, the bank-site nests are not uncommon, for during a trip by houseboat from Baramullah to Shadipur on 3 May a number of nests were noted in cavities in the river banks, one such containing six fresh eggs. The nest is usually a collection of sticks, often thorny ones, lined with hair, wool, rags, or other refuse. At times the sticks are dispensed with and the lining covers the accumulated rubbish at the bottom of the hole.

Egg-laying starts in the first week of May and fresh eggs may be taken throughout the greater part of that month. The Bandipur Nullah nest mentioned above, however, contained four young ones on 24 May, which points to the eggs having been laid at the end of April. Davidson (*Ibis*, 1898, 7) saw a single pair of birds building on 27 June at Ganderbal, possibly prior to producing a second brood. Rattray in 1903 took eggs at Sonamarg at an elevation of 8,000 ft, but the Jackdaw does not normally breed at such an elevation. The eggs, from 4 to 6 in number, vary a great deal both in coloration and size, some being long ovals, others broad and more pointed at the smaller end. The ground colour ranges from a bluish white to a bluish green or pale blue, and the markings, which may leave considerable areas of the ground in view or may cover the greater part of the egg, consist of well-defined blotches and specks of dark brown and brownish black with a few underlying ones of lavender grey. The shell is well glossed. They average 35·1 by 24·8 mm. Colonel Unwin makes the amusing point that 'they . . . are capital eating. I have seen them served in mistake for plovers' eggs at a garden party in Srinagar'.

WESTERN YELLOW-BILLED BLUE MAGPIE

Urocissa flavirostris cucullata Gould

(See Plate 2 facing p. 5.)

KASHMIRI NAMES. *Lōt Rāza* (*Lolab*), *Literāz*.
BOOK REFERENCES. *F.B.I.*, I, 44 ; *Nidification*, I, 29 ; *Handbook*, 10.
FIELD IDENTIFICATION. (4) A noisy blue-grey bird with an abnormally long and highly graduated tail, of which each feather is tipped with white and has a sub-terminal black band. As the tail is often spread out in flight and always on alighting, the bird is then a striking and beautiful sight. The head and breast are black and there is a curious white patch in the centre of the black nape.
DISTRIBUTION. A bird of mixed forests and thickly wooded country from approximately 4,500 ft up to about 7,500 ft, very occasionally higher. It is fairly common in all the valleys around the main Vale, less so perhaps on the slopes of the Pir Panjal. The farthest point at which we have seen it up the Lidar Valley is at Aru. It is to be seen on the motor road before reaching Rampur and is common in the Kishenganga Valley between Salkalla and Dudnial. It is on the whole a resident species,

MAGPIE

only those birds of the higher zones coming a little lower in winter.

The species comprises two races both confined to the Himalayas. This race extends from the Hazara border to Western Nepal where it meets the Eastern form.

HABITS AND NESTING. These magpies are essentially woodland birds but by that we do not mean to imply that they will only be found in the depths of thick forests. On the contrary they are more prone to inhabit the outskirts whence they can visit more open wooded country and forest glades. Though mainly arboreal, they may not infrequently be seen on the ground where they progress in ungainly hops with the long tail raised aloft out of harm's way. Out of the nesting season they are usually seen in small bands or family parties which at times provide a striking sight by flying one by one high overhead in a rather laboured undulating manner from the forest on one face of a nullah to the shelter of the woods on the opposite side. Such bands are then extremely noisy, uttering a varied repertoire of harsh notes mingled with some of a more musical nature. When nesting the story is different, for each pair is then restricted to its own territory, which may be of considerable extent.

The nest, particularly when well-concealed in a leafy tree, is small, definitely smaller than that of a crow, and is often a fairly neat structure of sticks with an inner lining of roots. Grass is also said to be used at times, and we once saw a bird carry a couple of strips of cloth to a nest it was building about 30 ft up in a large-leafed tree standing well away from the forest's edge. The framework of sticks of one nest was certainly an untidy network, but it was built in a tangle of small twigs halfway up a small dead pine and was quite unconcealed. The nests are flimsy structures; in fact at times so thin that the eggs can be seen through them from beneath. They are built not infrequently in small trees and consequently low down, one seen at Rampur being not more than eight feet from the ground.

The eggs are up to 5 in number, occasionally only 3, and often 4. They are usually rather pointed with an olive or yellowish-stone background, but the flecks and streaks of brown or brownish-red are sometimes so profuse as to cover it up. Sometimes the markings are larger and not so close together, so that underlying spots and blotches of a neutral tint show through. They measure 33·8 by 23·1 mm. The principal breeding months are May and June. We took a nest of five fresh eggs between Pahlgam and Praslun at the very end of the latter month. The bird often sits very close,

PLATE 3

Eastern Jackdaws at nesting hole

Indian Brown Dipper

PLATE 4

Simla Green-backed Tit approaching nest in roof of forest hut

Kashmir Grey Tit leaving nesting hole

and when we sent a man up the maple in which this particular nest was situated, 27 feet from the ground, the female sat on until he was level with the nest and then hopped up to a branch not two feet above him where she clung upside down looking as if she might drop on to his head at any moment.

BLACK-THROATED JAY

Garrulus lanceolatus Vigors

(See Coloured Plate IV facing p. 232.)

BOOK REFERENCES. *F.B.I.*, I, 60 ; *Nidification*, I, 46 ; *Handbook*, 15.

FIELD IDENTIFICATION. (4—) This bird, which is only slightly smaller than a House Crow, cannot be confused with any other Kashmir bird. The black head and loose crest, the white-streaked throat on a black ground and the close blue barring on the wings and tail are characteristic. Silent and found only in pairs in the nesting season, at other times it is a noisy bird going about in small bands in forest or well-wooded areas.

DISTRIBUTION. This jay only just succeeds in finding a place in these pages. We are told in every work that both the Black-throated and Himalayan Jays are common on the outer ranges of the Himalayas throughout their whole length, and in fact the jays for some reason unknown to us seem to have a definite objection to penetrating beyond them. We can trace no authentic record of a jay having been seen in the forests of those many valleys taking off from the northerly rim of the Vale of Kashmir or even on the inner side of the Pir Panjal range, except for one pair seen by Osmaston mobbing a Long-eared Owl in the Lolab Valley, and a bird reputed to have been seen by Ludlow at Achabal on 13 December 1945. We have, however, met with the Black-throated Jay in the lower reaches of the Kishenganga, that is around Salkalla, and in the Jhelum Valley only a few miles short of Baramullah where on 18 May Whistler also saw a pair. It is, of course, a common bird in the Murree hills, so there is no reason why it should not occur in suitable country throughout the length of the so-called cart road from say Garhi to Baramullah.

The Black-throated Jay is a non-migratory species with no races. Its full range includes the Sulaiman hills and the stretch of the Himalayas from Chitral and Gilgit to Nepal.

JAY

HABITS AND NESTING. The requirements of the Black-throated Jay are light forests of mixed type and even well-wooded country bordering them. Its distribution appears to be correlated with the commencement of the pine zone, but this may not be so. At any rate it is certainly to be found in the pine forests, in the mixed forests of pines and deciduous trees, and in the breeding season even in the valley bottoms away from the forests provided there are plenty of trees and shrubs about the area. When mating the noisy parties break up and each bird of a pair then becomes rather secretive and silent in its own particular territory, conversing with its fellow in subdued though harsh tones, occasionally giving tongue to a sharp loud squark not unlike the note of an alarmed woodpecker. When the nest is threatened, however, they become excited and revert to all the clamorous abuse of which they are outstandingly capable. Being arrant egg-thieves and baby-snatchers like most of the larger *Corvidæ*, they come in for considerable mobbing by smaller birds, which may often be seen following them about, marking their disapproval with considerable emotion.

During a trip up the Kishenganga Valley we came upon two or three nests, unfortunately at that time empty, which can be described as typical. They were one and all between 10 and 15 ft from the ground in young trees, one in a small tree in what amounted to a hedgerow a couple of hundred yards from the forest edge. Two others were within mixed forest in the crests of small saplings growing amongst the larger trees. One was very ill-concealed, the others well protected in the foliage at the tops of the saplings. All were about eight inches across, made of light sticks with a grass and root lining. The fine black moss roots are often used and we saw one bird carrying pine needles. This was in April and the birds are said to breed in that month, May and June.

To quote Stuart Baker : ' The eggs are typical Jays' eggs and the range of variation is small. The ground colour is olive brown or olive green. The surface is stippled all over with freckles, generally most minute, less often becoming fairly well defined small blotches of brown which nearly always coalesce at the larger end to form a ring or small cap. Nearly always at this end also are added a few twisted lines of black.' They are rather broad eggs somewhat pointed at the small end and average 28·8 by 22·0 mm. Three to 5 are laid, usually 3 or 4.

LARGER SPOTTED NUTCRACKER

Nucifraga multipunctata Gould
(See Coloured Plate IV facing p. 232.)

BOOK REFERENCES. *F.B.I.*, I, 67 ; *Nidification*, I, 54 ; *Handbook*, 16.

FIELD IDENTIFICATION. (4) A chocolate-brown bird heavily spotted with white, in size about that of a House Crow. The wings are rather short and rounded and in flight the tail appears somewhat graduated due in reality to the increasing amount of white on the ends of the outer tail-feathers, which when spread forms a white semicircle. The bird is a resident in the heavier forests, where its harsh squealings are often to be heard from the concealing tops of the tall pine trees.

DISTRIBUTION. The Spotted Nutcracker is first and foremost an inhabitant of thick pine and fir forests wherever these occur in our area, more particularly those containing the Blue Pine growing between 7,000 ft and 10,000 to 11,000 ft. It may be seen to a lesser extent in the mixed forests and comes out into the open to visit walnut trees and perrottia scrub in search of nuts. In the Kishenganga Valley we saw a bird at 4,500 ft at Keran. This hardy bird is a resident species and probably moves down only a little distance even in the severest winters.

This bird is confined mainly to the North-West Himalayas from Chitral and Gilgit to Kashmir and Bhadarwar, possibly considerably farther east. There are said to be specimens in the British Museum from Kumaon although it does not seem to occur in Garhwal. It is undoubtedly commonest in the pine forests of Kashmir. Stuart Baker states that he has received specimens from the Chumbi Valley in Tibet and Sikkim. A similar jump to the west takes it into the Sufed Koh where Whitehead records it as being found in small numbers. A nutcracker, probably also of the Kashmir race, is not uncommon in Baluchistan in the Fort Sandeman area.

HABITS AND NESTING. The nutcracker is essentially an arboreal species living and nesting in the pine trees from 20 or 30 ft up to the summits of the tallest, amongst the close upper foliage of which its members seem to delight in ensconcing themselves while complaining to the world in general in anything but pleasing tones. Most accounts agree that their calls are harsh ; but in fact they are not calls but rather squeals, somewhat reminiscent of the squealings of little pigs, and a bird or birds will indulge

NUTCRACKER

in these squealings for many minutes at a time. As to whether these complaints emanate in the main from young birds eager for food we are uncertain, but particularly in May and early June these birds are at their noisiest. It is then that one sees so many young birds about and this fact leads us to doubt the accuracy of statements to the effect that the Spotted Nutcracker's breeding season is May, June, and early July and further that they are double brooded.

There is little authentic on record about their nidification. Ward states that they breed from May to July at altitudes varying from 8,000 to 10,000 ft, but in a further note he says, ' The young of this bird were hatched out on 29 April, it is probable that this bird has two clutches during the summer '. Was this note by any chance based on an observation of Buchanan's, who stated that in the Astor District at an elevation of about 10,000 ft on 29 April he saw three young nutcrackers hardly able to fly ? He then added : ' I think, too, it probably has two broods in the year as I found a nest containing young in Sonamarg at the end of July.' Osmaston in his *Notes on the Birds of Kashmir* says eggs are laid chiefly early in May, but as he gives no details of the eggs presumably he took none himself. Between 16 and 21 May he also records seeing a family of fully-fledged young ' probably at least 2 months old '. In the Kazinag range in early June we saw many immature birds, and on 11 June at Nildori recorded seeing two birds in attendance on a young one not long out of the nest. Surely if eggs were obtainable from May to July, the many first-class oologists who have visited Kashmir in those months would have found them and recorded the facts. Added to this, two adults shot in Gilgit in the middle of May by Scully were moulting, the body feathers and the primaries being equally in process of renewal. No, we are convinced that March, possibly even earlier (Buchanan's young of 29 April must have been in the egg in March), and perhaps the beginning of April will prove to be the period in which the Larger Spotted Nutcracker normally deposits its eggs. Double broods are probably exceptional, if indeed the bird is ever double brooded ; Buchanan's July nest was more likely to have been the result of one or even two previously unsuccessful attempts to raise a family. We have visited Kashmir many times from halfway through May and have inspected a number of nests which agree well with the description of the structure said to be made by this bird, but never yet have we had the luck to find one still occupied.

These nests are rather compact stick affairs about 10 in. across.

but lichen seems to figure prominently in their composition, not only clothing many of the sticks but separate bits seem to be pulled off the trees and mixed with them. The lining of the cup, which is comparatively deep, is usually of roots and pine-needles. Those we have seen have been at least 30 ft up and it is probable that many are placed at more than double this height. They are sited in the pines and fir trees, usually close up against the main trunk.

The Nutcracker certainly lives up to its name where its feeding habits are concerned, although it partakes mainly of the seeds of the Blue Pine which, according to Osmaston, it obtains from the ripe cone by hanging inverted from the cone and prising open the scales with its bill, so that the ripe seeds fall into the open mouth. In 'Miscellaneous Notes' in the *J.B.N.H.S.* is an account of a bird being shot with its crop full of hazel-nuts (or rather nuts of the perrottia) which had been swallowed complete with the shell. The writer goes on to say that he later shot some birds with their crops replete with walnuts. In this case the nuts had not only been shelled but carefully divested of their bitter skins. On three consecutive days we surprised a couple of birds in a walnut tree near Gund village in the Sind Valley when the nuts must have been nearly ripe.

The only description we can find of the eggs is that given by Stuart Baker, in his *Nidification*, of 10 eggs from three clutches sent to him from the extreme north of the Chumbi Valley. He writes : ' Two clutches have the ground pale blue . . ., but the marks are all small blots and specks of brown with others, similar in shape and size, underlying them of pale grey.' The third clutch differed in having no blue in the ground colour and is more densely marked. The 10 eggs average 32·8 by 25·4 mm.

EASTERN RED-BILLED CHOUGH

Pyrrhocorax pyrrhocorax himalayanus (Gould)

(See Coloured Plate V facing p. 328.)

KASHMIRI NAMES. *Wān Kavīn, Wozij Tōnti Kavīn.*
BOOK REFERENCES. *F.B.I.*, I, 68 ; *Nidification*, I, 55 ; *Handbook*, 17.
FIELD IDENTIFICATION. (4) The choughs are somewhat slender, all-black crows, in summer only to be met with, usually

in flocks, in the wide open lands near to and above the tree line. In this species the vermilion-red bill is conspicuously long and curved. The legs and feet are also of this bright hue. They alight on the ground or on the rocks, seldom if ever on trees. The Red-billed Chough is less gregarious than the Yellow-billed species, particularly in the nesting season.

DISTRIBUTION. Widely but unevenly distributed throughout the uplands both on the Pir Panjal mountains and on the hills of the main Himalayan range. It is not, it seems, anything like so numerous in our area as the next bird, but this position is reversed once the main barrier is crossed into Ladakh and into the districts bordering it. In January choughs may be seen around Srinagar, but they soon disappear to high levels with any retreat of the snow-line. They are always to be seen around Sonamarg, at the heads of the Sind and Lidar Valleys and on the routes between, but they must move about a good deal, for one observer will record seeing numbers of them in a certain area while another of the same place at a different time will report none or only members of the next species. The clue to this is possibly to be found in their nidification, for, as soon as they commence to nest, the Red-billed Choughs are apt to become solitary or at least the larger flocks split up, whereas the Yellow-billed parties depart to their breeding-grounds *en masse* and nest in colonies often of great size, e.g. the colony at Sekwas, 11,500 ft, between the Sind and the Lidar Valleys.

The Red-billed Chough of the Himalayas is separable from the typical race of Europe, North Africa, and Central Asia by virtue of its slightly larger size. The habitat of this race is in the higher mountains enclosing India from North-East Baluchistan to Chitral and thence along the Himalayas to Southern Tibet and Bhutan. It is a resident bird, leaving its elevated habitat only temporarily under stress of weather.

HABITS AND NESTING. Choughs are gregarious birds feeding and flying about in flocks, often of considerable size. In the breeding season, with which we are mainly concerned here, the Red-billed Chough is apt to be more independent in its habits, single or but a couple of pairs nesting together being perhaps the rule rather than the exception, the flocks again collecting after breeding is over. It is not unusual for them to join the parties of Alpine Choughs, and the combined flocks may be seen feeding in company on the ground rather after the manner of rooks, marching about this way and that with a sharp eye for any particular seed or succulent insect. They are not averse to picking up scraps from a camp site, but are apt to be shy in their approach to it and to wait until it is evacuated. Nor does carrion come amiss. They are extremely wary birds and seem just as much at home on deep snow, and on almost barren plains, as on the grassy upland meadows. Their powers of flight are fine, but they do not appear to take part in those superb mass exhibitions of carefree flying indulged in by the next bird when feeding on winged insects. A. E. Osmaston remarks upon the close similarity of their call to that of the Jackdaw.

In Kashmir the nest, as we have already said, is often placed singly, or a few pairs may nest in proximity to one another, in caves and crevices and on well-sheltered ledges of inaccessible cliffs. We have seen a nest on a ledge in what one might term an open cave in the abrupt river bank of the stream near Rangmarg. It was not more than 15 or 20 ft above the water and an agile climber might have reached it. Although the birds were seen standing by it on two or three occasions, no attempt was made to reach it as it was then the beginning of August. The nest is usually a structure of sticks with a wool lining and in it generally 3, sometimes 4 and rarely 5, eggs are deposited. The ground colour is white, or white tinged with pink or green, and the markings consist of many blotches of umber-brown and brown with secondary markings of grey and lavender-grey. The markings as a rule cover the egg fairly thickly and evenly, at times being more closely packed round the larger end. They average 39·2 by 27·6 mm. There is little to go by as to dates within our area. In Tibet they are early breeders, some beginning in March, though the majority lay in April and May. Ward took eggs at 12,000 ft in the Lidar Valley but does not record the date, and we are inclined to think that the breeding season, so far as we are concerned, most likely dates from May. They are said often to have two broods.

YELLOW-BILLED CHOUGH

Pyrrhocorax graculus forsythi Stoliczka

KASHMIRI NAME. *Wān Kavīn*.
BOOK REFERENCES. *F.B.I.*, I, 70; *Nidification*, I, 56; *Handbook*, 18.
FIELD IDENTIFICATION. (4) Distinguishable in the field from the last bird only by virtue of its much shorter yellow bill, although in point of fact it is also a slightly smaller bird. Takes

a considerable portion of its food on the wing, the dexterous turns and twists of the members of a flock so engaged being most striking.

DISTRIBUTION. Like the last bird the Yellow-billed Chough resides at high elevations throughout our area. It may in summer be met with in parties feeding down to about 8,500 ft, but its main stronghold is much higher, from about 11,000 or 12,000 ft up to at least 16,000 ft. In winter it of course descends considerably lower, in most years being seen in the Vale whenever the snow is low on the surrounding hills.

Previously looked upon as not strictly separable from the typical race resident in Southern Europe and Central Asia, our Indian bird is now accepted as being of a different race, *forsythi*. It is distributed as a resident bird from the Kurram Valley, thence to the Himalayas, Central and South-East Tibet south to Sikkim.

HABITS AND NESTING. In Kashmir it is not the bold scavenger and hanger-on of man that it is in Ladakh, and little can be added here to what we have already said concerning the Red-billed Chough. The habits of the two birds are much the same, but the Yellow-billed Choughs undoubtedly take a high proportion of their insect food on the wing though they appear often to delight in flight purely for flying's sake. It is indeed a delightful sight to watch the evolutions of a flock which, suddenly taking advantage of the air currents in the vicinity of some towering cliff, sweeps upwards to become but a series of black dots, presently to come hurtling earthwards, swerving, rolling, tumbling, and sometimes diving headlong with wings closed almost to the body in abandoned carefree dashes, marking their pleasure with not unmusical calls which can be heard even when the birds are but little specks in the firmament.

The Yellow-billed Chough is the more sociable, and, although to be seen not infrequently singly and in pairs, it carries the habit of flocking even into its nidification, so that when a nesting site is discovered it is usually tenanted by a congregation of considerable size. Such for instance is the site near Sekwas on that well-known route between the Lidar and Sind Valleys which commences at Lidarwat and terminates at Koolan. The cliffs and crags, lying at 11,500 to 12,000 ft on the north side of the trail, provide ledges, crevices, and caves for the nests of many pairs. This is perhaps the site where Ward succeeded in getting at a nest. To quote from the *Nidification*, it was taken ' from a great colony breeding in holes in a high cliff in the Lidar Valley on the 22nd of May. In the latter instance the one nest taken was all that could be got at '.

PLATE 5

Simla Black Tit at nesting hole in grassy slope

White-cheeked Bulbul on nest

PLATE 6

Himalayan Tree-Creeper at nesting site

Brooks' Nuthatch at nesting hole

The only difference in the eggs is that on the whole the ground colour is paler and, according to Stuart Baker, they average a little smaller, that is 39·0 by 28·5 mm. It will be noticed, however, that these measurements are larger than those given for the last bird, which we took from Osmaston's *Notes on the Birds of Kashmir* as being the average size of Kashmir-taken eggs. The breeding season is in general later than in the case of the Red-billed Chough and is said to be May and June.

KASHMIR GREY TIT

Parus major caschmirensis Hartert

(See Plate 4 facing p. 13.)

KASHMIRI NAMES. *Ranga Tsar, Dantiwu.*
BOOK REFERENCES. *F.B.I.*, I, 76; *Nidification*, I, 60; *Handbook*, 19.
FIELD IDENTIFICATION. (2 –) The grey plumage, black head with white cheeks and the broad black line all the way from the breast to the abdomen render this familiar little bird quite unmistakable. Young birds are greenish above and have a tinge of yellow beneath, but the Green-backed Tit with which they might possibly be confused is of much deeper coloration, of heavier build, and is definitely a forest bird.
DISTRIBUTION. The Grey Tit is a comparatively common cheerful little bird spread all over the more open parts of Kashmir as well as throughout the lighter mixed forests. It ascends the side valleys to round about 7,500 ft, possibly somewhat higher. For instance, it is by no means uncommon at Pahlgam and we both photographed a bird which had its nest in the wall of the Ahlan Forest Rest House which is situated just on 7,500 ft. We noted it as common in the Kishenganga Valley all the way from Domel to Sharda, where its numbers rapidly decreased, the last one being seen at Janwai at 7,200 ft. It will, of course, be met with commonly all along the Jhelum Valley road from Kohala bridge to Baramullah. It is common in the Vale throughout the winter, being one of Kashmir's permanent residents. These are probably birds summering higher up the side valleys which have taken the place of emigrants to the Punjab plains.

This race of the Great Tit is more widespread than its name implies as it extends eastwards to Garhwal. The species as a whole covers Europe, North-West Africa, and the greater part of Asia including Japan, and is consequently divided into a great many races forming a European group with green backs and yellow underparts (viz. the Great Tit of the British Isles), and an Asiatic group with grey backs and whitish underparts.

HABITS AND NESTING. This sprightly representative of the Asiatic group is found in gardens and the haunts of man as well as in the open country ; in fact wherever gardens, groves, orchards, and hedgerows are there for it to explore. Although by no means gregarious and more often than not content with its own company and that of its spouse, it will also join the roving mixed bands which move through the lighter forests. On one occasion, while walking down the Lidar Valley in July, we saw some Brooks' Nuthatches, a pair of Tree-Creepers, Cinnamon Sparrows, and a number of Grey Tits feeding together in a noisy party. It is just as restless and energetic as other members of the family and indulges in all sorts of acrobatics in its intense search for grubs and insects, inspecting the undersides of the leaves, peering into crannies, and upending itself to get at an otherwise inaccessible hideout, while intermittently emitting its cheerful call-notes of *wurr-wichee-wich*, *wich-wichee wich*, or a harsh little churr if alarmed by one's presence. Like many of the tits, its innate boldness and lack of fear of mankind is intensified in defence of its nest, in which it will remain till removed forcibly, showing its anger by really alarming hissings and spittings. On one occasion we thought we were digging out an angry snake from a kingfisher's disused nest-hole, only to lay bare a poor little Grey Tit still sitting squarely upon seven fresh eggs, although three-quarters buried by lumps of earth and sand which had been dislodged from the roof of the cavity. After photographing the nest and eggs we carefully roofed in the tunnel and had the satisfaction of seeing the little heroine retake possession without further ado.

The use of kingfishers' old tunnels is a favourite one in those parts of the Vale where the canals run through the orchards and willow plantations, but holes in the sandy banks of the wells, in earthen or stone walls, in the walls of buildings, in crevices and holes in trees where only those are available, are likewise patronized. The site chosen is not as a rule far from the ground, the great majority of nests being only 3 or 4 ft up, but there are of course exceptions to every rule. The nest is a soft pad of various materials which generally fills the base of the cavity, being composed usually of moss with an inner thick lining of fur often mixed with wool and a few feathers.

The number of eggs varies considerably. We have seen as few as 4 ; 6 or 7 are quite usual, while a nest inspected on 4 May contained 8 ; Osmaston records up to 9. This last nest of ours was in a curious position—at the bottom of a considerable vertical tunnel, in the red-brown mossy fungoid growth which forms on the trunks of willows which have been standing in water. The breeding season in Kashmir does not appear to be very extended. It commences early in May and most young birds are disposed of by the beginning of July. We remember having noted many family parties in the lower Sind round about 26 June. The eggs are typical of the genus, white with reddish-brown spots and a few underlying purplish marks which are often collected into a zone round the larger end. They average 17·5 by 13·4 mm.

SIMLA GREEN-BACKED TIT

Parus monticolus monticolus Vigors

(See Plate 4 facing p. 13.)

BOOK REFERENCES. *F.B.I.*, I, 80 ; *Nidification*, I, 62 ; *Handbook*, 21.

FIELD IDENTIFICATION. (2) Differs from the Grey Tit in that the slate of the back is replaced by green and the greyish-white of the underparts by bright yellow. The immature Grey Tit has yellowish underparts but not the green back. Notice should also be taken of the prominent double wingbars. It is of a slightly heavier build.

DISTRIBUTION. The Green-backed Tit is widespread but rather uncommon in the forest areas of Kashmir. It is, we consider, a little more numerous in the extreme western parts, i.e. particularly in the country drained by the Pohru River. It is no doubt common on the outer slopes of the Pir Panjal range and in the foothills of Poonch and Jammu, but that is outside our purview. The Green-backed Tit is a local migrant, moving into the lower foothills in winter, while small numbers reach the Punjab Salt range.

This tit ranges through the outer Himalayas, parts of Burma, and Yunnan to Formosa, the typical race being found as far east as Eastern Nepal. It is resident throughout except for a seasonal vertical movement down to 3,000 or 4,000 ft, a few birds approaching but not entering the Plains.

HABITS AND NESTING. The habits as well as the nesting of the Green-backed Tit are in the main those of the Grey Tit, but there are certain noticeable differences. The Grey Tit is undoubtedly as much at home in the vicinity of man as it is away in the open country and in the mixed forests. In the outer Himalayas the Green-backed Tit, too, is a common inhabitant of the gardens, orchards, and roadside foliage of every hill-station, but in Kashmir it leaves such haunts to the Grey Tits and leads a secluded life in the forests from Valley level to some 9,000 ft. Davidson, writing in the *Ibis* of 1898, states: 'This Tit was decidedly rare and noticed by us only on a few occasions in the neighbourhood of Gund and *then only in the denser forests*' (the italics are ours). In the vicinity of Gulmarg Osmaston saw it on two or three occasions only. We have seen it infrequently, almost always in fairly heavy mixed forest, in most of the small valleys around the Vale of Kashmir but more often around the Lolab and in the forests of the Kazinag foothills.

Although it possesses loud call-notes which most people syllabize as *te-te-tee-it*, it appears in Kashmir to shed some of its outer hill-station boisterousness and becomes of a quieter, more retiring disposition, preferring to feed quietly in the sole company of its mate and often spending considerable periods on the ground, where it methodically searches the fallen leaves, or grubs about in the pine-needles for insects. It seems to be the general consensus of opinion that holes in walls are the sites most favoured by Green-backed Tits for their nests. This is, of course, quite true of the normal station but, alas, walls seldom exist in the areas most favoured by the bird in Kashmir, with the result that they must console themselves with the next best ready-made shelters, that is with crevices and knot-holes of usually quite small trees. The only nest Davidson took was some dozen feet from the ground deep down in a hole in a small tree, and on 29 May it contained six well-grown young. Three nests found by us, likewise at the end of May and likewise all containing young, two of them above the Lolab and one near Pahlgam, were also in small tree-trunks in holes between 4 and 7 ft from the ground. The nests consist of a large pad of wool on a lesser or greater foundation of moss and feathers. From the above, which are the only records we can trace, it would appear to breed in Kashmir as early as it does elsewhere, that is, laying in April or early May. The eggs, generally 6 to 8, but occasionally as few as 4, are like those of the Grey Tit but inclined to be longer ovals and on the whole more richly marked. They average 16·7 by 13·1 mm.

CRESTED BLACK TIT

Lophophanes melanolophus (Vigors)

KASHMIRI NAMES. *Pintsakon, Tājdār Tsar.*
BOOK REFERENCES. *F.B.I.*, I, 83; *Nidification*, I, 65; *Handbook*, 24.

FIELD IDENTIFICATION. (2−) Kashmir boasts of two Black Tits, to differentiate between which presents some difficulty in the field. The present bird is, however, the smaller of the two and of a slighter build. It also possesses a double row of spots on the wing-coverts which unfortunately are not very conspicuous except at short range. It is a fussy little bird at the nest, which ninety-nine times out of a hundred is in some natural hollow in a tree, whereas the Simla Black Tit's is invariably, so far as we can make out, in a hole in the ground.

DISTRIBUTION. The Crested Black Tit is widely distributed throughout the pine and fir forests of Kashmir right up to 10,000 or 11,000 ft, where the pines give way to birches. In the Kishenganga we first met with it at Keran as low as 5,000 ft and have no doubt that it will be found to occur in the Jhelum Valley from below Rampur, where the deodars commence. Both of the black tits are among the regular winter visitors to the main Valley, where they may be seen from November onwards in parties amongst the willows. Osmaston says they are always to be found in winter in the small Blue Pine wood on the Takht. They are, however, very hardy birds and many must remain at considerable elevations even in the depths of winter. Meinertzhagen obtained it at Baltal at 9,700 ft in early April when the ground was completely covered in snow. During the winter of 1932 A. E. Jones found it not uncommon in Rawalpindi.

This is a purely Western Himalayan species extending east only to Kumaon. Its farthest west appears to be the Kurram Valley (Sufed Koh).

HABITS AND NESTING. Although primarily a bird of the heavier forests, it does not confine its activities exclusively to them and is to be found in the lower mixed forests such as one finds near Pahlgam, and even in comparatively open wooded glades. It will not, however, be come across in what may be termed open parkland. It is undoubtedly the commonest tit in Kashmir, and, barring the Grey Tit, the one which most brings itself to notice, for it is a bold as well as an exceedingly energetic and fussy little creature, be it a member of a mixed gang of other tits, warblers, and tree-creepers feeding their way through the forests, or fussing around an intruder at its nest-site. Here again is another factor to help distinguish it from the Simla Black Tit : its call-notes are an oft-repeated *te-tewy te-tewy*, with variations of course, whereas the Simla bird has a much louder monosyllabic plaintive call. The Crested Black Tit also possesses some cheerful short phrases by way of a song.

When not engaged in nesting, the Black Tits like to join the roving bands of warblers and other little birds which are always such a feature of the fir forests. One moment the world is drear, the sound of birds absent and the forest a land of seeming death, a vista of lifelessness between the darkling tree-trunks, the embodiment of gloom and eeriness. Suddenly all is changed : the sun shines once more, for the foliage is alive with the green plumage of willow-wrens ; on the dark trunks the jerky movements of tree-creepers are discernible, and the flashing scarlet or yellow of minivets is proclaimed as they trickle from tree-top to tree-top, while the Black Tits are here, there and everywhere, hanging upside-down on the pine-cones, peering into crevices in the bark and searching the foliage, the undergrowth, and at times even the ground, in their ceaseless quest for insects, all the time adding their cheerful twitterings to the general chorus. In a few minutes the band has passed, leaving one almost wondering whether the stillness of the trees and the all-pervading silence was in truth ever broken.

The nest is placed in a tree or stump inside any natural hollow with a sufficiently small entrance and is generally anywhere from ground level to say ten feet up. On occasion nests may be 30 or 40 ft in the air, but these are exceptions rather than the rule and the majority will be round about 4 to 8 ft from the general ground level. Very occasionally a hole in a wall or bank may also be used. The shape and size of the cavity is a minor consideration, it being at times possible to touch the rim of the nest with the finger while at other times the nest is so far down that the bird has to travel a couple of feet to reach it ; in other words the site and the size of the entrance are the deciding factors and not the capacity of the cavity.

A photograph we possess shows a nest placed in a hole formed by the wood rotting away on either side of a fold in the trunk of a small tree, and was revealed merely by peeling away the bark. This nest was a collection of moss and oddments well lined with a felted pad of wool and hair. The amount of material depends entirely upon the shape and size of the floor of the cavity, but the felted pad is always present. It is the amount of moss which varies and which may at times dwindle to vanishing-point, while Hume, on the other hand, recorded a nest containing nearly half a cubic foot of dry green moss.

The parents exhibit great courage in the defence of their nests and young. Rather than leave them to their fate they will snuggle right down into the furry rim of the nest, at the same time emitting powerful snakelike hisses which, until one knows their real origin, cause a hasty withdrawal of the offending fingers from the cavity. We have even lifted a bird from its nest before it has condescended to take the hint and fly away. For some days we kept a nest under observation by removing a large piece of loose bark, looking at the outraged bundle of irateness on the nest and replacing the bark—with no ill effects upon her domestic economy. The clutch varies considerably, though 5 or 6 eggs is perhaps the most usual number laid. We have seen a full clutch of 4, while as many as 10 are recorded and yet another nest contained 8. The eggs are pure white, spotted with red or reddish-brown, the spots varying considerably both in size and density in eggs of the same clutch. Sometimes they are very small and very numerous, at others larger, but they can never claim to be called anything but small spots, while their usual appellation must be 'specks'. In the main they are densest at the large end where they often form a ring. In shape and size the eggs are rather blunt ovals averaging 15·8 by 11·5 mm., possessing a slight gloss and a comparatively strong shell. We consider May and June to be the breeding months in Kashmir, and doubt that eggs are likely to be found in April though the birds may commence building in that month at the lower limit of their range. Likewise very few nests will be found in even the first half of July which do not contain young.

SIMLA BLACK TIT

Lophophanes rufonuchalis rufonuchalis (Blyth)

(See Plate 5 facing p. 20.)

KASHMIRI NAMES. *Pintsakon, Tājdār Tsar.*
BOOK REFERENCES. *F.B.I.*, I, 85 ; *Nidification*, I, 66 ; *Handbook*, 24.
FIELD IDENTIFICATION. (2−) A slightly larger, more thickset bird than the last, and of a darker hue unrelieved by spots

or lighter markings on the wings. The nest is always at ground level and the danger-call monosyllabic and loud.

DISTRIBUTION, HABITS AND NESTING. Much of what we have written of the preceding bird applies to this species, except that it is spread in two races over a much greater area, that is from Turkestan to Western China, the typical race being found in Turkestan, north-east Baluchistan, eastern Afghanistan and the north-west Himalayas to Garhwal. Its distribution in our area, and general habits, are similar to *melanolophus*, and where one is to be found the other will probably be present. It, too, occasionally finds its way to the Plains, for A. E. Jones obtained it in 1927 at Rawalpindi, and Sang-jani. This bird is perhaps not quite so energetic and voluble, its notes being louder but less fussy and excitable. It is alleged to be rarer in Kashmir than the Crested Black Tit, but with this we are inclined to disagree, although it appears in the Vale in lesser numbers in the winter and is disinclined to drop at that season to such low elevations. As already stated, in the Kishenganga we found it the commoner of the two and we are inclined to the opinion that elsewhere it fairly well holds its own in numbers. Many observers, we are convinced, have recorded this bird as being the Crested Black Tit, thereby earning for the latter the credit of being much the commoner of the two.

In nesting habits there is indeed a difference, for the nest is invariably within some hole or cavity in the ground, often in the least suspected place and consequently overlooked. We have seen a nest within a hole running vertically downwards, so that to enter the parents used to drop head foremost from a twig about a foot directly above it. Others may be in rat-holes running into the faces of steep banks, while a favoured site is in a hole under the spreading roots of a tree. The entrance is often inconspicuous and hard to see, especially if in a grassy bank, but at times we have watched a bird enter a wide tunnel into which it has been possible to insert the whole arm. The nest is the usual collection of moss and fur or wool, but the materials are usually somewhat scanty in quantity. It is frequently some distance from the entrance; seldom less than a foot and often considerably more.

There is really very little on record about its eggs. The general consensus of opinion, however, is that it lays from 4 to 6 eggs which are rather more sparsely marked than those of the Crested Black Tit. The second week of May probably marks the commencement of building. At Bagtor in the Kishenganga Valley we watched a pair prospecting for a nest-hole on 18 May, but we have found small

PLATE 7

Western Variegated Laughing-Thrush at nest

Simla Streaked Laughing-Thrush at nest

PLATE 8

Rufous-backed Shrike

Indian Magpie-Robin, female, leaving nest

young in the nest towards the end of that month and throughout June and early July. Whymper mentions that it is difficult to hit off the right time to obtain fresh eggs as the birds have a habit of carrying in wool long after the eggs have been laid.

WHITE-THROATED TIT

Ægithaliscus niveogularis (Gould)
(See Coloured Plate II facing p. 40.)

BOOK REFERENCES. *F.B.I.*, I, 98 ; *Nidification*, I, 76.

FIELD IDENTIFICATION. (1) A forest tit only sparingly to be met with from about 8,000 ft. Quite different from the commoner Kashmir tits in that its coloration is conspicuously pale. The upper plumage is mainly ashy-grey. The black eye-band is conspicuous and so is a wide brown band separating the white of the throat from the pinkish buff underparts.

DISTRIBUTION. This little tit is rare in Kashmir and indeed uncommon throughout its known range, which extends along the north-west Himalayas from Chitral eastwards to Garhwal and Kumaon. The following is the sum total of its recorded occurrences within our area. Oates mentions having examined specimens obtained at Gulmarg. Osmaston saw one on 20 September at 9,500 ft in the fir forest below Killenmarg. He also obtained specimens from a mixed hunting party near Tragbal on 31 July and recorded seeing others during the subsequent three days. Meinertzhagen noted a pair on 5 April at 8,400 ft near Gund, and another pair on 21 March at 9,200 ft in the hills west of the Wular Lake. Ward records that on 25 August one was obtained at 11,000 ft, and finally he mentions a winter record of three having been obtained at 6,000 ft in some willow trees—in neither case, unfortunately, does he give the locality. The only nest record is that of one found by Captain Livesey on 31 May at about 8,400 ft, also on the ridge west of the Wular Lake about midway between Nagmarg and the point where it is crossed by the Lolab Pass. It would appear that at most this bird is a local migrant, descending somewhat during the winter months owing to stress of weather but probably never being found much below 6,000 ft.

HABITS AND NESTING. Meinertzhagen describes this lovely little tit as ' this exquisite little bird, perhaps the doyen of the group, a gentleman among gentlemen ', and then goes on to say that on both occasions his pairs were not in trees but were feeding on low brambles and withered flower-stalks in forest clearings, frequently uttering a high-pitched, almost Goldcrest note. Our personal knowledge of this bird's habits and nidification is confined to Captain Livesey's nest, as one of the authors was encamped with him on the Nagmarg ridge at the time of its discovery in 1921.

This nest and its site merit a full description. The ridge at this point consists of a narrow grass strip varying from about 50 yards to less than 10 yards in width. On the west, thick deodar forest commences at the crest and slopes away abruptly in an unbroken sweep for more than 3,000 ft to the Lolab Valley below. The eastern slopes, overlooking a corner of the Rampur-Rajpur basin and the Wular Lake, are nothing like so heavily afforested. Opposite the nest, which was in the outermost twigs of a lower branch of a deodar at the very top of the western sweep, the growth consists of only scattered trees, weeds, and bushes, country which strikes us as probably being of the type in which Meinertzhagen saw his birds feeding. The nest was about seven feet from the ground and caught the full force of a cold wind tearing across the ridge to such effect that our efforts to obtain photographs were rendered abortive. Incidentally, large patches of snow were still lying among the trees on the Lolab side. The nest was a pear-shaped lichen-covered contraption with a small opening near the top, and was most reminiscent of the nest of the British Long-tailed Tit, and of much the same size and shape. The parent birds were not particularly shy and visited the nest, which contained large young ones, while we stood a few yards off. Their approach, however, was from the thick deodar forest and not from the more open eastern slopes.

Whymper, in Garhwal, where this bird appears to be less rare, took 4 fresh eggs from one of his nests. These he describes as being white, rather feebly spotted with pink at the larger end. They varied in size from 14·3 by 11·2 mm. to 14·0 by 10·8 mm. He took these eggs on 26 June, but at an elevation between 11,000 ft and 12,000 ft. It is evident that the White-throated Tit is a hardy little bird which in summer is only to be found in the higher forest zone, and in winter but little lower.

BROOKS' NUTHATCH

Sitta cæsia cashmirensis Brooks
(See Plate 6 facing p. 21.)

BOOK REFERENCES. *F.B.I.*, I, 128; *Nidification*, I, 95; *Handbook*, 28.

FIELD IDENTIFICATION. (2) The nuthatches don't perch across the branches but move along them like the woodpeckers. Unlike the members of that family, however, they are equally at home moving head downwards and at any angle. Brooks' Nuthatch is easily separable from *leucopsis* by its darker plumage, the fulvous of the cheeks and throat changing rapidly to deep chestnut on the underparts. The upper surface is slaty-blue and there is a conspicuous black line through the eye to the shoulder. It is a forest bird.

DISTRIBUTION. A bird of the mixed forests and well wooded country mainly between 7,000 and 9,000 ft. We have always found it particularly numerous in the woods bordering the Wular Lake and surrounding the Lolab Valley, but it is well spaced throughout our area in the Pir Panjal mountains, the side valleys of the main range, Gurais, and the Kazinag hills, where we came upon it as low as 6,500 ft. Around Gulmarg Osmaston found it less numerous than the White-cheeked species.

This race of the European nuthatches, which extend in many forms from the British Isles down to north-west Africa, and across Europe and Asia to China, is to be found at least as far west as the Zhob Valley and then in the higher hills to Chitral and eastwards along the Himalayas as far as Kumaon. It is not migratory and is probably resident throughout the year at all but its highest elevations.

HABITS AND NESTING. This, in our opinion, is the commoner of the two nuthatches which breed in our area, and as it by no means sticks to the summits of the trees, as does the White-cheeked Nuthatch, it is the one more frequently noticed. Outside the breeding season these nuthatches are often members of the feeding parties of tits, warblers, tree-creepers and other birds which so suddenly appear upon the scene to lend an air of liveliness and bustle to a dull patch of forest which but a moment previously has seemed devoid of all living creatures, and which a few minutes later lapses into its unbroken silence as the chattering little band passes out of hearing.

In habits, and coloration, Brooks' Nuthatch agrees closely with the English bird, so that it is most often to be seen moving about the trunks and larger branches of trees searching the bark for insects and larvæ. It is exceedingly agile, unlike the woodpeckers and tree-creepers being quite independent of its tail as a means of support. It can, and in fact does, move on a vertical surface up and down, sideways and diagonally, with the body in any position and with the utmost facility. On being disturbed it has the curious habit of freezing, that is of remaining quite motionless, often hanging head downwards in the rather grotesque attitude shown in the illustration, the sharp slightly up-trending bill pushed stiffly outwards.

It possesses a harsh alarm-note which it uses freely when displeased with one's proximity to its nest. It has also a very loud call which can be likened to the bleating of a kid. It cannot be termed a shy bird and will enter the nest-hole quite openly if one remains still and not too close. It is evident, in fact, that it considers its masonry a sufficient protection for its treasures. For it reduces the size of its nest-hole to the diameter of an eight-anna piece with a fine cement of its own manufacture. The hole may be a disused woodpecker's or other cavity in the trunk or on the underside of a sloping branch of a tree, at any height above ground from 7 or 8 ft to 30 or more. The cement work is not as a rule prominent, but at times it may take the form of a short cone. It occasionally extends for some distance around the hole, filling up any little depressions round it and being smoothed off, even as a mason smooths off his mortar when pointing a wall. The eggs, which are laid in April and May and possibly early June, are from 5 to 7 in number, but as many as 8 have at times been taken. They are typical of the nuthatches as a whole, being white, well-marked with spots and small blotches of red. At times they have a distinct gloss. They average 19·7 by 14·4 mm.

WHITE-CHEEKED NUTHATCH

Sitta leucopsis leucopsis Gould
(See Coloured Plate II facing p. 40.)

BOOK REFERENCES. *F.B.I.*, I, 130; *Nidification*, I, 97; *Handbook*, 28.

FIELD IDENTIFICATION. (2) Easily distinguished from Brooks' Nuthatch by the white cheeks and white underparts. Remains almost exclusively in the upper branches of firs and pines. Has a loud call which has often been likened to the bleating of a kid.

DISTRIBUTION. Generally distributed in somewhat small numbers throughout the forests of our area from some 7,500 to 10,000 ft. Its habitat overlaps that of Brooks' Nuthatch and Osmaston found it the commoner of the two in the vicinity of Gulmarg, but on the whole it is a bird of higher levels. Meinertzhagen, however, states that he found it there between 6,500 and 9,800 ft.

In the *Nidification* Stuart Baker gives this bird the peculiar distribution ' Throughout the north-west Himalayas from Afghanistan and Baluchistan to Garhwal '. Ticehurst, however, does not list it amongst the birds of Baluchistan, its farthest west certain records appearing to be those relating to the Kurram Valley and Sufed Koh (Whitehead and Wardlaw-Ramsay).

HABITS AND NESTING. The habits of this nuthatch are in many respects similar to those of Brooks', but there are certain well-defined differences between the two. The latter is found equally in mixed forests and in the pine woods, whereas the White-cheeked Nuthatch confines its activities almost exclusively to conifers. It feeds largely in the upper parts of the trees and seldom comes down into the undergrowth. The bird's presence is often given away by a loud bleating call uttered from the crown of a fir tree, to be followed shortly by its stubby silhouette against the sky as it flits on short rounded wings to the next tree. In nesting it does not follow the usual nuthatch practice of reducing the size of the entrance-hole with plaster, but makes its nest and places its eggs in a natural hollow in a tree, usually at a considerable height from the ground. On three occasions we have noticed nests in long crevices at the summits of gaunt dead trees which had been split by lightning. Another pair was nesting in an old woodpecker boring which had also been split open. Osmaston took two nests about 18 ft from the ground, one in a big silver fir, the other in a hole about 2 in. by 3 in. in a yew. The quantity and type of nesting material vary considerably. In some cavities nothing but a few strips of grass and other oddments are to be found, but more compact nests of moss with a felted lining of fur or hairs have also been recorded.

Eggs may be taken in May and June, some at least being deposited very early in the former month, if not in April, as we have noticed feeding operations in progress at the end of May. The eggs are rather variable in shape with the small end rounded. They are white with a slight gloss and are spotted—which marks often form an ill-defined mottled cap—with small spots and specks of bright reddish-brown and purplish-grey. Usually 6 to 8, but as few as 4, are laid, and they measure 18·1 by 13·8 mm.

LAUGHING-THRUSHES

WESTERN VARIEGATED LAUGHING-THRUSH

Trochalopteron variegatum simile Hume
(See Plate 7 facing p. 28.)

BOOK REFERENCES. *F.B.I.*, I, 174; *Nidification*, I, 135; *Handbook*, 35.

FIELD IDENTIFICATION. (3) On our first encounter with it our shikari likened this bird to the White-cheeked Bulbul so common round his home in the Vale. Later a friend informed us he had seen a jay where we feel sure no jays exist. To begin with, this Laughing-Thrush has no semblance of a crest, but on seeing it in flight the points of resemblance which had impressed themselves upon both these observers at once become evident. The Variegated Laughing-Thrush is of course considerably larger than the bulbul and smaller than a jay, but in flight the tail is spread, showing up boldly the dark rectrices with their white tips. The wing pattern is perhaps more jaylike when the bird is at rest. The contrasting dark and whitish portions of the head are also conspicuous. The general tone of both the upper and lower plumage is a rich ochraceous brown, the lower becoming a bright pale buff posteriorly. The flight is unsteady, sharply reminiscent of that of the Large Grey Babbler of the Plains. Lastly, its loud calls, *weet-a-woo-werr*, are quite unmistakable.

DISTRIBUTION. In so far as we are concerned almost entirely confined to the belt between 7,500 and 11,000 ft on the inner slopes of the Pir Panjal mountains encircling the southern and western sides of the Valley. Here it is a widespread and comparatively common bird. It is perhaps more numerous in the breeding season around 9,000 and 10,000 ft amongst extensive patches of viburnum and other bushes. Osmaston says it is not found in the vicinity of Gulmarg, but is not rare in the birches and rhododendrons on the slopes of Aphawat between 10,000 and 11,000 ft. We have a record of its occurrence at Pahlgam, and Osmaston recorded it at Praslun on 9 August, but it is uncommon on the northern side of the Vale. We came across it in some numbers on the slopes leading up to the Margan Pass into Kishtwar at the head of the Naubug Valley at around 10,000 ft. It is certainly far from uncommon in the nullahs between the Sinthan and Banihal Passes where the Pir Panjal mountains sweep round in a wide curve. A resident bird, moving down to lower levels under stress of weather.

This species consists of a Western and an Eastern typical form, restricted in the *Fauna* to the Western Himalayas as far as Nepal.

The Western race, however, also occurs on the Samana (North-West Frontier Province), and so probably in between. The dividing line between the races is about Chamba.

HABITS AND NESTING. The Variegated Laughing-Thrush is a highly interesting bird, being the bigger and by far the more arresting of the only two members to be found in Kashmir of that large and widely-spread genus *Trochalopteron*. In the rather restricted portions of Kashmir which it inhabits it is not uncommon, but it is an arrant skulker and much prefers to hop and creep through the undergrowth by devious paths than to take to wing. For this reason it is a poor flyer, tottering unsteadily from some thicket to alight quite often short of its objective, whence it finishes the journey with a ratlike run. During these apparently difficult excursions the tail is spread wide, the short wings beat spasmodically, and the flight is interspersed with short sailings during which the bird sways visibly as if finding difficulty in keeping its balance. The loud calls serve to draw attention to its presence in places where its secretive habits might otherwise render it liable to escape notice. In our opinion, in Kashmir it is not a bird of forest interiors but a lover of extensive patches of bushes, particularly the patches of viburnum at the forest edges, around the margs, and in thickets in the valley bottoms. Where adjacent trees occur, it is sometimes to be seen mounting in rapid runs and hops from branch to branch before plunging once more into the undergrowth beneath.

Like most of its family it is a somewhat noisy bird, conversing loudly with its own kith and kin over wide distances, *weet-a-weer* or *weet-a-woo-weer*; or again, when agitated, muttering and squealing in the undergrowth in subdued tones which sound like a nestful of young chicks clamouring for food. Before and after the breeding season these birds collect into small parties when they often call with the same *pte-weer* note of the Streaked Laughing-Thrush, albeit louder and clearer in tone.

The altitudinal range in the breeding season is fairly wide. We have seen it at a little over 7,000 ft and it certainly goes up to 11,000 ft, but most birds are to be met with in the zone between 8,000 and 10,000 ft.

The nest is a large and somewhat untidy cup of coarse grasses thinning down towards the centre where it always contains other materials in the form of stuffing, such as strips of birch bark and leaves, or a little moss and other rubbish below the inner lining of finer grass and rootlets or pine-needles. Two adjacent nests, possibly built by the same architect, contained laminated strips of birch bark, layer upon layer, some measuring as much as 8 in.

by 3 in. In diameter the nests may reach quite 9 in. This structure is usually placed in the viburnum bushes about 3 to 4 ft from the ground, or at higher levels when in the forks of small trees growing amongst the bushes. The two nests mentioned above were well concealed at the extremities of low fir branches. Whistler mentions that the nest is also placed in trees at a considerable height from the ground and that both birds take part in incubation.

The eggs are large and quite handsome, having a pale greenish-blue background with not very numerous but conspicuous irregular blotches and spots of reddish and varying shades of brown and dull purple. They are noticeably long but blunt-ended ovals measuring 27·8 by 21·0 mm. Three is said to be the normal clutch, but we have more frequently noted only 2. Larger numbers are abnormal, Stuart Baker recording that Rattray once took a clutch of 5. They are said to breed throughout May and June. In 1904 Ward, however, took a nest of fresh eggs in Kashmir as late as 8 August, while the owner of a nest found by us at 9,000 ft on the way to the Marbal Pass laid its 3 eggs between 7 and 9 July.

SIMLA STREAKED LAUGHING-THRUSH

Trochalopteron lineatum griseicentior Hartert

(See Plate 7 facing p. 28.)

KASHMIRI NAME. *Sheen-a-pī-pin.*
BOOK REFERENCES. *F.B.I.*, I, 181; *Nidification*, I, 142; *Handbook*, 38.
FIELD IDENTIFICATION. (3—) A warm-brown bird of the undergrowth unrelieved by any other colour until it flies to the next patch, when the pale tips of the feathers of the rounded tail become visible. Only at close range do the black or white shafts of the body feathers become evident. Creeps about the undergrowth in a ratlike fashion, often running unsteadily from one patch to another and giving vent to a plaintive whistle, *pte-weer pte-weer*.
DISTRIBUTION. A common bird in undergrowth and patches of scrub, eschewing only the thickest forest. It is found from Domel to Baramullah, up the side valleys off the main Vale, and on their enclosing slopes to about 8,000 or 8,500 ft. We noted it in the Kishenganga Valley from the river's mouth up to Gurais. Although found around its rim, these thrushes appear to be absent from the greater part of the Vale of Kashmir until they descend on the

PLATE 9

Indian Bluechat at nest. Male on left

PLATE 10

Indian Stonechat, male, at nest

Western Dark-grey Bushchat, male

approach of winter even into the environs of Srinagar. They are, however, to be found breeding the moment the flats and marshes of the Vale are left behind, while a few remain to nest in the thick scrub of the Sind River delta between Ganderbal and Shalabug. They are as numerous in the Wardwan Valley as elsewhere. They do not migrate in the true sense of the word, but have a seasonal vertical movement under stress of weather.

The species is divided into five races which inhabit the hills along the frontiers of India from Baluchistan to Bhutan. The race found in our area extends along the Himalayas from the Murree Hills south-east to Kumaon.

HABITS AND NESTING. The Streaked Laughing-Thrush is a sneaky secretive bird which loves to creep about the thickets and undergrowth and to use its wings only to get as unobtrusively as possible from one patch of cover to another. If it can attain this object by a ratlike run across a path or up a bank, it will do so in preference to flying and may also at times be spotted mounting a tree in a mixture of runs and hops, its tail appearing to get in the way and being flicked first to one side and then to the other. When it thinks itself unobserved it will mount a boulder or creep and hop to a twig above the general level of the surrounding bushes, at times even mounting to the lower branches of the larger trees. Here it will call to the world in general, employing a couple of shrill whistling notes, *pte-weer, pte-weer*, repeated until it finds something better to do or is disturbed, whereupon it dives headlong back to cover. Like all laughing-thrushes it has a number of other and varied noises in its repertoire, one a weak chirp nearly always accompanied by a flick of the tail. It also sports a definite song, loud in tone, but feeble in quality, comprised of but a single bar ending with some reeling notes. Lastly, one must not forget the plaintive *twee-twee-twee* whence it derives its Kashmiri name of *Sheen-a-pī-pin* and which we have always connected with anxiety and alarm.

Although a typical laughing-thrush in its habits and so given to furtive ways, it often disconcerts one by appearing in the least expected places. For instance, we have seen one on a roof-top in the middle of a village, and also at times well away from cover out in an open field. It is, of course, well known as a garden bird in every hill-station in the Himalayas and the very last thing that could be said of it is that it shuns the haunts of man. In fact, except in the heaviest forest, it will be found wherever low cover exists, be it in a garden, in the hedgerows, in a thicket on a piece of waste ground, or in isolated scrub on an open hillside, or

LAUGHING-THRUSHES

yet again in a patch of bushes in an opening in the forest; and from our lowest levels about Dulai or Domel up to some 8,000 ft. Occurrences higher than that are doubtful. We have seen it in Gurais where it is common enough. It is numerous around Pahlgam; up the East Lidar certainly as far as Praslun, but we cannot remember ever having seen or heard it at Tanin. It occurs on the hills around the Lolab, also at about 8,400 ft; in Gulmarg, and in the Ahlan Nullah around the Forest Rest House, but by moving up the Marbal Glen a thousand or so feet we seemed to leave it behind. It is in fact a bird of the lower and medium levels, where its rather bulky nests are to be found in plenty in spring and early summer.

It commences building towards the end of April, concealing its untidy grass nest generally in bushes and brambles, in perrottia scrub, at times at the ends of low fir branches, and even, as Osmaston reiterates, in thick grass on a sloping bank. It is seldom above 3 or 4 ft from the ground and sometimes lower. Although large for the size of the bird, measuring at times 8 or even 10 in. across, it is always tucked well out of sight. We can recall finding only one nest open to view; this was about 3 ft from the ground in a dwarf willow, but it was screened by taller bushes growing closely around the site. The nest on the outside is made of coarse grass, the ends often sticking out untidily and loosely, adding to the general bulk. As the centre is approached, the material is more compactly woven until the deep cup, about $2\frac{1}{2}$ in. in diameter, is reached: this is usually neat and tidy and finished off with thin grass or grass roots. Pine-needles may also be found incorporated in the inner lining, and dead leaves often underneath it.

The eggs, 2 to 4, seldom the latter number, and usually 3, are a velvety unspotted blue varying rather in shape and averaging 25·8 by 18·7 mm. They have a slight sheen. May and June are the most popular months for eggs, but fresh eggs may be taken well on into July and Osmaston took a nest of four eggs at Praslun on 9 August. Some eggs must be laid as early as the third week of April for near Ganderbal we have seen the fledglings leave a nest on 13 May.

This bird, incidentally, is said by local shikaris of the Anchar and Ganderbal areas, on whose statements in this respect it is safe to rely, to be the victim of the Pied Crested Cuckoos which penetrate into the Kashmir Valley in very limited numbers during the latter half of June. We notice that Stuart Baker's 167 eggs of this Cuckoo include 3 from nests of this species.

HIMALAYAN BLACK BULBUL

Microscelis psaroides psaroides (Vigors)
(See Coloured Plate IV facing p. 232.)

KASHMIRI NAMES. *Wān Bulbul, Khruhun Bulbul.*
BOOK REFERENCES. *F.B.I.*, I, 369 ; *Nidification*, I, 338 ; *Handbook*, 66.

FIELD IDENTIFICATION. (2+) A noisy arboreal species, often in small flocks which move in irregular driblets from tree summit to tree summit giving vent to harsh squeaks and squealings by way of continuous conversation. A dark grey bird with a longish slightly forked tail, a tousled crest and a red bill.

DISTRIBUTION. In the early spring Black Bulbuls are common in the Vale, particularly in and around Srinagar where they feed in flocks, principally on the fruit of the Persian lilacs. Later they move into the lower reaches of the side valleys where in late May the flocks break up for breeding purposes. They do not penetrate the valleys deeply, remaining in the well-wooded lower margins up to about 7,000 ft. They are then common in such valleys as the Bandipur Nullah, and up the Sind as far as Koolan or possibly Gagangair : we do not remember ever seeing one as far up the Lidar as Pahlgam. In the Kishenganga they are to be met with from the mouth of the river up to at least Reshna, 13 miles above Keran. In the Jhelum Valley we have only seen them for certain at Domel—where in early May they were present in large numbers preparatory to moving up to the Vale—and at Rampur, but no doubt they are present the whole way from Baramullah downwards. In the winter they drop to lower levels and may even be seen a short distance out from the foothills. Ludlow, however, noted them on 13 December in flocks at Achabal so they do not entirely leave the Vale.

The Black Bulbuls are a widespread genus in the hills of India extending thence to Japan. In India there is only one species divided into four races, the race which we are concerned with ranging from eastern Afghanistan and Chitral along the Himalayas to Bhutan.

HABITS AND NESTING. The Himalayan Black Bulbul is the street arab of the bird world, invading the trees in ragged noisy bands which move boisterously up the road, as it were, soon passing on to annoy the neighbours in the next district. In point of fact, their wheezy squarks and piglike squealings are in some ways not unpleasing, for they so quickly and thoroughly bring a feeling of life to any gloomy patch of forest they deign to traverse. Seldom is it that these birds come anywhere near the ground, for as a general rule they pass along the tree-tops rather in follow-my-leader fashion, a straggler or two suddenly darting after the main party complaining loudly at having been left behind. Even the mid-levels of the foliage are usually left untouched in their restless travelling search for berries and insects. Although largely vegetarian in diet they may sometimes be seen making sallies from an elevated perch after flies and hymenoptera, somewhat after the fashion of a flycatcher.

In spite of their great preference for the tree-tops, when the parties break up for nesting purposes the site chosen for the home is often low down towards the outer end of some horizontal branch. The last nest we found, on 28 May, was in process of construction on a horizontal fork about four feet from the extremity of a walnut branch. This tree was overhanging the path in a bit of park-like country in the Erin Nullah, near the point where the valley begins to narrow down. The nest was not above ten feet from the ground. Davidson also records that the only two nests he and his companion found were about fifteen feet up small trees at Kangan, Sind Valley, where on 20 and 21 June they held respectively 3 fresh eggs and 3 young. Incidentally, Davidson adds, ' It does not seem to come higher up the river [than Gagangair], nor did we find it in any case higher up the hills '. We too must admit to never having heard or seen it in the heavier flanking forests of the valleys, but always at the forest edges and in the well-wooded flats and parklands of the valley bottoms. A typical locality, where it is always to be found, is that delightful stretch of country up the Bandipur Nullah, at and beyond Sonarwain. It must not be taken for granted that the nest is always, or even usually, placed at such low elevations above ground level for it is generally accepted that the nest is often built, as one would expect, at a considerable height up. As Stuart Baker writes (*Nidification*, I, 339), Gammie's nest at 50 ft from the ground, and Marshall's in a bush in a steep bank at the forest edge and only 8 ft up, probably form the two extremes and most nests will be found some 25 to 35 ft up.

In Kashmir the breeding season is probably confined to May and June, but records are too few to be definite on that point. When breeding the pairs are very devoted to one another, the male never being far away from his spouse whether she be seated on the nest or taking time off for feeding purposes, when he will usually be seen trailing after her conversing in not unmusical tones. One call at this time is a comparatively melodious excited *pip-per-tree*,

PLATE II

1. Ticehurst's Willow-Warbler. 2. Long-billed Horned Lark. 3. White-cheeked Nuthatch. 4. Blue-headed Robin (adult male). 5. Scully's Red-rumped Swallow. 6. White-throated Tit. 7. Hodgson's Shortwing (adult male). 8. Indian Short-billed Minivet (adult male). 9. Little Forktail. 10. Himalayan Goldcrest. 11. Wall-Creeper. 12. Hodgson's Yellow-headed Wagtail (adult male)

sometimes uttered on the wing with the body held stiff and slightly bowed as the bird flutters from one tree to the next.

The nest is a typical bulbul structure rather like that of the White-cheeked Bulbul, but neater and not quite so flimsy, nearly always having a foundation of dead leaves and dried grass with walls of woody stalks or fine twigs well plastered with spiders' web. Moss may be used to a greater or lesser extent and the Erin Nullah nest had some pine-needles in the lining.

The number of eggs varies, occasionally only 2 being laid, usually 3 and sometimes 4. The markings are prone to considerable variation, but the average egg has a pale pinky-white ground which is fairly thickly sprinkled with medium-sized spots and small blotches of red-brown and purple-brown with other small blotches of lavender and inky-grey dispersed amongst them. The markings are clear and well defined. In shape they are rather long eggs, measuring 26·2 by 19·1 mm.

WHITE-CHEEKED BULBUL

Molpastes leucogenys leucogenys (Gray)

(See Plate 5 facing p. 20.)

KASHMIRI NAME. *Bil-bi-chūr.*
BOOK REFERENCES. *F.B.I.*, I, 389 ; *Nidification*, I, 364 ; *Handbook*, 71.
FIELD IDENTIFICATION. (2+) The perky cap-and-bells crest curving right over the bill, the white cheeks, dark olive-green upper parts and the yellow patch under the tail render this confiding and cheerful bird with the pleasant tinkling notes quite unmistakable. A common visitor to the houseboats and to picnic parties in the Mogul gardens.
DISTRIBUTION. Common from Kohala to Baramullah. In the Kishenganga Valley at the end of April we saw it only as far up as Salkalla (4,500 ft). As that valley warms up, however, it probably extends its range to at least Keran. Throughout the length and breadth of the Vale of Kashmir it is an exceedingly common bird. Particularly is it numerous in and around the towns and villages, in the gardens and groves of Srinagar, and even in the bazaars. It enters the side valleys for but a short distance, decreasing rapidly once they assume an upward trend. Descending the Sind Valley, for instance, it is invariably met with in the neighbourhood

of Kangan, at which point it can be said that the avifauna of the hills meets that of the Vale. Although so obviously at home with the human race, of which it has little or no fear, the White-cheeked Bulbul will also be found in light forest and secondary growth and amongst the hedgerows and bushes well away from habitations. It is, however, essentially a bird of the bushes rather than the trees. In winter large numbers literally invade the houseboats and houses, in the warmth and security of which they live on the best of terms with their human occupants. A resident bird.

The species extends in a number of races from Iraq to the hills of Assam north of the Brahmaputra, and in the peninsula down to Central India. The form in our area, with its highly developed crest, is purely Himalayan, occurring from about 2,000 ft in the foothills up to, in parts of its range, as high as 9,000 ft.

HABITS AND NESTING. So tame and confiding is this little bird of pleasant form and cheerful ways that it is bound to bring itself to notice within a short time of one's arrival in the Vale. Should you be in a houseboat on the Dal lakes, it will probably not be long before one flutters down from the willows hard by to peep in at the window, or if on the Club terrace, overlooking the wide smooth-flowing river, one tarries awhile to indulge in coffee and biscuits, it is almost a certainty that a bulbul will alight on the edge of the sugar-basin to scatter its contents hither and thither, or flutter about the table in search of crumbs, whilst a picnic in the Nishat or Shalimar Gardens would hardly be complete without a couple of pairs soliciting a share of the provender. It must not be concluded that their whole fare consists of sugar and crumbs, for their diet in the main is a mixed one of berries and insects, such as ants, beetles, and the creeping things amongst the leaves, whilst we have seen them make fluttering attempts to catch elusive butterflies. Further light is also thrown on their food-supply by Sir Walter Lawrence, who writes in his *Valley of Kashmir*: 'The mistletoe (*ahalu*) attaches itself to the walnut and the people make no attempt to remove it. They say it would be cruel to rob the bulbul of its favourite food.' Although so companionable and ornamental about the garden it can be somewhat of a pest, particularly among the vegetables and fruit, where peas, currants or raspberries are apt to be devastated by its depredations. Nevertheless, it is a gay, energetic and fussy bird, which is an adornment to one's garden and a tonic to any bird-lover's spirits. Usually found in pairs, these occasionally band together for some particular purpose into a small party.

They nest in Kashmir mainly in May, some in April and a good number in June, placing their flimsy structures in varied situations from 2 to about 8 or 10 ft from the ground. The majority of outdoor nests are built in bushes; wild roses seem popular in the country, but in the gardens creepers or any convenient plant giving a modicum of cover may be utilized. The lower branches of chenar trees are by no means neglected. Clusters of leaves often spring direct from the larger branches providing not only support but good concealment for nests of this bulbul, Tickell's Thrush, and the Ring Dove. In the Dal lakes approaches we have seen numbers of nests in low pollarded willows growing out of the water and flanking the channels. We once found one on the veranda of the Hyan Forest Rest House, but a quaint though commonly utilized site in the Vale is tucked into the thatch of the roof of a dwelling-house, not only inside the lofts but frequently inside a room which is occupied by the members of the household.

The nest is the usual flimsy structure indulged in by so many of the *Pycnonotidæ*; a thin-walled cup of interwoven woody plant stems and springy bents lined with fine roots and fibres or a little thin dried grass. The nests vary a good deal, however, both in bulk and depth, sometimes being fairly substantial and as deep as or deeper than their width, but often frail to look at, although tougher than they seem. On the whole they are by no means ill-concealed, often being tucked away in the interior of a bush, but at times quite obviously no attention whatsoever has been paid to concealment. With such nests the crows play havoc, and even the shrikes take toll, but the bereaved parents quickly console themselves with renewed efforts to establish a family. The eggs, numbering 3 or 4—though we have seen occasional clutches of 5— are white or pinky-white, handsomely and closely mottled, speckled, or blotched with reddish-browns and underlying flecks of inky purple. They have little in the way of gloss and measure $22 \cdot 8$ by $16 \cdot 7$ mm.

HIMALAYAN TREE-CREEPER

Certhia himalayana limes Meinertzhagen

(See Plate 6 facing p. 21.)

KASHMIRI NAME. *Koel Dider*.
BOOK REFERENCES. *F.B.I.*, I, 430, and footnote to page 431; *Nidification*, I, 412; *Handbook*, 78.
FIELD IDENTIFICATION. (1+) An inconspicuous slender little

TREE-CREEPER

bird with the upper plumage mottled dark brown and fulvous to tone with the bark of the fir trees up which it progresses in little mouselike runs when searching the crevices for spiders and insects with its curved attenuated bill. At close quarters it will be noticed that the tail is cross-rayed with narrow blackish bars.

Hodgson's Tree-Creeper (*Certhia familiaris hodgsoni* Brooks) also occurs in the area and has the same habits. It is not nearly so common as *limes*, from which it can be told by the absence of cross-barring on the tail, a characteristic which cannot be spotted satisfactorily in the field at even the closest range. In endeavours to trace this species we have on occasion shot tree-creepers which we thought lacked the tail bars only to find the victim an undoubted *limes*. According to Meinertzhagen, Hodgson's Tree-Creeper does not occur in the Vale in winter, though he obtained it at 8,000 ft in March, west of the Wular Lake.

DISTRIBUTION. A widely distributed and common bird in the breeding season from about 6,500 ft to the tops of the forests. It does not eschew the mixed forests, but is much more numerous in the heavier pine woods and amongst the firs, where it breeds in the largest numbers at between 8,000 and 9,000 ft. We found it common throughout the entire length of the Kishenganga Valley, coming upon it towards the end of April at Pateka where the chir pines first grow down to the river's edge. In the spring and autumn numbers may be seen working the trees of the willow groves in the Vale. In the winter it comes down lower still, being found even some distance into the plains of the Punjab and north-western United Provinces, but large numbers still remain throughout that season in and around the Vale and in the Jhelum Valley below it.

The species, so far as India is concerned, is divided into three races occurring in Turkestan, Afghanistan, and the hills of the North-West Frontier from Baluchistan to the Himalayas, along which it extends certainly as far as Kumaon. Our race is found westwards of Simla to Afghanistan and Baluchistan.

HABITS AND NESTING. An energetic forest bird which will probably first come to notice through its habit of fluttering down lightly but rapidly from the summit of one tree to land with a slight scratching of claws against the bark low down on the trunk of the next. From here it works its way energetically upwards in short mouselike runs, using its narrow tail as a support after the manner of a woodpecker. Occasionally it slithers sideways to investigate the treasures of some crevice in the bark, but to search a fresh area it will usually fly down a few feet and commence its upward

PLATE 11

Western Dark-grey Bushchat. Male on right

PLATE 12

Western Spotted Forktail on nest

Western Spotted Forktail, female with worn tail-feathers, approaching nest

passage over again. It never works with the body at an angle or head downwards like the nuthatches, nor does it perch on the upper surface of a branch. It is an exceedingly active and sociable little bird, always helping to swell the numbers in those feeding bands of tits, warblers, and other birds which every now and then work their lively way through the trees. As soon as nesting is finished, bands or family parties of tree-creepers will be found feeding in unison, keeping in touch with one another by emitting penetrating little squeaks as they trickle through the forest. They cannot be termed particularly shy birds; in fact they will often appear on a tree close by, but so quick and restless are they in their movements that one seldom has a chance of observing them for more than a few seconds at a time. Quite often, too, they appear to ignore one's presence when entering the nest. On the other hand we have at times, when attempting photography, found a bird most shy of the camera lens.

The nests of these tree-creepers are most interesting. In the first place they usually choose some natural slit in the trunk of a tree or a cavity behind a loose flap of bark. At times a woodpecker boring may be appropriated, but as a general rule the entrance is a narrow slit little wider than the bird. As regards the height of the site from the ground we have come to the conclusion that, provided the cavity suits in other respects, it matters little whether it is 4 or 40 ft from the ground. As can well be imagined, where the bark is peeling away from the wood in large strips, the cavities so formed are apt to be bottomless pits. The tree-creeper ingeniously gets over this by first blocking the hole by dropping in sticks which wedge themselves across the cavity and upon which the nest proper may then be constructed. At times the mass of sticks reaches amazing proportions, depending presumably on where the first ones condescend to lodge. Basil-Edwardes records a nest in which the stick formation was some 18 in. deep. He described the sticks as being deodar twigs mostly about half the thickness of a lead pencil. We have seen many such sticks not only about this width, but also about the length of a pencil, to carry which must have taxed the strength of the birds to the utmost. In Kashmir the main nest usually consists of a pad of wool, straw, bits of bark and other rubbish, with an upper loose layer of feathers in which the 4 to 6 eggs are deposited. Where the hole is shallow, the stick foundation is entirely dispensed with and the nest may dwindle to a few scraps of rotten wood and a feather pad. The bird sits closely and the first intimation of a nest is often when the female flutters out in one's face as the result of a sharp rap on the tree or of pushing in a stick to see the result. We once noticed a bird clinging to the back wall of the cavity while we were measuring its depth. On withdrawal of the stick she retreated once more to the bottom.

Most eggs are deposited, we should say, in the month of May, but fresh ones may be found in the first ten days or so of June. Davidson took many nests with eggs round about Sonamarg in the last week of May and first fortnight of June. This agrees well with our own discoveries, but we once watched building operations between 18 and 23 April, in which only one bird, presumably the female, took part, and have seen nests with young in May. A pair was also seen only building in mid-June. As a general rule the post-breeding bands are going about in large numbers by the beginning of July. Incidentally, the newly-fledged young are somewhat darker than their parents. The eggs have a dull white or whitish-pink ground freely marked with tiny freckles and small blotches of red-brown, the latter often forming a zone round the larger end. They are apt to be rather oval eggs not much compressed at the smaller end and average 15·8 by 12·2 mm.

WALL-CREEPER

Tichodroma muraria (Linnæus)
(See Coloured Plate II facing p. 40.)

KASHMIRI NAME. *Lamba Dider.*
BOOK REFERENCES. *F.B.I.*, I, 441 ; *Nidification*, I, 417 ; *Handbook*, 79.
FIELD IDENTIFICATION. (2+) Moves about the surfaces of boulders and cliffs after the manner of a tree-creeper on the bark of a tree. Over its wide range from Switzerland and Southern Europe to Tibet, it is universally dubbed the Butterfly-bird from its habit of showing off the bright crimson patches and white spots of the wings against its otherwise french-grey plumage while fluttering about the cracks and crevices when unearthing its insect food. Has a narrow, long, and slightly curved bill, rounded wings, and an indirect hoopoe-like flight.
DISTRIBUTION. The Wall-Creeper is a common winter visitor to the Kashmir Valley ; indeed numbers penetrate to the plains of the Punjab and the North-West Frontier Province adjacent to the Himalayas and Frontier hills, and we once watched a bird

working the walls of Jodh Bai's palace at Fatehpur Sikri, 22 miles from Agra. In March and early April they are particularly numerous in the Vale on passage to Ladakh and beyond, but some at least remain on the south side of the range to breed at elevations of 11,000 ft upwards.

The Wall-Creeper is a widespread species with no races, breeding under alpine conditions from Central and Southern Europe to Turkestan, Mongolia, and down to the higher ranges of the North-West Frontier of India, where Whitehead found it breeding in the Sufed Koh. It also nests in and north of the Himalayan range.

HABITS AND NESTING. In winter the Wall-Creeper lives up to its name, as it is to be seen searching the crevices and chinks in the walls of buildings as well as rocks, boulders, and cliff faces ; anything in fact which does not encroach upon the tree-creepers' typical preserves. Osmaston notes that it is not uncommon from November to March on the rocky slopes of the Takht and elsewhere in and around Srinagar. He goes on to write that when the snow came down in January, a Wall-Creeper used to visit his houseboat almost daily to search over the wooden walls and mosquito-proof netting for insects. This was a heaven-sent opportunity to observe this beautiful creature at close quarters, for as a general rule they are somewhat shy or at any rate so restless and energetic that they soon move on with that flickering hoopoe-like flight affording only a fleeting glimpse of their handsome vestments. As soon as spring is in the air the vast majority cross the great Himalayan range to the wastes which lie beyond, but those which remain within our limits betake themselves to the beetling cliffs and precipices above tree-level, penetrating even to the barest stony uplands nearest the permanent snows.

For this reason few of their nests have been noted, though if one went up specially to search for them they might turn out to be not so very uncommon. Be that as it may, the fact remains that the only records we can quote for our area are that Osmaston reported seeing a pair in July at 12,000 ft near the headwaters of the Lidar, and on the 30th of that month we observed a newly-fledged young one being fed at about 11,500 ft under circumstances which left no doubt in our minds that the nest had been near by. A route from the Upper Wardwan Valley crosses the Gulol Gali at about 14,500 ft and drops down into the East Lidar Valley near Shishram Nag. It is a bare rocky route for the most part, which starts from a small open valley at Rangmarg climbing steeply for about the first thousand feet between vertical precipitous cliffs whose towering walls converge to close quarters as one nears the summit of the first shale slope. A Wall-Creeper with its slender bill packed with food flew across some 20 ft above our heads from one face and landed by a small shelf on the opposite wall, where it proceeded to slither up to a young bird whose head and shoulders could be plainly seen just above the line of the shelf. We watched the young one being fed, after which the parent returned to the cliff immediately above the path and commenced searching for a further supply of insects. Either the young one was still incapable of flight or disinclined to move for it made no attempt to follow its parent. Young birds, incidentally, do not possess the black chin and throat which the adults attain in the breeding season and the spots on the wing are rufous instead of white.

The nest is usually tucked well inside some narrow crevice in the cliff face and consequently more often than not is quite inaccessible. It is a pad of moss, wool and feathers, fur and hairs, feathers forming the bulk of the lining. In it 4 to 6 broad pointed eggs are laid, dull white sparsely freckled with dark red-brown, mostly round the larger end. They measure 21·3 by 14·9 mm. Osmaston points out that the nests are often near streams. The usual breeding season in the Himalayas is said to be May and June, possibly earlier.

KASHMIR WREN

Troglodytes troglodytes neglectus Brooks

BOOK REFERENCES. *F.B.I.*, I, 446; *Nidification*, I, 419; *Handbook*, 81.

FIELD IDENTIFICATION. (1+) The Kashmir Wren is a bird of broken ground in the forests from some 8,000 ft upwards. Boulder-strewn hillsides above the tree line also have their quota. With its dark brown close-barred dumpy little body it is very mouselike as it moves about the rocks and fallen trees. It keeps on disappearing into crevices and little caves, popping up again with its short tail stiffly erect, only pausing in its ceaseless activity to pour forth a rapid sweet song, strikingly loud and vibrant for so small a bird.

DISTRIBUTION. Occurs commonly from about 8,000 ft upwards, particularly in the forests wherever fallen trees and patches of rock occur. It is also to be found at times well above the tree-level on boulder-strewn hillsides. In winter it descends somewhat and may then be seen on the lower rocky slopes in and around the

Vale, not uncommonly entering the gardens in Srinagar. Meinertzhagen obtained specimens in the Vale in the second week of March, but by the beginning of April he noted them in the Sind Valley up to 8,100 ft.

Two races of this wren, which is conspecific with the English bird, occur commonly within Indian limits, from the Sufed Koh along the Frontier hills to Kashmir and thence in the Himalayas to Bhutan, the Kashmir Wren meeting the Eastern form about Garhwal.

HABITS AND NESTING. This rotund little wren is just as energetic as its English cousin and has very similar habits. Its cheerful happy song, tumbling forth so rapidly and so spontaneously, will give away its presence just as surely as its restless activity. Forever on the move, it will be seen flitting from log to log, creeping under stones and through cavities and tunnels under the boulders, or sneaking through undergrowth only to appear on the top of some rock, bush or stricken forest giant where it will suddenly burst into loud song, bobbing this way and that with its short tail flicked stiffly upright.

The nest is an exceedingly compact sphere with an inconspicuous entrance-hole. Favourite situations seem to be in the roots of fallen trees or wedged into the roof of a Gujar's unoccupied hut. Often it is well concealed, sometimes pushed so far in between the boulders or into cracks and crevices in fallen logs as to be quite out of sight and even inaccessible. At other times it merely escapes notice through its similarity to its surroundings. The materials used are mostly moss and grass with an internal lining of hair or feathers and an external layer of lichen, moss, or other camouflaging material. What may be termed cock nests, unlined, are frequently met with, and these of course never hold eggs. At times, should it be placed far inside a crevice in the rock or in a deep hole in a tree, the usual domed nest is dispensed with. Of two such nests, one, at Tanin at nearly 11,000 ft, was in a hole in a broken birch stump; the other was at Lidarwat in a hole in a tree a good 15 ft from the ground. Both these nests were merely loose untidy collections of a little grass and some feathers and more or less filled up the backs of their cavities. Davidson also noted this type of nest in Kashmir, for he wrote : ' Others were in holes in banks or dead trees and consisted merely of a few feathers separating the eggs from the rotten wood.'

Up to 5 eggs are laid though as few as 3 only may at times be found. These seem rather large considering the size of the bird and are usually fairly long ovals. The markings, which are seldom numerous and at times wanting, are small red spots generally collected around the large end. They average 17·0 by 12·5 mm. Although Stuart Baker in the *Nidification* says they nest in Kashmir from 6,000 to 10,000 ft, we have never seen the bird below 8,000 ft in the breeding season, while they certainly nest up to at least 12,000 ft. The lowest we have seen them in summer is in the Gagangair gorge below Sonamarg, while we watched a Wren building at 12,500 ft at Sona Sar in the East Lidar Valley on 2 August, but whether this nest was ever intended to hold eggs is another matter. Meinertzhagen, however, took a nestful of 5 newly-hatched young in a building at Burzil Chauki at 11,300 ft on 24 August. We have taken fresh eggs, 3 in a nest at Zojpal at 11,000 ft, in the second week of July, but there is no doubt that the great majority of Wrens lay in June.

KASHMIR WHITE-BREASTED DIPPER

Cinclus cinclus cashmeriensis Gould

KASHMIRI NAME. *Dungal.*
BOOK REFERENCES. *F.B.I.*, II, 2 ; *Nidification*, I, 436 ; *Handbook*, 82.
FIELD IDENTIFICATION. (3) Has the rotund figure and short up-tilted tail of the Brown Dipper, but is easily separated from it by the darker plumage relieved by a pure white breast. Only found at high levels on glacier- and snow-fed streams. Nestlings are grey above speckled with white, and largely white below with the dark brown bases of the feathers showing through here and there.
DISTRIBUTION. The White-breasted Dipper is not common anywhere in our area, being really a trans-Himalayan form which replaces the Brown Dipper on the Tibetan plateau. It occurs, however, sparingly at high elevations on the Pir Panjal mountains and a little more frequently on the southern face of the main range. We have seen one or two pairs in the Kolahoi area, and it is these glacier-fed turbulent streams, and particularly the clearer waters descending rapidly from snow beds and upland catchment areas, which would appear to be the most frequented by it, though it does at times visit the mountain tarns. It is a resident species which, so far as we know, never descends so low as the Vale, even in the most severe weather.

It is, of course, simply a race of the European bird whose forms are very widespread in both Europe and Asia, and its main habitat is the Tibetan plateau and the Inner Himalayas as far east as the Dibang or the Brahmaputra.

HABITS AND NESTING. This bird of the streams, so closely allied to the dipper of the British Isles, is only to be found at high elevations in the treeless glacier-worn valleys where rushing streams of ice-cold water tumble and hiss their way past open banks. Those losing elevation somewhat rapidly with consequent frequent small falls and fast-flowing races are most to their liking, and Osmaston states that they seem to prefer the clearer swiftly-moving torrents. The Rewil stream is a good instance, crossed as one reaches the scored-out Lidar Valley just short of Kolahoi. Here, on three separate visits, we met with a pair of White-breasted Dippers near the stream's junction with the main river. On the other hand they are at times to be encountered on the mountain tarns as the following extract from *The Valley of Kashmir* shows. It reads : ' Once, on reaching the edge of a small tarn lying at the eastern foot of the Kotwal peak and 14,000 ft elevation, I saw the water circling as if a fish had just risen, and while I was watching for another rise, a Dipper emerged and, seeing me, flew off down the little stream which runs from the lake.'

This bird enjoys the same restless habits as the commoner Brown Dipper, bowing jerkily on the spray-covered boulders in midstream, running forward to disappear completely in the frothing water to pop up again after quite an appreciable time like a buoyant cork. But it is unnecessary to repeat what is written about the Brown Dipper, as the remarks thereon apply almost equally to this bird.

The nest, however, has its differences. That of the White-breasted Dipper is more difficult to find for it is often an ill-shapen mass on a boulder, or log if there be any in midstream, which, built into piled-up refuse, looks more like a haphazard collection of rubbish than a fabricated nest. Its lining of moss or leaves filled in with fine grass is, however, always thick, warm, and quite dry. Conspicuously placed domed or flattened oven-shaped affairs of moss and leaves on ledges of small cliffs, and other sites over-looking the water, are at times to be come across, but the nests are seldom easy to spot, or rather one should say easy to recognize as nests. Whymper in a note in the *Journal* credits them with building ' two fairly distinct types of nest ; one kind is placed on the ground among short grass by the water's edge, an oven-shaped nest thatched with grass and with the entrance very low down, looking like a tiny kaffir hut ; the other kind is a round ball (much rounder than any of the *Cinclus asiaticus*' [*tenuirostris*] nests that I have

DIPPERS

seen) as big as a football and placed on a boulder in midstream without any attempt at concealment although sometimes the boulder can be easily got at; it is made of grass and leaves and has the entrance in the middle '.

The eggs are white, rather rounded ovals with a smooth but glossless shell and measure 25·9 by 18·5 mm. Up to 5, usually 3 or 4, are laid. In spite of the high elevations at which this bird is found, nesting is said to terminate usually by the end of June, but Stuart Baker states in the *Nidification* that he had two nests from Sekwas taken at 11,500 ft and 12,000 ft on 16 and 17 July. Unfortunately he does not divulge their contents. He also records that the lowest level in Kashmir from which he has received eggs—on 5 April—is 5,000 ft, and remarks that it is still quite possible that the nests will be found in winter at very low levels. We can trace no records of the White-breasted Dipper ever having been seen in or below the Vale and feel there may be some mistake over this 5,000 ft record. Meinertzhagen relates that even in winter it remains between 8,000 and 12,000 ft, in summer breeding down to the former level. So far, however, we have never met with it in our area below 10,000 ft.

INDIAN BROWN DIPPER

Cinclus pallasii tenuirostris Bonaparte

(See Plate 3 facing p. 12.)

KASHMIRI NAME. *Dungal.*
BOOK REFERENCES. *F.B.I.*, II, 4; *Nidification*, I, 438; *Handbook*, 82.
FIELD IDENTIFICATION. (3) A uniform dark-brown plump bird with short tail and wings which flies up and down the side rivers and torrents close to the surface, alighting on stones in midstream, where it bobs and jerks itself from side to side, often running into the water and even disappearing completely beneath the surface. The squamated plumage of the young bird gives it the appearance of being a speckled grey and white in general coloration.
DISTRIBUTION. A common bird on all rivers and side streams of Kashmir up to at least 10,000 ft, no matter whether they run through forest or open country. It is numerous along the entire course of the Kishenganga River through to Gurais, and along the Jhelum from Kohala to Baramullah. It is absent, however, from the placid waters of the Vale. This hardy little bird is a resident

PLATE 13

Blue-fronted Redstart. Male on right

PLATE 14

Plumbeous Redstart at nest. Male above

species, only moving down from its highest levels when forced by the severity of the weather.

The Brown Dippers, comprising several races, are widespread from Northern and Eastern Asia to Japan and down to the Himalayas. The Indian race is found in Turkestan and Afghanistan, and along the Himalayas to Eastern Assam north of the Brahmaputra.

HABITS AND NESTING. Although first place in numbers undoubtedly goes to the Plumbeous Redstarts, the Brown Dipper must come next on the list of the common birds of the Kashmir rivers and streams. Those watercourses which are reduced to mere trickles alone will lack their Dipper population. For the Dipper loves fast-flowing water with falls and pools in which he can at times immerse himself completely in search of his aquatic insect food. His flight is swift and direct, the short wings working rapidly as he skims the surface, swerving from the direct line with a protesting *tsit-tsit* only to avoid some protruding boulder. Seldom it is that he ever flies over the land, for if it is physically possible he will follow the torrent, and such an offence as cutting his corners he never contemplates. The Dipper seems always in a hurry; he can seldom be still. The moment he alights, which as often as not he does actually in, rather than by, the water, he bobs this way and that and then commences to feed by running into the water, swimming a few feet up against the current only to allow himself to be washed back to his starting-point. He then probably walks straight in again in a quite deliberate manner and disappears from view, bobbing up some seconds later perhaps in the same place, perhaps on another stone on the far side of what would appear to be such a boiling, swiftly moving mass of water that no bird or even fish could possibly withstand its force. In these diving expeditions the wings are used as paddles to assist the bird in accomplishing these difficult feats, and it has also been established that the Dipper can actually walk on the beds of even fast-moving streams. This accounts for the apparent deliberation with which he disappears from view.

The Dipper is not really shy; a little quiet watching will usually suffice to disclose the nest. Two nests out of three are placed in inaccessible situations wedged into crevices in water-logged caverns by the sides of falls or roaring torrents so that to gain access to the nest the bird has to fly or swim through an opening often but a few inches above the general surface. One such nest, which by chance we were able to investigate, was in a small cave roofed in by a fallen log. The nest was fixed into a recess in its rotting underside and was an untidy collection of moss more or less following the walls of the recess but projected upwards at the rear into a sort of incomplete canopy. It could not be called a domed nest by any stretch of the imagination. Fairly often the nest is placed in full view on a ledge of a boulder in midstream or low down on a cliff-face overlooking it. It is then a massive oven-shaped affair of moss with a wide entrance-hole in the side. The moss is usually wet from the dashing spray of the torrent or from drippings from herbs and ferns surrounding it. The lining, which miraculously contrives to remain dry, consists of roots and leaves, and in it from 3 to 5 eggs, usually 4, of a satiny texture and pure white in colour, are laid. They are rather elongated eggs and average about 25·4 by 18·8 mm.

The breeding season is most prolonged, due in part to the bird's wide altitudinal range. Osmaston says it starts in February and certainly at the lower levels we have seen young birds out of the nest early in April. Numbers of birds are still nesting in May and June, while on two separate occasions in July we have watched building operations at Pahlgam, and once photographed a bird still feeding its young in a nest near Lidarwat at 8,500 ft in the first week of August. The highest nest we have seen was one just ready for eggs in early July at an elevation of 10,000 ft on the route to Astanmarg. While taking photographs at a nest in the Wardwan Valley, where all the Dippers encountered had young by the end of May, it was noticed that the smallest of fishes were fed frequently to the young as an item of diet.

INDIAN BLUECHAT

Luscinia brunnea brunnea (Hodgson)

(See Plate 9 facing p. 36.)

BOOK REFERENCES. *F.B.I.*, II, 14 ; *Nidification*, II, 4 ; *Handbook*, 83.

FIELD IDENTIFICATION. (2) A skulker in the bushes which is seldom seen, but whose presence is at once apparent from the call of three sharp notes, each louder than the last, which precedes a short song. The male is dark blue above, chestnut below, and has a white supercilium. The female is a plain olive-brown robinlike bird.

DISTRIBUTION. Occurs alike in close or open forest from about 6,000 to 9,000 ft provided there is a sufficiency of thick low cover.

A common summer visitor to Kashmir, not arriving on its breeding grounds until the middle of May. It winters mainly in the hills of Southern India and Ceylon. In India we are concerned chiefly with this Bluechat though another species of the genus visits Assam in winter. It nests from the Sufed Koh on the Afghan border and throughout the Himalayas to Bhutan ; also in North Yunnan. A smaller race is apparently resident above 5,000 ft in the Chin Hills in Burma.

HABITS AND NESTING. The Bluechat is such an arrant skulker that, although we have seen a number of nests and know the bird's habitat well from its unmistakable notes, we have each only once had a good look at close quarters at both male and female, and that, we should add, was through the peep-hole of the hiding-tent. The male is a beautiful bird, but unfortunately so bashful that he can seldom bring himself to leave the thickest of cover. It is said that in the breeding season he often mounts to the top of a bush to proclaim his little song. We have heard him do so from the bare branch of a pine, but in our experience he utters far more often from the very heart of the bushes he so hates to leave. And as if in an attempt to make sure that the passer-by will pay some attention to his feeble but rather sweet lay, he invariably precedes it with those three shrill notes. The Bluechat's most favoured habitat is undoubtedly amongst the wealth of viburnum bushes which grow in extensive patches in every valley in Kashmir, but, provided there is, too, forest of some kind to provide shadow and gloom along banks and nullahs, there he will be found, no matter of what the brushwood may consist.

This shade would appear essential, for the nest is by no means always amongst the cover he most affects but often nearby in a hole in the bank of a shady ravine, in the buttress roots of a tree or under a projecting stone, but the shade is just as important as the proximity of cover. It seems to be agreed that the nest is always well concealed. This appears to require qualification. Its immediate precincts are not infrequently such that it is plainly visible without any clearing being necessary, but it is placed in such sheltered and gloomy surroundings that its presence is easily passed over. Two instances will perhaps serve to explain what is meant. First, a nest in the roots of a huge overturned pine. It was placed in full view in a little earth-filled depression ; a nest such as a robin might build in England. The tree was at the bottom of a steep slope on the very fringe of the thick forest. Around the fallen tree viburnum grew closely, but where the roots had been torn from the earth, the ground was bare. The moment, therefore, the bushes were parted, an open space was disclosed perhaps 8 ft each way with the nest 4 ft from the ground staring us in the face. The slope faced north; the bushes were tall; the locality beautifully sheltered. The second nest was even more open than this. It was in the roots of a hazel growing on the side of a tiny ravine, but so shut in by the overhanging foliage above and the thick forest around, that half a minute's exposure at no small stop was necessary to take a photograph of the nest. It is, of course, true that many nests will be found well hidden right in amongst the roots of the bushes and in depressions in a bankside shielded by overhanging ferns, weeds, or grass.

The female is as cautious as her mate. One we were photographing at the nest not once sat for five minutes at a stretch. At any suspicious noise, even fifty yards away at the top of the bank behind the nest, she quietly sneaked off to the cover of a bush by the side of our hiding-tent. Once, however, a sitting female did allow us to have a good look at her during a visit late in the evening.

The nest is generally an affair of moss, dead leaves and roots lined with hairs and a few feathers, but in one nest it was noticed that pine-needles had been freely used. Either 3 or 4 eggs are laid, nearly perfect ovals, the so-called small end being very blunt. The texture is silky and the colour a pure hedge-sparrow blue, very even in tone and never marked in any way. They average 20·0 by 14·6 mm. The breeding season is June and July, the latest record we have of a nest with eggs being the 17th of the latter month, but they probably lay after that as the bird is a late migrant. When the young are hatched, the parents become very fussy, but, though they will approach quite close to the intruder, *tack-tacking* the while just like a perturbed stonechat, it is seldom that they will even then venture into view.

HODGSON'S SHORTWING

Hodgsonius phœnicuroides (Gray)
(See Coloured Plate II facing p. 40.)

BOOK REFERENCES. *F.B.I.*, II, 21; *Nidification*, II, 12.
FIELD IDENTIFICATION. (2) A retiring bird which keeps exclusively to thick bushes and scrub jungle and to undergrowth along the forest edges. It is little seen for it dodges about and

skulks restlessly within the cover which it seldom leaves. Only when unobserved is the male in the habit of mounting to a twig transcending the general level of the bush to practise his short but pleasant song. It will then be noticed that he carries his rather long tail over the back in a typical robinlike manner. The male is a handsome bird, largely bright but fairly deep slaty-blue with a white abdomen. The tail-feathers are black with a patch of red on the basal half of all but the central pair. The female is more soberly arrayed, being a fairly warm olive-brown all over.

DISTRIBUTION. Occurs in thick bush and scrub both in the Pir Panjal mountains and along all the valleys of the main range from about 8,000 ft up to at least 10,000 ft. It is not a forest bird and will not be found far within the trees, but it shows a distinct preference for undergrowth immediately bordering the forest edge. We noted it in Gurais and found it common around Suknes in the Wardwan Valley. Shortwings are also found on the Pir Panjal where Osmaston reports them as common in and above Gulmarg up to about 11,000 ft, frequenting the low dense scrub of viburnum, skimmia and juniper. At the other end of the range, we noted them in the Marbal Glen. We have no information about their winter movements, but judging from the remarks in the first edition of the *Fauna*, they may drop into the foothills.

The known breeding range of this species extends from the North-West Frontier, the Kurram Valley, throughout the Himalayas and on to Yunnan. The Yunnan race is that found east of Myitkyina.

HABITS AND NESTING. It is unfortunate but true that many of India's highly colourful birds are most self-effacing and loth to show off the beauty of their plumage. Hodgson's Shortwing is by no means uncommon, but dozens of bird-lovers have surely passed through its haunts on their way to Sonamarg and Baltal, or between Pahlgam and Lidarwat, around Gulmarg, up the Marbal Glen, and along a hundred and one bush-flanked paths without once suspecting its presence. It does, however, possess a triple melancholy call which is quite distinctive, the middle note of the three being the highest, and the last the lowest. This must not be confused with those three sharp notes of the Bluechat with which that bird precedes its short thin song, particularly as both these arrant skulkers often occupy the same areas. This retiring nature cannot be due to lack of courage, for any threat to the nest invariably brings the owners to the fore, following the searcher closely albeit still keeping as far as possible out of his direct vision. While opening up one nest to view, we had the male fussing and chacking within three feet of us. We levelled the reflex in his direction but, try as we might, he would not remain in uninterrupted view of the lens even for a fraction of a second. The female was just as bold and, once ensconsed in the hide, we had no difficulty in obtaining her picture. They are said to be very pugnacious in the event of other birds trespassing upon their territory.

The nest is placed close to the ground or up to about 2 ft from it and is usually tucked well into the roots or lower parts of a bush with rank grass and herbage hiding it, at other times with no concealing surroundings other than the cover of its own bush. Considering the size of the bird the nest is deep and very bulky, having thick untidy walls of coarse grass generally with some dead leaves intermixed and an inner lining of fine grass, sometimes incorporating a few hairs and a feather or two. One nest we saw was a good 8 in. across with a cup about $2\frac{1}{2}$ in. deep, but we would say that the diameter is more often about 6 in. At times 2 eggs only are laid, but 3 is undoubtedly the normal full clutch. Whitehead once took a nest on the North-West Frontier containing 4 eggs, and others with a like number have been taken in Garhwal. The colour of the egg is a deep unspotted green-blue and the texture is smooth and satinlike, the combination being most pleasing. They are inclined to be very long, slightly compressed ovals measuring on an average 22·7 by 16·1 mm.

The birds do not arrive on their breeding grounds from their as yet still imperfectly known winter quarters until well on into May so that most eggs are to be found in June. Davidson took nine or ten nests in that month round about Sonamarg. At Suknes, at 10,000 ft in the Wardwan Valley, on 19 July we saw young ones only recently out of the nest being fed by their parents. This shortwing is said to be a popular dupe of the Asiatic Cuckoo.

NORTHERN INDIAN PIED BUSHCHAT

Saxicola caprata bicolor Sykes
(See Coloured Plate III facing p. 136.)

BOOK REFERENCES. *F.B.I.*, II, 26 ; *Nidification*, II, 19 ; *Handbook*, 85.

FIELD IDENTIFICATION. (2) To be seen in waste land and mixed cultivation perched on a low bush, weed-stalk or even a stone from which it makes sallies to the ground to pick up insects.

Along the Jhelum Valley road a favourite perch is on the telegraph wires, even when these are slung high above the ground. The male when at rest sometimes appears all black except for the white abdomen, although the large white kidney-shaped patches on the wings always show up well in flight. The female is a nondescript brown.

DISTRIBUTION. The Pied Bushchat seems as much a bird of the Plains as of the hills, although in the outer ranges it has extended its sway up to as high as 8,000 ft. It does not, however, seem inclined to push beyond the foothills and in consequence, although it extends throughout the Jhelum Valley from the Plains to the vicinity of Baramullah, which after all is hardly 5,000 ft above sea level, it has not even turned the corner, as it were, to colonize the lower slopes around the main Vale. In point of fact it is common only up to Chenari, but we have seen occasional birds to within five miles of Baramullah. Its penetration of the Kishenganga Valley is yet more feeble, for even towards the end of April we did not see it beyond Pateka (2,700 ft). This Northern race is migratory, the hill birds leaving them for the Plains while numbers of birds which breed in North India find their way in the cold weather to the borders of the Madras Province. Hugh Whistler in the *Handbook* gives the distribution of the species, which consists of many races, as Transcaspia, Afghanistan, Persia, India, Burma, the Philippines and Java, and says that our race, *bicolor*, breeds from the extreme north-west, Baluchistan and Sind, along the outer Himalayas and in the neighbouring plains.

HABITS AND NESTING. The habits of this chat are little different from those of the Indian Stonechat. It affects the same type of bush-dotted stony hillside and cultivated and semi-cultivated country, in which it hunts for its food in like manner by descending from a low perch to seize the creeping insects which catch its eye. It will also make fluttering sallies into the air in attempts to catch a moth or tempting fly. It has a pretty habit of showing off to its mate by launching itself a few feet upwards from its perch, to sink slowly to the ground with wings aspread and white axillaries and rump-feathers conspicuously fluffed out, singing as it falls to posture for a few moments beside its, we hope, duly impressed spouse. Its danger-calls are much the same harsh grating sounds, and it also possesses a sweet but subdued little song.

Its nesting shows certain well-defined characteristics, however, for particularly in the Plains it makes its nest in a hollowed grass tuft on level ground, although holes in walls and cavities in earthy banks are also made use of. Which type of site is most patronized

STONECHAT

between Kohala and Baramullah we are unfortunately unable to state, as Davidson, who took a number of their nests in the more leisurely days when Kashmir had to be reached by tonga, has omitted to say anything about their situations. In these days of faster travel, ornithologists seem always to have been imbued with an urge to reach the Vale as soon as possible and have apparently never bothered to tarry on the way to make the necessary investigation of the avifauna of the Jhelum gorge. The nest itself is generally a pad, more or less fitting the shape of the cavity if in a bank or wall, of bents and grasses and roots, often with a few leaves or other rubbish intermixed, and a neat cup lined with finer grass, hair, fur or wool.

Either the birds commence operations later than one would expect or else they have more than one brood, for at the end of June Davidson's nests mostly contained fresh eggs and we have noticed them carrying nest materials at the beginning of that month. The usual nesting season is somewhat prolonged, as eggs have been taken elsewhere from March to August. It is, of course, likely that well within the hills the latter months are the more popular ones. Stuart Baker in the *Fauna* asserts that both sexes take part in all nesting activities; building, incubation, and, of course, tending the young. We must confess, however, never to having found the male incubating and it is also the view of that astute observer Mr Sálim Ali that the female alone incubates. In colour the eggs, which vary from 3 to 5, have a pale bluish-white or sometimes pale stone background with fairly well pronounced reddish-brown freckles and small blotches, most numerous in a zone round the larger end. They average 17·6 by 13·9 mm.

INDIAN STONECHAT

Saxicola torquata indica (Blyth)

(See Plate 10 facing p. 37.)

KASHMIRI NAME. *Dofa Tiriv.*
BOOK REFERENCES. *F.B.I.*, II, 28; *Nidification*, II, 21; *Handbook*, 87.
FIELD IDENTIFICATION. (2) A common bird at the lower and medium levels on the barer hillsides and on waste land and open cultivation, where it is particularly fond of perching very upright on the tip of a weed or low bush from which it makes

PLATE 15

Blue-headed Robin, female, on nest. Note the pale rim round the eye

White-capped Redstart taking food to young

PLATE 16

Nest and eggs of Black-backed Robin

Nest and eggs of Indian Blue Rock Thrush in quarry face

descents to pick up insects from the ground. The male is largely black above with a conspicuous white collar and white patches on the wings. At close quarters the pale rufous breast shows up well. The female has the same restless habit of flicking the wings and waving the tail, but is a soberly arrayed, nondescript brownish bird with no distinctive coloration. The courtship display of the male is also helpful in identification, for he flings himself into the air to a height of fifty feet or more, singing all the time, and then planes down with wings outspread. During this performance he appears entirely black and white.

DISTRIBUTION. This stonechat is a fairly common bird throughout the Jhelum Valley from Kohala to Baramullah, also along the Kishenganga up to and including Gurais. In the Wardwan Valley we found it in fair numbers between Inshan and Basman, becoming rare between there and Suknes. Around the Vale of Kashmir it is exceedingly numerous the moment the rice-fields cease and the rather bare slopes around the rim of the valley are reached. It occurs commonly up to 8,000 or 8,500 ft, becoming scarce above that elevation though we have met with it at times in the neighbourhood of 10,000 ft, at Baltal for instance. It is to be seen on both sides of the Vale, but we notice that it is not included by Osmaston in his list of the birds of Gulmarg. By the early autumn it has found its way throughout the whole of India except Southern Madras and is a conspicuous bird amongst the crops and on waste land until the following April.

The stonechat is a widely distributed species in Europe, Africa, and Asia, from the British Isles to Japan. The Indian race not only summers in the hills enclosing India from North Baluchistan along the Frontier and thence throughout the Himalayas, but also in a large slice of Western and Northern Asia. Besides a number of winter visiting races there is also another resident race in the terai and swamps of the Indo-Gangetic plain. For fuller information we commend a perusal of the *Handbook*. According to the *Birds of Burma* the Indian race cannot be admitted to the Burma list.

HABITS AND NESTING. The Indian Stonechat is a bird of rather bare stony hillsides and open slopes provided there is a modicum of weeds, stones or little bushes to provide low perches, and of dry cultivation with a certain amount of bush, rank grass or weeds between the fields to provide it with nesting sites and vantage points. It is quite at home in village lanes and hedgerows, but not so much so as is the Dark Grey Bushchat ; nor does it patronize the forest edges to the extent that that bird does. It seldom ascends

STONECHAT

into what one may term real trees, preferring to potter about near ground level, perching upon some twig 2 or 3 ft up whence it pounces upon a luckless beetle or other insect, to return forthwith to the same or another perch. This preference for foraging from a low stance instead of moving about the ground is a very characteristic habit and is the normal method of obtaining food. The male has a pleasant but feeble little song, while both sexes indulge freely in a harsh danger-note, *pee-tack*, *pee-tack*, particularly when they consider the nest is threatened. This grating *tack* can be well simulated by tapping a couple of stones together. It is, of course, from this that the trivial name of stonechat is derived.

They are cautious birds in their approach to the nest; consequently in spite of their fussiness its position is not easily fixed as considerable pains are taken to conceal it under a stone, in a small cavity in a bankside, low in the roots of a little bush or amongst weeds. It is usually a small and neat structure, sometimes of coarse grass, sometimes comprising a quantity of softer materials in the form of moss or leaves, and it is lined with hairs, fur, fine rootlets, and often a few feathers. The nests are usually placed on slopes or in banks, but a few nests may be found on level ground. Four or 5 eggs are laid; on one occasion we saw 6. These have a very frail shell, little or no gloss, and are a delicate pale greenish blue with small not very pronounced freckles of reddish which are usually most numerous in a zone near the larger end. They average 17·3 by 13·5 mm.

Cuckoos not infrequently victimize these little birds. The breeding season is prolonged, so it is probable that many pairs have two broods. Osmaston says they commence to arrive from the Plains about the middle of March, Meinertzhagen noting males about the 6th and the first females on the 24th. They leave in September and October. They do not, however, commence building until the end of April. Large numbers of eggs are laid during May and early June while we have seen eggs in the third week of July at both ends of their altitudinal range.

WESTERN DARK-GREY BUSHCHAT

Rhodophila ferrea ferrea (Gray)
(See Plate 10 facing p. 37 and Plate 11 facing p. 44.)

KASHMIRI NAME. *Dofa Tiriv.*
BOOK REFERENCES. *F.B.I.*, II, 36 ; *Nidification*, II, 29 ; *Handbook*, 89.
FIELD IDENTIFICATION. (2) Although a typical bushchat in most ways, the rather long tail and the head and wing markings give to the male a shrikelike appearance. Prefers a higher vantage-point than the Indian Stonechat and also affects the forest edges. The rasping danger-call, like the winding of a clock, is characteristic. The female, as the plate shows, is an inconspicuous bird with little distinctive about her.
DISTRIBUTION. Rather erratically distributed throughout our area, mainly from about 6,000 to 8,500 ft chiefly in open forest, scrub, and pasture land well provided with bushes and broken ground. It may often be seen perched on the hedges around villages. It is commonest at approximately 7,000 ft. We found it exceptionally common at this elevation in the little forest-encircled valleys lying between the Wular Lake and the Lolab, where it is more common than even the Indian Stonechat. At Domel and in the lowest reaches of the Kishenganga it was noted as common in April. It may not then have been on its breeding ground, but it will probably turn out to be a permanent resident both in the Jhelum and lower Kishenganga valleys. It is equally common on the Pir Panjal slopes, more so perhaps than on the north side of the Vale. In winter it evacuates the higher areas, but few reach beyond the fringes of the Himalayan foothills.

As a species this bird extends from the borders of Afghanistan and Chitral throughout the Himalayas to Assam and Burma and on into Yunnan and China. It is throughout but a local migrant in winter, from higher to lower elevations.
HABITS AND NESTING. The resemblance to a small shrike ends chiefly in its appearance, except that, being an insect-feeder, it has something of the shrikelike manner of hunting from a fixed perch to which it returns after short sallies to the ground to seize a beetle, grasshopper or other luckless creeping thing. In other habits and in nidification it is a typical chat, and what has been said of the Indian Stonechat certainly applies to this bird, although it undoubtedly prefers rather more wooded country and extensive patches of bushes, and often occurs just within the outer fringes

of the mixed forests. It has a harsher alarm-call and is equally loth to give away the position of its nest, but the male becomes more and more fussy the nearer one gets to it, rasping out the hoarse danger-note in quite a frenzied manner. Magrath aptly terms this call a *geezing* noise and likens it to the winding up of a watch. The males have quite a pretty song which they proclaim from a conspicuous vantage-point, such as the top of a hedgerow, or the pinnacle of the highest bush in the vicinity. It is quite canarylike in sweetness and tone but unfortunately somewhat short and unvaried.

They do ascend high into the trees but prefer an existence within 12 ft of the ground. The nest is always on the ground, concealed in a cavity in a bank, in the roots of a bush, or in a depression in amongst the roots of a tree on a sloping hillside, and is usually a lightly-built structure of grass and small bents, and often a little moss and rootlets, lined scantily with thin grass, a few hairs and a feather or two. We have also found pine-needles used for this purpose. The eggs are not unlike those of the Indian Stonechat in appearance, but are slightly larger and their ground colour is rather more blue. The faint markings are more profuse as a rule; in fact, at times they almost cover the egg, giving it a reddish hue, but they are frequently collected in a zone around the large end. The shell is distinctly hard and has not much gloss. The eggs average 17·9 by 14·2 mm.

The breeding season in Kashmir would appear to start towards the beginning of May and continue throughout June. We have seen young in the nest at the end of May and have taken eggs, but not fresh ones, in the first week in July. This chat is certainly cuckolded by the Asiatic Cuckoo and probably by the Himalayan Cuckoo also, although we have no direct proof of the latter assertion.

WESTERN SPOTTED FORKTAIL

Enicurus maculatus maculatus Vigors

(See Plate 12 facing p. 45.)

KASHMIRI NAME. *Shakhel Lōt.*
BOOK REFERENCES. *F.B.I.*, II, 57; *Nidification*, II, 51; *Handbook*, 95.
FIELD IDENTIFICATION. (3) A bird of contrasting black and white plumage of which the upper parts are spotted and barred,

the lower pure white. Even more distinctive features, however, are the long forked tail drooping slightly at the end, the two halves of which are often spread wide apart; and secondly the white legs and feet. The white forehead, the feathers of which are erectile, is also conspicuous. A bird of the shady streams.

DISTRIBUTION. The Spotted Forktail breeds in Kashmir certainly from Valley level up to about 8,000 ft. In the Kishenganga Valley we first met with it early in May in the vicinity of Keran. It occurs and probably also breeds from this level, about 4,500 ft, in the Jhelum Valley; there are plenty of suitable streams about Rampur, and we have seen it on the roadside between Chenari and Uri. It is not uncommon around the edge of the Vale, and becomes increasingly numerous up to about 7,500 or 8,000 ft on all the streams of the side valleys of both the main Himalayan and Pir Panjal ranges. It does not occur in the Vale, where there are no shady streams rippling over stony beds; we have seen it, however, at Manasbal and at Achabal. It is said to be resident, but the birds of the higher levels undoubtedly descend somewhat in winter. We have in fact seen it at that season in the Himalayan foothills as low as 2,000 ft.

The Spotted Forktails, of which there are a number of races, extend from the extreme Western Himalayas to China and Siam. The typical race is spread from the North-West Frontier through Kashmir to Nepal, where it meets the Eastern form.

HABITS AND NESTING. Of all the birds of the Himalayan torrents there is none so dainty as this forktail. In the main it shuns the larger rivers, except as an easy means of passage from one side stream to another, preferring the smaller torrents running through forest and well wooded country. It is in fact probably commonest on those streams with little water whose course follows a boulder-strewn shaded nullah provided with tiny pools, tinkling waterfalls, and shadowed banks with overhanging grass and bushes. One first becomes aware of its presence on hearing its exceedingly shrill call-note, reminiscent of the penetrating alarm-notes of the other torrent birds, yet readily distinguishable by its higher pitch. Drawing alongside the point where it is feeding round the margin of a little pool, one gets the impression of dainty white feet tripping lightly over the mossy stones and a curving tail spread wide and lifted out of harm's way like a sedate Victorian lady holding up her crinoline to avoid the wet. With another high pitched *pe-eet* it launches itself into darting flight, and flits easily up the stream bed to commence feeding again fifty to a hundred yards farther on. Each time it is disturbed, the same performance is repeated, until at last, at the limits of its territory, it unexpectedly veers off into the forest and flies away, gleaming and twisting through the tree-trunks back to the point where it was first disturbed. And all the time it is moving over the stones, the long forked tail waves gently up and down.

Steeply falling mountain streams are not essential to its well-being, for there is at Manasbal a cut conveying water from the Sind River to irrigation channels around and beyond Manasbal village. Although somewhat lacking in the shade of trees, we found a pair of forktails occupying a stretch of this conduit where it rounds the hill of Ahateng. At Sonarwain, too, in the Bandipur nullah, a pair had a nest under the bank of just such another cut behind the camping-ground near the bridge. At Achabal we have seen it at the trout hatcheries.

The nest is a beautiful piece of workmanship; we have never yet found an untidy one. Of course it is made of materials which bed down well, living wet moss and moss roots with the clay intermixed, with a lining of skeleton leaves, but it is so compactly put together and the rim so well rounded off. The birds are loth to give away its position and indeed it is often extremely well hidden, being placed generally in a cavity or on a ledge under an overhanging bank. Where ferns and grasses droop over it, it is quite out of sight and very difficult to locate. As a general rule it is placed directly over the water, although occasionally it may be a few yards away up the bank on a steep nullah-side. We have at times found nests quite conspicuously lodged on a ledge of a boulder or in the roots of a large tree, but even then, being made of living mosses which tone in with adjacent herbage, they are apt to escape notice. The full clutch is 3, but 4 eggs are not uncommonly found. The ground colour of the egg is usually a very pale greenish or stone colour, freckled all over with yellowish and reddish brown. The markings are generally small and evenly distributed, but at times boldly spotted eggs may be found in which the underlying inky-grey marks are easily discernible. They are rather long, have little or no gloss, and measure on the average 24·8 by 17·6 mm.

We have found fresh eggs in both May and June, but numbers of birds must commence laying towards the latter end of April. The sexes are alike, but we have observed a bird feeding its mate on the nest, which may be an indication that the female alone incubates, although writing of the Eastern form, Stuart Baker records that both birds incubate and both assist in the construction of the nest.

LITTLE FORKTAIL

Microcichla scouleri scouleri (Vigors)
(See Coloured Plate II facing p. 40.)

BOOK REFERENCES. *F.B.I.*, II, 65 ; *Nidification*, II, 61 ; *Handbook*, 95.

FIELD IDENTIFICATION. (2) This is the rarer and smaller of the two forktails found in Kashmir. Patterned in black and white like the Spotted Forktail it is not only far smaller but has a short tail. It haunts precipitous shady streams with dashing waterfalls.

DISTRIBUTION. A fairly rare bird, but widely distributed in our area from 4,000 ft up to about 10,000 ft.

This little bird is a Himalayan species found as far East as the Upper Chindwin and North-East Burma, but not, according to the *Birds of Burma*, in the Shan States as is recorded in the *Fauna* and in the *Nidification*.

HABITS AND NESTING. Little Forktails are probably more numerous in Kashmir than at first sight they appear to be, for they are silent, unobtrusive little birds which love small but steeply-falling torrents where waterfalls are many and the action of the rushing water has cut deep down to form shady little ravines and shrouded rocky cuts with dripping fern-bedecked vertical banks. Only in winter do they occasionally forsake the streams for the banks of the larger rivers. They have not the restless energy of their larger cousin, and may frequently be seen standing almost motionless on some slippery rock at the foot of a roaring cascade. We once lay on a bank above Aru, at about 9,000 ft, with a little fellow perhaps 30 ft directly below feeding on the insects in the water splashing round its feet. We dropped pebble after pebble all round it, but it remained quite unperturbed, taking no notice of our presence whatsoever. Magrath in an article in the *Journal* gives an excellent description of the habits. ' Although these Forktails ', he writes, ' are constantly entering the water to bathe, I did not see them do so . . . in search of food. They, however, constantly stand on boulders over which the water flows at a depth of an inch or so, picking up the insects that flow towards them. Often they make dashes under the spray of falling water and sometimes pick up their food out of a foaming rush of water by hovering just above it.'

The nest is an exact replica of that of the Spotted Forktail, being made of living moss and moss roots, lined with skeleton leaves. It is, of course, much smaller, about $3\frac{1}{2}$ or 4 in. across. It is a most difficult nest to find, always matching its surroundings to perfection, and being placed at times very high up on some tiny ledge of a dripping cliff face or even behind a waterfall. Records are consequently few, and for Kashmir we can only unearth one. Ward records finding a nest on 19 June containing hard-set eggs. This must be the nest referred to by Stuart Baker in the *Nidification* as being at between 8,000 and 9,000 ft near Sonamarg. The breeding season in other parts of the Himalayas is apparently April, May, and June, and this is probably also correct for Kashmir. The eggs number up to 3, and there is one record of 4 (Ward's). In colour they are a dull white sparingly marked with reddish, sometimes very faintly and chiefly at the larger end. They measure 20·1 by 15·0 mm.

BLUE-FRONTED REDSTART

Phœnicurus frontalis Vigors

(See Plate 13 facing p. 52.)

BOOK REFERENCES. *F.B.I.*, II, 69; *Nidification*, II, 64; *Handbook*, 97.

FIELD IDENTIFICATION. (2) A typical redstart inhabiting broken country above the tree line. The male has wings and head, breast and upper back, black washed with blue, but the cobalt-blue forehead, whence it derives its name, is very inconspicuous. The remainder of the plumage is a bright chestnut-red except for a black band at the end of the tail. This band readily distinguishes it in the field from the Kashmir Redstart which inhabits the same ground, but is less common. The females of both species are very similar, being earthy brown birds with chestnut tails and a fulvous tinge to the plumage. According to Osmaston this species lacks the characteristic redstart habit of intermittently shivering the tail, but this is a point upon which we think further observation is required.

DISTRIBUTION. In the breeding season these birds are found on the margs and hillsides above general tree-level wherever the ground is sufficiently broken and rocky. Open grassy slopes are not patronized ; there must be a plentiful sprinkling of rocks and boulders, or broken stony ground interspersed with juniper bushes and perhaps a few odd birch trees ; or better still, a combination of both these types of ground. The Blue-fronted Redstart

Western Himalayan Rubythroat at nest. Male above

Kashmir Red-flanked Bush-Robin, female, leaving nest

Central Asian Blackbird approaching nest

is rarer on the Pir Panjal slopes than on the Himalayan range, but may be found on both up to 14,000 ft or even higher. In winter it does not descend to the Plains with the Kashmir Redstart, but contents itself with moving down to the outer Himalayas and foothills, where it becomes a common bird about the hill-stations and particularly amongst cultivation.

The Blue-fronted Redstart nests at high altitudes throughout the entire length of the Himalayas and extends into Southern Tibet, the Northern Shan States, North-East Burma and Western China.

HABITS AND NESTING. The Blue-fronted Redstart is typical of the genus. It may be seen on a rock, low bush, or even the bare branch of a gaunt birch struggling bravely for existence in the neighbourhood of 11,000 or 12,000 ft. The bright tail with its black terminal band readily identifies it as it turns and bobs this way and that, or returns to its perch after a sally to the ground to snap up an insect whose movements have caught its watchful eye. They are shy little birds, at once setting up a mournful *swe-up, swe-up* when an intruder approaches the territory wherein the nest is situated. This is often placed upon the ground, well hidden in a hole under a log, in amongst stones, or in a deep crevice between boulders. At times it is more open to discovery, and we found three nests in one day, two of which were placed in cavities in the ends of fallen trees, and the third, providing the subjects of our illustrations, in the top of a birch stump about 5 ft off the ground. Magrath once found a nest 20 ft up in a shallow hole in a birch in a small grove of those trees growing at an elevation of 11,300 ft; while Whymper actually goes one better by recording a nest 30 ft up a tree.

The nests are a pad of dry moss lying on a foundation of coarse grass and bents with an inner lining of hair or thin roots, and a few feathers. Those we found were not very tidy structures and were lined with fine white rootlets. Usually 3 eggs are laid and occasionally 4. These have a ground colour varying from pale grey to stone with a pinkish tinge, while the markings are profuse and minute of a reddish or pinkish brown colour. The shell is smooth and without gloss and the eggs average 19·4 by 14·75 mm. June is probably the favourite breeding month, but Ward obtained three clutches at 11,000 ft on the 5th and 11th of August. There is no doubt that in spite of the high altitudes at which it lives, this redstart is freely victimized by numbers of cuckoos which ascend beyond the tree line to carry out their parasitic breeding habits. The young Asiatic Cuckoo in the photograph

REDSTARTS

facing p. 173 had Blue-fronted Redstarts for its foster-parents, but it was noticeable that in this area, in which the main bird population consists of this and the Kashmir Redstart, accentors, rubythroats, a few pipits and many Stoliczka's Mountain-Finches, the Small Cuckoo (*Cuculus poliocephalus poliocephalus*) was also much in evidence.

KASHMIR REDSTART

Phœnicurus ochruros phœnicuroides (Horsfield & Moore)

BOOK REFERENCES. *F.B.I.*, II, 76 ; *Nidification*, II, 69 ; *Handbook*, 97.

FIELD IDENTIFICATION. (2) A mainly black bird with orange-chestnut tail and abdomen, seen sparingly above the tree line in broken country. Its most characteristic habit is the manner in which it frequently vibrates or shivers the tail. To be recognized from the preceding species, which occupies the same terrain, by the lack of a black band at the end of the tail, and by the blacker plumage with no blue wash about it.

DISTRIBUTION. This redstart is a summer visitor to the highlands of Kashmir, large numbers wintering in India, where they penetrate as far south as the Godavari. ' Kashmir Redstart ' appears rather a misnomer, for this race has a very wide range and the majority of its members which pass our way go well beyond the central Himalayan range to breed in Ladakh and Western Tibet, leaving comparatively few of their number within our area. As would be expected, therefore, fewer birds are to be found nesting on the Pir Panjal mountains than on the Himalayan range. It will not be found in the breeding season below 10,000 ft. The most accessible of its haunts, where it is to be found in some numbers, is between Burzilkot and Shishram Nag in the East Lidar Valley, and at Astanmarg.

The Black Redstarts are extremely widespread in a number of races breeding at high elevations in Europe, Asia, and parts of Africa. In India the North-West frontiers from Baluchistan to the Himalayas are also to be included in the breeding range. The species winters from the British Isles to South-West China, southwards into India and Burma.

HABITS AND NESTING. During the winter months this representative of the Black Redstarts is a familiar garden bird, hopping about the ground between the shrubs and flower-beds in its search for grubs and insects and clinging to the shady patches as if allergic to the direct sun. It is not fearful of the approach of mankind and allows close observation of the oft-repeated rapid quivering of the tail. Once in its breeding haunts, however, it becomes a shy suspicious creature, setting up a plaintive *pe-up* as soon as its nesting site is approached. Here it is only to be found well above tree-level on broken boulder-strewn ground, where it will be seen perching upon a rock, or other low vantage-point, frequently flickering the tail and bobbing this way and that before dropping to the ground to seize its insect food. 'The male', says Meinertzhagen, ' has rather a pretty short song of six or seven notes followed by a curious husky sound, resembling that made by pouring shot into a bottle.' When the young are hatched, it is apt to become bolder, and on one occasion a pair with newly hatched chicks entered their nest freely when we were seated a few feet distant. Their fear, however, was only held in check so long as we sat still with our backs to the crevice, a foot inside which the nest was placed.

The usual sites chosen by this bird are cavities under stones, inside crevices between rocks, or on ledges of boulders sheltered by overhanging weeds or bushes. The nest is made of a variety of oddments, dry grasses, thin twigs and roots, lined with scraps of grass, short hairs and feathers. The eggs, from 3 to 6, have no markings, and are either very pale blue or almost white. They average 19·9 by 14·3 mm. They are laid in June and July.

WHITE-CAPPED REDSTART

Chaimarrhornis leucocephalus (Vigors)

(See Plate 15 facing p. 60.)

KASHMIRI NAMES. *Chets Tāl, Kumīdi (Kolahoi).*
BOOK REFERENCES. *F.B.I.*, II, 79; *Nidification*, II, 74; *Handbook*, 98.
FIELD IDENTIFICATION. (2) A frequenter of torrents and river beds, most adequately described by its name. The black foreparts relieved by the glistening white cap in conjunction with the fiery back and tail, which is terminated by a black band, render it quite unmistakable.
DISTRIBUTION. From May onwards, when this bird is breeding

throughout a part if not the whole of its range, it may be met with along all the streams with open banks and along the rivers almost as soon as one leaves the Vale, certainly from the 6,000 ft level. We are inclined to think, however, that many of the birds noticed at these lower levels in May and early June move upwards and provide the later breeding birds at the higher elevations. In June from 8,000 ft upwards they frequent every river and torrent up to at least 12,000 ft, except perhaps those running through deep forest which have no open banks. The White-capped Redstart cannot be termed a true migrant within its Indian limits, but this gradual altitudinal movement is most marked, ending practically at the snow-line in summer and little short of the Plains in winter. It is, in fact, an uncommon winter visitor to the Punjab Salt range, and has occurred in February and March in Rawalpindi.

It breeds from the hills of Northern Baluchistan and the frontiers with Afghanistan, along the Himalayas, and thence to Western China. In Burma its winter movements are more extensive for it penetrates to most of the hill systems with the exception of the Pegu Yomas and into Tenasserim.

HABITS AND NESTING. This well-groomed little redstart is never to be found far removed from running water. It may visit rock-strewn ground and cliffs in search of food, but sooner or later it will return to its mountain torrent to flit from stone to stone, dipping and fanning out its black-tipped chestnut tail, then tilting it right over the back, every now and then giving vent to its plaintive far-reaching call-notes. Each pair has its own territory, which appears to be rather extensive, a fact which tends to make the marking down of a nest most difficult. For the nests are always hard to find and the birds themselves give one no assistance. In fact they are remarkably cunning birds. The nest is habitually well out of sight in a cavity under some overhanging bank—riverside paths seem to provide many such sites—in amongst stones or in a deep crevice between rocks, or again well inside the hollow so often to be found at the extremities of large broken-off horizontal branches, or of tree-trunks which have crashed across the torrent. It is usually, but by no means always, close to the water. One found in Kishtwar was in a cavity in the bank of a path running high up on the side of a dry side-nullah, the river being at least 300 yards away; another, at Lidarwat, was in a broken stump in the forest a good 50 yards from the water.

The nest is a compact one of bents, roots, grass, and other bits of odd materials picked up along the streams, with a thick lining of wool and hair. Three, 4, or sometimes 5, eggs are laid, very pale sea-green marked rather profusely with streaks of rufous-brown and underlying blotches of lavender and neutral. They measure 24·6 by 16·8 mm. The limits of the breeding season are somewhat doubtful, but according to Stuart Baker June and early July form the main period. Personally we feel that late May—we have taken eggs in the third week—and June may be favoured by those birds remaining between 8,000 and 9,000 ft, but that many birds go to much higher elevations and probably breed considerably later, right on into August.

PLUMBEOUS REDSTART

Rhyacornis fuliginosa (Vigors)

(See Plate 14 facing p. 53.)

KASHMIRI NAME. *Kola Tiriv.*
BOOK REFERENCES. *F.B.I.*, II, 81 ; *Nidification*, II, 75 ; *Handbook*, 100.
FIELD IDENTIFICATION. (2) These plump little birds, the size of a robin, are never seen away from broken water, be it that of a mountain stream or the thundering cascade of the larger rivers. The sluggish waters of the Jhelum in its passage through the Vale and the glassy surfaces of the canals and jheels have no attraction for them. The male is dark leaden-grey with a short tail of dull chestnut ; the female much lighter in tone having a white tail with a dark brown wedge running into it. Both sexes have a habit of fanning out the tail, at the same time waving it up and down. In the case of the female the resultant manner in which the white intermittently catches the eye has been aptly likened to the scintillations of light off water gently stirring.
DISTRIBUTION. Numerous on every river and torrent in Kashmir. One meets with it at Kohala and it extends up to at least 9,000 ft, but stragglers may be noted considerably higher. They are absent only from the bed of the main Vale, where the waters of the Jhelum and its tributaries run placidly between smooth banks. In the winter they descend to the foothills, while stragglers find their way along the banks of suitable streams for a few miles into the Plains.

The Plumbeous Redstarts extend as a species from the extreme Western Himalayas right across Northern Burma, Central and Southern China to Formosa, and south to Northern Siam. In

REDSTARTS

winter they are common on the hill streams in Burma and Tenasserim, but their status in all but Northern Burma is not yet fully known.

HABITS AND NESTING. It is quite impossible to leave the main Valley in any direction without at once coming in contact with these soberly-arrayed yet striking little birds. We like the name Water-Robin as applied to them in early works ; it suits them. For robinlike they are in plump build, confiding ways, and watchful eye. Be one walking along the banks of a wide river or climbing laboriously over the rocks and boulders of a mountain torrent, the fluttering ascents and trilling little song of the male Water-Robins will catch the eye and ear on every stretch. At times there must be a pair to each 50 yards, and it is seldom that they are spaced farther apart than a quarter of a mile. They are very jealous of this territory, too, no matter what its limits, and the males will chase away viciously trespassers of their own species of either sex. In that they show no chivalry ! They perch indiscriminately on the low boughs of trees near water, on the stones at its edge, or the tops of spray-bedewed boulders protruding from the churned-up waters of the fiercest torrents, picking up an insect here, a bedraggled fly there, and often after the manner of flycatchers they will make little sallies into the air after fleeter game. And all the time the short tails are rhythmically fanned out and depressed, a distinctive and deliberate combined movement.

During the prolonged breeding season the males often indulge in a tinkling wheezy song. This is uttered generally as the bird flutters from a perch at the water's edge, rising a few feet into the air on vibrating wings to sink slowly down on to some stone a few yards out in the torrent. The females seem to take little notice ; they are perhaps too preoccupied, as upon them devolves the whole onus of incubation.

The male at nesting time becomes a most efficient sentinel. He seldom visits his nest and certainly never feeds his spouse upon it, but he remains within warning distance, at once breaking out into a complaining call-note, which carries well over the turbulent waters, the moment he considers that danger threatens. As like as not the female will slip unobtrusively from her treasures in response to his summons, so that on arrival at the water's edge one will but behold a happy pair of redstarts playing about the stream as if they had no cares in the world. Once the eggs are hatched their caution is apt to evaporate, and it is then an easy matter to watch both birds to the nest with beaks full of insects, for the male now takes on his due share.

The situations chosen for the nests are many and varied, but are usually overlooking or close to the water. We have at times come across nests as many as 50 yards in from the banks, but such are definite exceptions to the rule. Perhaps the most favoured sites are those actually overhanging the water's edge, and comprise any well-hidden cavity or depression capable of holding the nest; such, for example, as in the split end of a broken stump, a depression in the bark of an old tree, a cavity in its roots, a small ledge on a mossy fern-bedecked rock, a narrow cleft between two boulders, in the pollarded end of a willow branch, and even in a hole in a sloping grassy bank; in fact, anywhere which provides in the bird's opinion safe disposal and is not more than say 12 to 15 ft from the ground and generally not more than as many yards from the water. As a rule the nest is barely out of sight, the front rim in fact often showing but toning in quite well with its surroundings. On a few occasions we have found what might be termed normal cup-shaped nests in small upright forks, and Whymper also writes of them building nests like flycatchers, 10 or 12 ft up, against the trunks of trees. We once found one in the decaying top of a post by a stile. It was well-nigh impossible not to put a hand in it in getting over the fence. Needless to say it was soon destroyed.

The nest is a small compact structure of moss, roots, and often dead leaves, shaped according to its site, the egg cavity, about 2 in. across and somewhat deep, being lined with fine roots and hairs. The eggs, usually 4 and occasionally 5, although 3 only may constitute a full clutch, are very robinlike, a greenish-white ground blotched and spotted with reddish-brown, at times with well-defined secondary markings of inky-grey. The markings are heaviest about the large end. They measure 19·8 by 14·6 mm. The breeding season is unduly prolonged and points to two broods being raised in the year for we have found fresh eggs in every month from the latter half of April to the end of August.

RUBYTHROAT

WESTERN HIMALAYAN RUBYTHROAT

Calliope pectoralis pectoralis Gould

(See Plate 17 facing p. 68.)

KASHMIRI NAME. *Yāquat Hōt*.
BOOK REFERENCES. *F.B.I.*, II, 92 ; *Nidification*, II, 81 ; *Handbook*, 102.
FIELD IDENTIFICATION. (2) A dark retiring bird which skulks in the weeds and bushes on the hillsides above tree-level, the male at times mounting a boulder or the topmost twig of a straggling patch of juniper to indulge in his not unpleasing lay, cocking the tail in a thoroughly robinlike manner and showing off a blue-black breast-band which encloses the crimson throat. When flushed off the nest, the duller brown female, who lacks these colourful adornments, prefers to run like a vole through the stems rather than take to flight, making it very difficult to catch even a glimpse of her.
DISTRIBUTION. A bird of the uplands throughout our area, where it is to be found wherever bushes and broken ground occur from the limits of the trees to at least 13,000 ft. It is particularly partial to the extensive patches of juniper in which have been found the majority of the nests which have fallen to our lot. In the winter it drops to lower levels, being found in the outer ranges and foothills normally down to about 3,000 ft, and a few may reach the adjacent plains. We have no information as to whether any then remain within our area.

A Western and an Eastern form, *confusa*, breed throughout the Himalayas to North-East Assam. Whitehead found the Western race nesting on the North-West Frontier at the head of the Kurram Valley.
HABITS AND NESTING. Rubythroats are rather shy birds, preferring to keep out of sight by running about between the bushes and rocks, but the beautifully coloured males may often be seen singing on the summit of a bush or moving about the top of a boulder, when the tail is frequently cocked up at an angle like that of a robin. Of this song, Magrath says that it is loud, continuous, and shrill, but compasses some very pleasing notes. That we consider a fair description of its vocal attainments. Rubythroats nest on the ground, generally on a slope, the nest being concealed in a tuft of coarse grass, or in thick weeds, but more frequently than in any other situation it is tucked well into the roots of juniper, usually on the outskirts of a patch where it is not too tall and thick.

The nest is as distinctive as its owner and cannot be confused

PLATE 19

Tickell's Thrush at nest

Grey-headed Thrush

PLATE 20

Nest and eggs of Himalayan Jungle Crow, built on a clump of Himalayan Mistle-Thrush

with that of any other Kashmir bird, being a domed structure of grass with the entrance-hole large and inclined towards the top, rather like a coconut with a good slice taken diagonally off a top quarter. It is loosely constructed of grass only and with no lining other than slightly finer grass. We have never taken any other type of nest. The extent of the dome of course varies with the type of situation, but we cannot help feeling that Buchanan made a faulty observation when he described the first nests he took in Kashmir as ' cup-shaped '. On the other hand Whymper also talks of finding undomed nests with young. So far we have not found a nest as low even as 10,000 ft, but in the *Nidification* Stuart Baker writes as follows : ' In Kashmir, Ward, Buchanan, and others have found them breeding at considerably lower levels [than 12,000 to 13,000 ft], such as near Aphawat at 9,000 ft.' The lower slopes of Aphawat are, however, some 2,000 ft higher than this, so Stuart Baker may possibly have made some mistake in the elevation he gives. The eggs are not unlike large editions of the Indian Stonechat's, being a washed-out blue, often with a slight greenish tinge, but the pale red markings are few and at times practically absent. Three to 4 are laid in June and in July, in the first half of which month we have come across numbers of fresh eggs. They are broad ovals inclined to be rather pointed at the small end and average 21·6 by 15·4 mm.

KASHMIR RED-FLANKED BUSH-ROBIN

Ianthia cyanura pallidiora Stuart Baker

(See Plate 18 facing p. 69.)

BOOK REFERENCES. *F.B.I.*, II, 101 ; *Nidification*, II, 86 ; *Handbook*, 103.

FIELD IDENTIFICATION. (2—) As so many males are found breeding in immature plumage, in which they somewhat resemble the females, we feel that the impression formed on our first meeting with these bush-robins fits the case well. That impression, to quote from a diary written at the time, was of ' a little earthy-coloured fellow with a white throat, dusky breast and abdomen, red flanks, and a large suspicious eye '. The red flanks spring into prominence the moment the wings are raised and there is a definite rufous tinge in the upper parts. In full dress the male is very handsome. His upper plumage is a deep purplish-blue, a band of it extending right across the breast, while a broad supercilium, the rump and upper tail-coverts are bright ultramarine.

DISTRIBUTION. Common over our whole area in the heavier forests, mainly those on the steeper slopes lying between 8,500 and 11,000 ft, and somewhat partial to those containing birch trees, they are also to be found at the higher elevations in much thinner forest where the tree growth is almost exclusively confined to birches. In the winter they are said to drop to lower elevations, 5,000 to 6,000 ft. There would, however, appear to be a more extensive movement, probably to the outer ranges, for Meinertzhagen noted them as abounding in Gulmarg between 6,400 and 8,600 ft in September and as fairly common in the lower Sind Valley in early April.

A widespread species whose three races are found from the Urals and Siberia to China and Japan, its southern nesting limits being in the higher mountains of the North-West Frontier, the Himalayas, and thence to Yunnan, though its status in Burma is uncertain.

HABITS AND NESTING. Stuart Baker in the *Nidification* states that these birds are not at all shy, but that they are so quiet and unobtrusive in their habits during the breeding season that they do not attract attention. The males breed in immature plumage, not more than one out of every three nesting males having the fully adult coloration. True it is that normally they are quiet and self-effacing birds, even the male in his brightest array being liable to pass unnoticed, but personally we should have been inclined to omit those words ' during the breeding season '; for once a pair have a nest to guard they become fussy and suspicious in the last degree. It is impossible to step into the confines of their territory without having one or other of the birds continually in attendance, proclaiming ceaselessly to all and sundry in that rasping danger-call that a bold bad man is in the offing. Not that this behaviour is really an aid to the searcher after nests, for the nest may be well over a hundred yards away in any direction, and so long as the bird thinks that it is under observation it is far too wary to go anywhere near it. In such an area there are invariably many possible sites, but there is no doubt that cavities under fallen logs, those that have been lying about for some time, have a great fascination for them. Other possibilities are in holes under surface roots and at times even in a hole in a plain bank provided its entrance is well screened. One nest we found was plainly visible to the eye, being in a narrow deep slit in a very rotten log off which all the bark had long since fallen. It was screened to a certain extent by a couple of podophyllum plants, but looking down at an angle half the cup was plainly visible. Such a site is quite exceptional and the nest is often placed so far in that it is difficult to reach. A nest found on a steep bank at Tanin was 2 ft from its entrance-hole. We got at it by rolling away the log, thereby laying bare a tortuous rat-hole with the nest in a cavity under the middle of the log. The nearer rim of the nest from which the female with the large inquiring eye is just emerging, was some 5 in. under the projecting surface root. Yet another nest, under a fallen birch, had separate entrance and exit holes. It certainly seems a fact, as pointed out by Davidson, that the steepest of hillsides are the most patronized, but there are of course exceptions to this rule. ·

The nest is a typical robin's nest of dried grass, generally with a few dead leaves and often other matter such as a little moss, and it is lined with finer grass and hair or fur. Osmaston noted that musk-deer hair is a favourite lining material. The eggs number from 3 to 5, the majority of nests holding 4, and are a pale greenish-white; Stuart Baker says a pure chilly white rarely faintly tinged with pink. They have inconspicuous freckles of pinkish-red, often in a zone at the larger end. A few eggs lack markings of any kind. They average 17·8 by 13·5 mm. Eggs are laid throughout May, the great majority being hatched by the end of the first week of June, though a few eggs may still be found in the second week.

The Red-flanked Bush-Robin possesses little in the way of a song, merely a three-noted call in which the middle note is a tone lower than the other two. It is not so much a ground bird as one of the lower foliage of the trees, but when on *terra firma* it hops about in the approved robinlike manner, often cocking up the tail in the air.

BLUE-HEADED ROBIN

Adelura cœruleocephala (Vigors)

(See Coloured Plate II facing p. 40, Plate 15 facing p. 60 and Plate 16 facing p. 61.)

BOOK REFERENCES. *F.B.I.*, II, 104; *Nidification*, II, 89.
FIELD IDENTIFICATION. (2) The male is adequately represented in the coloured plate. The female is almost wholly brown with a ferruginous tinge to much of the plumage, particularly the upper tail-coverts, but she has one distinctive feature which shows up well in the photograph, namely the ring of pale feathers

around the eye. Haunts bushes, scrub, and broken rocky ground at medium to high altitudes and is rather a skulker. Nowhere common.

DISTRIBUTION. There are few records for Kashmir, but, according to Osmaston, it occurs sparingly on steep rocky hillsides on the Himalayan range between 9,000 and 11,000 ft. Magrath caught a young female, which he thought must have dropped down the hillside from above, when he camped—in June it appears to have been—at 7,000 ft in the Surphrar Nullah, Sind Valley. Ward obtained specimens in April, but these are marked Kashmir, giving no actual locality. We saw a male amongst some rocks on the summit of the Margan Pass, and took the nest which provided the photograph of the female on 19 July at Suknes in the Wardwan Valley.

Stuart Baker gives the range of this species as Turkestan to Afghanistan and Baluchistan, and the Himalayas to Sikkim and Bhutan. We cannot trace Baluchistan records and it occurs in Eastern Afghanistan only, but Whitehead found it breeding on the Sufed Koh between 7,500 and 12,000 ft, where it is not uncommon. In Garhwal it is recorded as nesting up to 14,000 ft, but there is only one record from Bhutan and it appears to be rare in Sikkim. It is said to drop to lower levels in winter, and Whistler records it at that season at Banjar and along the Beas Valley at from 4,000 to 7,000 ft, and Jones says it arrives in Simla (7,000 ft) in late October, returning to higher elevations in March. On the other hand Stevens obtained his only Sikkim specimen at 11,500 ft on 19 February.

HABITS AND NESTING. Accounts of this striking little bird's habits, which have been variously likened to those of a chat, a redstart, and a robin, are somewhat conflicting. Osmaston says that it avoids forest and prefers open rocky country. The bird seen on the treeless summit of the Margan Pass, in June incidentally, was amongst some boulders not far from those extensive patches of dwarf rhododendrons which ascend from the southern flank of the Pass for some thousands of feet. Stuart Baker remarks in the *Fauna* that it certainly does feed on the ground sometimes, but it also haunts bushes and scrub and may not seldom be found hunting about for insects in the higher branches of the trees. Our nest near Suknes was in a short nullah lying at about 10,000 ft which had a few large trees on one flank and bushes on the other side and in the bottom. It was very well concealed some inches from the ground in the roots of a leafy bush and we never saw the birds away from its immediate vicinity. They were definite skulkers and loth to come out into full view, at any rate near the nest. Brooks, however, as long ago as 1875 likened them in manner and notes to the Red-flanked Bush-Robin, adding that the differences are so slight that they might very well stand in that genus. Their ways around the nest struck us as being akin to those of a shortwing—for which we at first mistook the male when we caught a glimpse of him through the bushes.

The nest was bulky and composed largely of coarse grass. We unfortunately omitted to record the lining, but it appears that most nests contain a considerable quantity of dry moss and roots, and are lined with some hairs and often a few feathers. It is always well hidden and the birds are very secretive about its position. Besides being concealed in bushes, it may be in cavities amongst the roots of trees, in a bank, or under some stone. Nests are usually on or near the ground, but Buchanan took one near Murree in a hole in a tree 5 ft up. As Osmaston suggests, the birds may breed on steep open hillsides, but Stuart Baker lays down that they habitually breed both in forest and in scrub-jungle. This is supported by Scully in his ' Contribution to the Ornithology of Gilgit ' (*Stray Feathers*, Vol. X), who states that it is common there in the forests from the third week in March to the end of September. We presume Stuart Baker means open forest with plenty of undergrowth. The cock has a pleasant song of fair quality. The female sits very close and deposits her eggs in late May and June, though the 2 eggs in our Suknes nest hatched out on or about 24 July. Up to 4 eggs are laid, which vary in tone from pale grey-green to dull creamy-buff, being microscopically speckled with faint light red. Occasionally the specks are bolder. The shell has a slight gloss and the eggs are broad somewhat pointed ovals. They average 19·3 by 14·4 mm.

INDIAN MAGPIE-ROBIN

Copsychus saularis saularis (Linnæus)

(See Plate 8 facing p. 29.)

BOOK REFERENCES. *F.B.I.*, II, 113 ; *Nidification*, II, 97 ; *Handbook*, 106.

FIELD IDENTIFICATION. (2+) A spritely denizen of groves and gardens, clothed in deep blue-black and white, which hops about the open ground and lawns or in the shade of the trees and plants. It holds itself very erect with the tail cocked over the back and

MAGPIE-ROBIN

presently mounts to a neighbouring bough or wall to pour forth its vibrant cheerful song. In the female, the subject of the illustration, the deep black of the breast and sides of the head is replaced by a more sober grey, the upper parts being a grey-brown. The central tail-feathers are black, the remainder pure white.

DISTRIBUTION. The Magpie-Robin occurs in any numbers only between Kohala and Garhi. Thence it becomes rapidly scarcer as Baramullah is approached, in which place we have so far seen it but once. There are no other records from the Vale of Kashmir, though in 1895 Davidson saw a pair on 20 June at Kangan in the Sind Valley.

The Magpie-Robin extends through a great part of the Oriental region from India and Ceylon to China and Malaya, the typical race occupying all India except Mysore and the Madras Province, and all Burma except the south of Tenasserim. It is rare or absent in many of the drier parts of North-West India.

HABITS AND NESTING. A great many travellers on their way to Kashmir make their first halt at one of the pleasant dak bungalows in Domel or Garhi. The shady compounds with their patches of deep shadow and pleasant sunlight, outbuildings and grassy lawns, are typical of the Magpie-Robin's favourite habitat, though it will, too, be found in plantations, groves, and shady hedgerows at a distance from the dwellings of mankind. It will not, however, be found in anything approaching heavy forest. It is, perhaps, more a bird of the plains than of the hills. Although it is numerous in the hill-stations of continental India and ascends the outer foothills of the Himalayas to some 6,000 ft, it has never penetrated far towards the inner ranges. For this reason, although not uncommon around Domel and Muzaffarabad, it has not assailed the gorges of the Kishenganga at all and is only seen with any certainty along the Jhelum Valley road as far as Uri. In addition to the one record for Baramullah, we have seen the bird at times between Rampur and Baramullah, but nearer to the former than to the latter place. The loud and really sweet song of the male is acquired in the breeding season and it is rather amusing to hear the faint and squeaky yet enthusiastic efforts he pours forth in the early spring. When he comes into full voice, there are few more pleasant sounds to listen to as one sips early-morning tea, for the Magpie-Robin is an early riser and delights to broadcast to the world in general his views on the splendours of the Indian dawn.

Being a familiar garden bird it makes the fullest use of the sites provided for it by man, tucking its nest into holes and niches in walls, in gables and on sheltered ledges, but it does not eschew a natural hollow in a tree, which kind of site is, of course, the normal one chosen by birds nesting in the country. The site is seldom more than a few feet from the ground, though in places we have seen nests extremely high up, one for instance in a South Indian hill-station in the summit of a very tall palm. The female alone appears to incubate, the male being a hard taskmaster who drives her back to the nest whenever she leaves it for more than a brief interval. The nest is not particularly tidy or symmetrical, being a pad, shaped according to its site, of dead grasses and roots, fibres and other oddments to be found lying about the ground, the materials becoming finer as the cup is reached, sometimes incorporating hairs and some feathers. The 4, or sometimes 5, handsome eggs have a ground of blue-green, rather variable in tone, well covered with blobs and blotches of various shades of reddish-brown. They are broad ovals averaging 21·9 by 17·1 mm. The normal nesting season for India as a whole is from the end of March right up to the end of July, but in our area March, at any rate, should be cut out.

CENTRAL ASIAN BLACKBIRD

Turdus merula maximus (Seebohm)

(See Plate 18 facing p. 69.)

BOOK REFERENCES. *F.B.I.*, II, 123; *Nidification*, II, 108.

FIELD IDENTIFICATION. (3+) Only to be seen moving about open rocky ground at high elevations above tree-level. The male is not unlike the typical English blackbird, but often a duller black. The female has a grey tinge about her, and if seen at sufficiently close quarters, distinct mottling of the lower plumage.

DISTRIBUTION. Generally, though not commonly, distributed throughout our area above the normal tree limit on both the main Himalayan and Pir Panjal ranges. In the breeding season it will not be met with below 11,000 ft. This large race of the Common Blackbird breeds at high elevations from the Kurram Valley and in the Himalayas eastwards to at least Bhutan.

HABITS AND NESTING. Although there is not much on record about this thrush, it is more numerous than is generally supposed. One seldom passes the tree line without running into these birds. Wild open hillsides with a sprinkling of boulders and bushy patches are most to their liking, where they can feed on the grassy slopes

BLACKBIRD

and hop about the rocks or occasionally fly up to perch on the bare branch of some isolated dead tree if any such be present.

We took no note of the song and found the birds, in fact, quiet and retiring, but not excessively shy, though not allowing of a very close approach. Buchanan, however, describes the song for us as ' having little of the flutelike tone one associates with the song of the Common Blackbird. Although containing some pleasing notes, the song of *M. maxima* is largely composed of wheezy drongo-like utterances with an occasional loud whistle not unlike that used by Kashmiri shepherds when herding their flocks.' When disturbed suddenly the shrill cackling alarm-notes are similar to those of their English cousin, but are not so freely indulged in and are perhaps not so loud.

We found them breeding at Astanmarg, off the East Lidar Valley, when camped at approximately 11,500 ft, and have no doubt that at this elevation they breed wherever they occur. Osmaston says they nest on the slopes below Aphawat, where he saw young at 11,000 ft ; and again he records young being fed at 12,500 ft at Sona Sar in the East Lidar Valley. In mid-June, on the slopes directly above the ski hut at Killenmarg where the extensive patches of rhododendrons begin, we saw a bird with material in its bill but were unable to trace its ultimate destination, which appeared to be some distance farther up. In the *Nidification* Stuart Baker refers to a nest taken by Buchanan containing 3 eggs on 5 July 1905, also near Aphawat. At Astanmarg we found three nests in the first five days of that month, while yet another pair undoubtedly had a nest on the steep hillside opposite our camp which was dotted with battered old birch trees and large rocks. These nests were on the ground ; two of them tucked into the buttress roots of gaunt birches in the extraordinarily broken-up ancient moraine towards the centre of the marg. The third was at the foot of a very large boulder on almost level ground. It was lightly screened from view by some podophyllum plants. This last nest had newly-hatched young ones in it. Of the others, one was empty, and one contained 3 nearly fresh eggs which the female was incubating.

The nests were like larger editions of those of the Common Blackbird, being compactly built, deep, and bulky. The eggs were of the blackbird type, but more richly coloured ; one, in fact, was blotched rather like a Mistle-Thrush's egg. In fact they agreed very well with Stuart Baker's description in the *Nidification* in which he says that the dominating colour is red-brown and not greenish as in the Common Blackbird. He gives the average size as 33·05 by 23·45 mm.

PLATE 21

Blue-headed Rock-Thrush. Male on right

PLATE 22

Himalayan Whistling-Thrush taking food to young

Nest and eggs of Himalayan Whistling-Thrush

GREY-HEADED THRUSH

Turdus rubrocanus rubrocanus Gray
(See Plate 19 facing p. 76.)

KASHMIRI NAME. *Wān Kastūr.*
BOOK REFERENCES. *F.B.I.*, II, 132 ; *Nidification*, II, 119.
FIELD IDENTIFICATION. (3) A bird of the mixed forests. The pale grey head and neck and the warm rufescent plumage at once single it out from the rest of the subfamily. It feeds largely on the ground, moving about quietly and unobtrusively, but the male loves to mount to the topmost pinnacle of a tall pine to proclaim his fine song.
DISTRIBUTION. A denizen of the thick Silver Fir forests and of mixed fir and deciduous woods lying between 7,000 and 10,000 ft. It is undoubtedly commonest at the lower edge of this zone in mixed forests such as are found around Pahlgam in the Lidar Valley. In the Kishenganga Valley we first met with it above Sharda (about 6,500 ft) in early May. It was then mating. Ludlow encountered it in the Wardwan Valley at 7,000 ft and we also saw it there at Uriwan, and again near Suknes at over 9,000 ft. We have no information about its winter distribution in Kashmir and it may well be a resident, though Stuart Baker states in the *Fauna* that it descends to the foothills in winter and may be seen for a short distance into the Plains. Writing of Garhwal, however, A. E. Osmaston says he only once met with the bird, and that was on 31 January at 5,000 ft, on the day following a heavy fall of snow. It seems probable that it is only a partial migrant and that bad weather alone will send it downhill. On the other hand, writing of Simla, Jones records that it arrives there in small flocks in October, leaving for higher elevations farther west in March.

The species, with two races, extends throughout the Himalayas and Eastern Tibet to Northern Burma, Kansu and Yunnan. Whitehead met with a family party in August at 8,000 ft on the slopes of the Sufed Koh.
HABITS AND NESTING. The Grey-headed Thrush is on the whole a retiring bird rather given to avoiding observation. It is, however, the finest songster in Kashmir so that where it occurs—and it is widely distributed—it will most certainly be heard even if it is not seen. Particularly in the mornings and evenings does the male love to mount to the topmost pinnacle of some magnificent fir standing above its fellows, whence it broadcasts its fine song far and wide. When not so employed it moves silently about the

bed of the forest hunting for worms and other insect food in a typically thrushlike manner. In addition to the song both sexes have the usual cackling alarm-notes of the genus.

The nidification of this bird has been somewhat misrepresented owing to the earlier records being almost entirely of nests found on the ground or low down against rocks and tree stumps. Unwin in his list in *The Valley of Kashmir* states that 'it breeds in May and June, often placing the nest in the lower branches of a (Budil) fir-tree (*Abies Webbiana*)'. Ground sites we find are rather the exception than the rule. The majority of nests are undoubtedly to be found from 8 to 12 ft from the ground, often well concealed, in the summits of fir saplings or close up against the trunks of birch, fir and other trees where two or more branches or a mass of concealing twiglets spring forth. At times the nests are conspicuous; we discovered a number at Pahlgam and Surphrar in perrottia scrub which were anything but well hidden. Nests are certainly to be noticed, and not infrequently, on ledges of rock in the forest, in the roots of a tree, and even right on the ground in a depression in a bank. We once found a nest so placed into which an evil-looking viper had thrust its head to scoff the three young birds it contained. It was the frenzied cackling of the agonized parents which had drawn our attention to the spot, but owing to their secretive nature the birds do not ordinarily give away their nest's position, even when they have young.

The nest is fairly well built but is not particularly bulky, having as a rule rather thin walls. The materials used are twigs, bents, roots and moss in varying proportions, the moss constituting the outer material. The lining is usually of grass, roots or pine-needles. The nests are not mud-lined in the manner we connect with the nest of the Song-Thrush at home, but a considerable quantity of mud is generally present immediately under the lining.

The eggs, from 2 to 4, 3 being the commonest number, are handsome, being not unlike eggs of the European Blackbirds in tone, but the brown markings are usually larger and more pronounced, sometimes being quite large blotches, so that the underlying grey markings also show up well. They measure 30·6 by 21·6 mm. The breeding season extends from the latter half of April, throughout May and June, and well on into July. We found a nest on 2 May containing two newly-hatched young and an egg, whilst of ten nests taken around Pahlgam between 18 May and 5 June in 1932, three contained young, five held eggs, and two were being built. Ludlow took three fresh eggs at Chengher in the Wardwan Valley as late as 24 July.

TICKELL'S THRUSH

Turdus unicolor Tickell

(See Plate 19 facing p. 76.)

KASHMIRI NAMES. *Kastūr, Kao Kumr.*
BOOK REFERENCES. *F.B.I.*, II, 138 ; *Nidification*, II, 122 ; *Handbook*, 113.
FIELD IDENTIFICATION. (3) An ashen-grey bird, paler below, which feeds on the ground mainly on worms and snails in a typically thrushlike manner. Very common in the gardens in Srinagar and round about the villages. A fine songster. At close quarters it may be noticed that in the female the throat is almost white, bordered by black stripes, and that she has a row of spots forming an inconspicuous gorget across the upper breast.
DISTRIBUTION. A common bird throughout the Vale, particularly so in the gardens and orchards. We have seen it in the Jhelum Valley from Domel upwards, and have noted a few birds in the Kishenganga Valley around Keran. It extends up the Kashmir side valleys to about 6,500 ft, possibly 7,000 ft, but we do not remember having seen it at Pahlgam. It is migratory, wintering in the northern half of India south to about a line drawn from Khandala to Vizagapatam. It returns to Kashmir at the end of March ; Meinertzhagen noted first arrivals on the 24th of that month.

Its full range in the Himalayas, to which it is confined in the breeding season, is from Chitral and Kashmir to East Nepal.
HABITS AND NESTING. That blessed feeling of content which steals over the Englishman in the Kashmir Valley on his escape from the grilling monotony of the hot weather in the Plains is undoubtedly enhanced by the accumulation of those many little impressions which help to bring about a sense of *home*. One of these impressions, which perhaps helps as much as any other, is the rediscovery of ' the blackbird on the lawn '. In this case, however, the blackbird is Tickell's Thrush, which, ashen-grey instead of black, is nevertheless reminiscent of Britain with its blackbirds and thrushes through that most characteristic habit of hopping about the lawns and through the flower-beds, every now and then stopping with head cocked on one side to gaze intently at the turf before extracting an unsuspecting blindworm. In point of fact Tickell's Thrush is not entirely carnivorous, as in the autumn the apples, that is chiefly the windfalls, come in for considerable attention. From April to July or August, this bird

THRUSHES

is a fine songster, although in our opinion his notes are far surpassed by that great songster of the forests, the Grey-headed Thrush. Still, especially in the mornings and at eventide, Tickell's Thrush delights the ear with his sustained singing ; a fine song but rather monotonous compared with the more varied cadences of the Song-Thrush.

Common throughout the Vale, and not alone in the gardens and groves of the villages and towns where it is a familiar and far from shy bird, it is to be found in any wooded area, and, once the Vale is left behind, to some extent even in the thicker forests which rim the hillsides. At Keran we came upon them in heavy mixed forest above the old Rest House at between 5,000 and 6,000 ft, yet on the opposite side of the river, in open scrub. It is, however, on the whole a bird of gardens and more open populated areas. Magrath, speaking of the variations in song of males of the same species in different areas, mentions the case of Tickell's Thrushes which he noted at about 7,000 ft in the Sind Valley. ' The song of Tickell's Ouzel,' he wrote, ' which in the Kashmir Valley below, is a monotonous repetition of a few notes, was here much more varied and melodious. So much was this the case that on occasions I was led to stalking a singing thrush not recognizing it from the song.'

As far as their nest-building is concerned they are confiding birds. Many nests may be seen somewhat ill-concealed in the mulberry trees, willows, and poplars within village limits and in gardens and groves. Most of them are at heights of between 8 and 20 ft. Occasionally nests may be found in the thicker bushes a few feet from the ground, but compared with the numbers in the forks of mulberry trees, in the heads of pollarded willows, and supported by clusters of leaves in the chenars or by shoots against the trunks of tall poplars, the bush nests are few. They are also at times found in cavities in banks. The nest is typical but as a rule neither large nor particularly neat. The main structure is, however, compact and the cup deep. It is built chiefly of dry grass with a lining of fine roots, but usually contains a quantity of moss, a few leaves and other material.

The eggs number 3 or 4. They are not unlike English Blackbirds' eggs, but are smaller and not so grey-looking, as the streaks, which are generally distributed closely over the surface, are reddish-brown or brown. Occasionally these markings are not quite so profuse and more in the nature of small blotches. They measure 27·8 by 19·5 mm. The breeding season is May and June, but fresh eggs may occasionally be taken in the first week of July.

HIMALAYAN MISTLE-THRUSH

Arceuthornis viscivorus bonapartei Cabanis

(See Plate 20 facing p. 77.)

KASHMIRI NAME. *Techal Kastūr*.
BOOK REFERENCES. *F.B.I.*, II, 154; *Nidification*, II, 134.
FIELD IDENTIFICATION. (3+) A large brown thrush with a distinct greyish tinge about the back and large black drops on the breast. In flight, which is swift and strong, the tail is noticeably rather long. These birds keep mainly to the larger trees and as soon as breeding is completed collect into small flocks which may be met with in July and August at high elevations up to and just beyond the limits of tree growth. A fine songster in the pine forests.

DISTRIBUTION. Not particularly common, but widely distributed throughout the fir and birch forests of Kashmir from about 7,000 ft upwards. In the Kishenganga Valley we met with them at Sharda, a little above 6,000 ft. After breeding they are to be seen in flocks which ascend to the birch woods at the highest elevations at which these trees grow, that is at about 12,000 ft. It would appear that this bird is a resident with but a seasonal movement, returning to its breeding areas or a little lower for the winter months. There is, however, a record of a female having been obtained in February by Waite at Kallar Kahar in the Punjab Salt range. Whistler found it in Rawalpindi in winter and A. E. Jones obtained one in Dehra Dun. We can find no records of its ever having been seen in the Vale.

This species is widespread in Europe and Asia, the Himalayan race being found from Transcaspia to Central Asia, south to North Baluchistan and the Himalayas as far east as Nepal.

HABITS AND NESTING. In Kashmir Mistle-Thrushes are to be found only in very well-wooded country, in the less dense portions of the fir forests, and amongst the birches, from say 7,000 ft up to the limits of the trees. This applies on both the Pir Panjal and main Himalayan ranges. They are independent hardy birds caring little for habitations and cultivated country. In the breeding season they are to be seen in pairs, but later, from July onwards, they become to some extent gregarious and may then be met with, usually at high elevations, in flocks of some size. For instance, at Astanmarg at 12,000 ft a flock was seen with an estimated strength of at least twenty birds. ' Flock ' is perhaps not the best term to use ; it tends to convey the idea of a collection of birds flying closely together, as starlings do, for instance, and feeding in close proximity to one another. These collections of Mistle-Thrushes, however, still retain their rather wild individuality, for their members are generally scattered about the trees and hillsides, but on being disturbed take the same general direction of flight. Magrath, of a trip from the Lidar to the Sind Valley in July and August, wrote that ' flocks of 20 birds or more were not uncommon, both here [Sonamarg] and in the Lidar Valley. . . . One flock in Sonamarg was in the habit of visiting a certain spot on a mountain stream running through forest, for bathing purposes, and for several days in succession one could always count on finding birds at their ablutions if the place was visited about 6 p.m.'

In May and June the Mistle-Thrush is no mean songster. Like the Grey-headed Thrush it prefers to take up its station on the pinnacle of some tall fir on a steep hillside from which to broadcast its loud notes. It builds in the fir forests in those two months, placing the nest, sometimes supported by the twigs at the extremity of a bough, but often rather conspicuously low down against the trunk of a fir resting on a stout branch springing out horizontally. Few are any great height from the ground ; in fact we have seen one or two so low that one could look directly into the cup. Osmaston records finding a nest of three fresh eggs near Gulmarg on 16 June about 6 ft from the ground. Incidentally, he seemed to consider this a late date for fresh eggs for he states that it was probably a second nest. Yet, on the other hand, he records, without comment, finding a nest at 9,000 ft on a lower branch of a silver fir on 23 June, also 6 ft up and also with three eggs in it, and another only 5 ft from the ground on 15 July with three young ones. The fourth and last of a clutch of four eggs was laid on 9 June in a nest found by us near Gogaldara Forest Rest House.

The nest is always a deep cup, varying somewhat in size but usually rather a massive affair, of the blackbird type with a mud core lined with fairly fine grasses. The outer part of the nest consists of fine twigs, coarse grass, leaves, ferns, lichen, etc., the grass forming the major part of the material. Three eggs appear to be laid more than any other number, but 4 are not uncommon and at times 5 are recorded. They are far more boldly marked than any other true thrushes' eggs to be found in Kashmir, the blotches of red-brown always revealing a considerable amount of the greenish-grey to pinkish ground colour between them. The markings, too, are generally more numerous towards the larger end. There are secondary blots of a lavender hue. They are rather oval eggs with a fair gloss and measure on an average 31·3 by 22·4 mm.

BLUE-HEADED ROCK-THRUSH

Monticola cinclorhyncha (Vigors)
(See Plate 21 facing p. 84.)

KASHMIRI NAME. *Pāla Tiriv.*
BOOK REFERENCES. *F.B.I.*, II, 171; *Nidification*, II, 149; *Handbook*, 116.

FIELD IDENTIFICATION. (3—) This is a species of pine woods and mixed forests, not being found in open country. The sexes are widely divergent in coloration, the male being deep blue above with cobalt crown and white wing-patch, and chestnut below. The female is mainly olive-brown with the underparts scaled, as each whitish feather of the breast and abdomen is tipped with black. The song of the male is loud and pleasant but monotonous. It is uttered morning and evening from the summit of a tall tree.

DISTRIBUTION. A not uncommon bird of the forests and well-wooded areas mainly from 6,000 to 9,000 ft, but sparingly found a thousand feet or so on either side of these limits. Judging by its habits elsewhere it can confidently be expected to breed in the Jhelum Valley from about Chenari or Uri and in the Kishenganga Valley from, say, Keran upwards, but there are no records. It winters in continental India, the majority preferring the hill tracts of Central India and the Western and Eastern Ghats to the actual plains.

Its full range in the breeding season is along the northern confines of India from the Sufed Koh eastwards to Assam. Records for Burma appear doubtful.

HABITS AND NESTING. The male is easily remarked, not alone by his bright plumage, but also by the loud monotonous song in which he freely indulges, sometimes singing on the wing as he planes from one tree to another. The Blue-headed Rock-Thrushes are perhaps the first birds to herald the awakening day with their loud notes, which have been accurately syllabized by Magrath as *tew-li-di, tew-li-di, tew-li-di, tew*. These rock-thrushes are birds of the trees and open pine forests rather than rocky country and are not often to be seen on the ground. The nest, however, is always thereon or very close to it, either in a depression in a bank protected by an overhanging stone or stump, at the foot of a briar or other bush on a steep bank, or in amongst the buttress roots of a pine, the first and the last positions being perhaps the most favoured. They are loosely-built structures, and in Kashmir we have never found a nest which did not contain large numbers of pine-needles, a few twigs, and at times one or two really large pieces of bark supporting the front edge. A few hairs may also be found in the lining. The nests are often quite unconcealed, but the sitting female with her dark closely-barred breast sits exceedingly tight and is apt to avoid discovery on this account. When finally dislodged, she goes off silently, flying directly away for some distance, and as the nest is often in open pine or other forest growing on the steepest of hillsides she makes off straight down the slope. They are exceedingly loth to return to a nest until quite certain that they are unobserved, and are in consequence difficult birds to photograph, for in addition to the usual objection to movement in the vicinity of the nest, they are intensely suspicious of any noises from within the hide.

The eggs, usually 3 in number, are thin-shelled ovals, white in ground colour, freckled all over with very pale red markings. At times faintly blotched eggs may be met with. They average 23·7 by 17·9 mm. The main breeding season throughout Kashmir would appear to be May and June, but Ward records taking eggs as early as 13 April and as late as 5 August. We found one nest containing fresh eggs at Pahlgam late in July, but the majority of birds lay much earlier than this. Stuart Baker says that both birds incubate, but we have so far never put the male off the nest.

INDIAN BLUE ROCK-THRUSH

Monticola solitaria pandoo (Sykes)

(See Coloured Plate IV facing p. 232 and Plate 16 facing p. 61.)

KASHMIRI NAME. *Pāla Tiriv.*
BOOK REFERENCES. *F.B.I.*, II, 174 ; *Nidification*, II, 152; *Handbook*, 117.
FIELD IDENTIFICATION. (3—) An all dull-blue bird haunting open stony hillsides and precipitous ground, where it perches on boulders and ledges of cliffs. Much addicted to quarries ; never seen in forest. The male, whose plumage is brighter than the female's, has a habit, when courting, of planing down a slope from one boulder to another, singing as he goes.
DISTRIBUTION. A solitary but not uncommon bird on the treeless stony slopes which enclose the northern and western boundaries of the Vale. Although it is more numerous about these low altitudes

PLATE 23

White-browed Blue Flycatcher, male, at nesting cavity. Note the conspicuous supercilium almost meeting on the nape

Jerdon's Accentor on lichen-covered nest in tree

PLATE 24

White-browed Blue Flycatcher. Male on right, showing the wide collar interrupted on the breast

between 5,000 and 6,000 ft, it occurs sparingly throughout our area wherever conditions are suitable up to at least 13,000 ft. We have never failed to meet with it on the scrubby, rock-strewn promontories bordering the Wular Lake, where on one occasion we noted as many as three pairs distributed around the same headland at Kyunus. The hill of Ahateng overlooking the Manasbal Lake also has its quota, but it is unnecessary to go farther than the Takht where it breeds even on the slopes overlooking the city. The Blue Rock-Thrush is a great wanderer, leaving Kashmir on the approach of winter to penetrate to the extreme south of India, and even to Ceylon, whence it returns by stages. On migration, though a silent unobtrusive bird, it does not shun contact with mankind and may often be seen on the roofs of houses even in the larger cities.

The rock-thrushes are extremely widespread with consequently many races. The species is native to Southern and Central Europe and North-West Africa, ranging across Asia to China and Japan, moving south in the winter. The Indian race breeds throughout the Himalayas to North Assam. Meinertzhagen considers the North Baluchistan breeding bird to be *pandoo*, and so also will be the Kurram bird.

HABITS AND NESTING. In Kashmir, happily for the ornithologist, few of the breeding birds resemble one another sufficiently closely not to be easily identified ; excepting of course amongst the large group of willow-warblers. This rule certainly applies to the species under review. A dingy blue bird somewhat similar to Tickell's Thrush, but rather more slender and slightly smaller, which haunts old buildings, quarries, and boulder-strewn treeless slopes instead of shady gardens, can be none other than the Blue Rock-Thrush. It is an evasive silent bird on the whole, cautious and retiring in the vicinity of its nest, whose whereabouts it is at pains to conceal. Probably one's first knowledge of its presence will be a glimpse of a dark silhouette against the skyline as it eyes you from the top of a small cliff or the summit of a boulder. More particularly at the beginning of its nesting time, the planing habit of the male is to be observed. In these excursions he is to be seen descending the hill, launching himself from one vantage-point down to another on slightly upturned wings so that the flight is somewhat slow and fluttering. He seldom ascends much during these runs, but merely volplanes to the next resting-place, and as he does so sings his pleasant subdued song. A. E. Osmaston describes this as a soft melodious rather short whistle, which at times reminded him of an English Blackbird.

WHISTLING-THRUSH

The female is much less blue and more grey in tone than her mate and is a silent bird, sitting very close in her well-hidden nest or seemingly quite content to perch indefinitely on some vantage-point near by until she is quite certain that she is unobserved before venturing to the nest with material for its construction, to incubate, or to take food for her young. Although Dodsworth says that the hen alone collects the building material, we have seen both birds do so. The female is said to do all the building and the incubating of the eggs, but both undoubtedly take part in feeding their young.

The nest is by no means pretentious, being a shallow affair of a little grass and roots in varying quantity and thickness, lined with finer roots and a little hair. The bulk and shape of the structure vary in accordance with its site, wedged well inside a vertical fissure in the rocks being a favourite one, but it may also be placed on a ledge well concealed from view. The nests are always hard to find, often difficult to reach, and sometimes quite inaccessible. The eggs are 3 or 4 in number, clear pale blue in colour. Many are quite unmarked. Occasionally a whole clutch, though not infrequently only one egg in it, may have a few faint reddish spots about the larger end. They average $26 \cdot 0$ by $19 \cdot 1$ mm. Buchanan took a nest of 4 eggs near Srinagar on 27 April, and our own finds comprise both eggs and young in May and a pair noted to be only building on 7 June, so the breeding season would appear to be from about the latter half of April to at least the first half of June, possibly later in the higher parts of their range.

HIMALAYAN WHISTLING-THRUSH

Myiophoneus cœruleus temminckii Vigors

(See Plate 22 facing p. 85.)

KASHMIRI NAMES. *Hazār Dastān, Kastūr, Kāva Kunūr.*
BOOK REFERENCES. *F.B.I.*, II, 180; *Nidification*, II, 155; *Handbook*, 119.
FIELD IDENTIFICATION. (3+) This well-groomed thrush, so agile on its rather long legs, is often to be seen hopping about the boulders of a mountain torrent or moving rapidly along some overhanging branch. The sexes are alike, and the plumage is a deep blue-black, the silver tips of the feathers about the head and shoulders only showing up at close range. The wild meandering song of the male and the sharp call-notes carry a considerable distance.

DISTRIBUTION. The most widely distributed thrush in Kashmir. It is absent from the middle of the Vale, but, once the marshes and slack waters are left behind, the Himalayan Whistling-Thrush makes its appearance, while an odd bird or two is generally to be seen amongst the rocks at the foot of the Takht on its north and west sides. From the lowest altitudes at Kohala to 10,000 ft or so it will be found breeding wherever rushing water is not far off. It does not perhaps enter the thickest forests except where torrents and rivers run through them. In winter it is seen more frequently in the Vale and around Srinagar, but numbers move down to lower levels in the foothills, a few being found well out into the Plains, and in the Punjab Salt range. This species has two races in the Indian Empire which inhabit the hills of the North-West Frontier from North-East Baluchistan, thence through the Himalayas to Burma, Yunnan, Siam and Indo-China. The western race penetrates into Burma as far as the Chindwin and the Arakan Yomas.

HABITS AND NESTING. A cheerful and energetic bird at all times. Be the sun shining in a cloudless sky or the rain lashing in furious gusts the boulder-strewn course of an upland torrent, if a Whistling-Thrush is there, its penetrating, meandering, but none the less pleasurable song will be heard. We have seen a bird on the nest with heavy rain-drops splashing off its plumage singing at the top of its voice while we took shelter from the storm under the overhanging boulder on which its nest was placed. Nor does it confine its love of music to the breeding season, for we have heard its lay in winter. And so bursting with surplus energy is it that at times it even starts its whistling when in flight. With its well-cared-for, sleek plumage of deep blue-black, slender black legs and yellow beak, few more pleasing sights can be imagined than that of a Whistling-Thrush hopping effortlessly from boulder to boulder in search of its insect food or moving lithely up the sloping trunk of a tree leaning over a boiling torrent, every now and then stopping work to burst into song. Like all birds which live in the vicinity of rushing water, the Whistling-Thrush possesses exceedingly strident call-notes, *tzeet tze-tze-tzeet*, which carry far above the roar of the waters. The flight is strong and swift, so that the bird may often be seen high up a gorge side or flying swiftly down through the steep forest to the stream at its foot. It is equally at home hopping on its long legs about the bed of the forest, turning over the leaves for insects, or pulling the wet moss off the stones at the water's edge to use as material for its nest.

The nest is a bulky structure redolent of strength like its owners. It is made largely of wet moss and moss roots with plenty of mud attached, so that as it dries it clings to its support, be this a narrow sloping ledge on a boulder in midstream or overhanging the bank, a hollow in the roots of a large tree leaning over the water, or even on a horizontal branch arching out into space. Other nests we have noticed have been on the rafters of unfrequented buildings, in the buttresses of the primitive log bridges, and, on one occasion, right inside a wide cavity in a dead tree, a site far more suited to an owl. It is a common occurrence to find nests tucked out of sight inside small caves in the beds of the streams and behind the tumbling waters of a small fall. The lining is generally of roots, more rarely of grass or ferns.

The breeding season is long, as two clutches are often laid. Usually a second nest, not far from the first, is built for the reception of the second brood, but occasionally only one nest is utilized. In the lower Kishenganga in the second half of April we noticed many pairs building, but even then a few appeared to be already engaged in incubation. In May and June we have taken a great many nests throughout the Kashmir valleys, and have come to the conclusion that it is at the end of the latter month, and in July, that the majority of the second nests are constructed. As late as August a few eggs are still to be seen, even at medium elevations. The number of eggs laid is often only 3, though 4 are not infrequently come upon. Five eggs are said to be laid at times, but so far we have never met with this number in any of the many nests we have taken in Kashmir, the majority of our nests holding 3 eggs or young. The eggs are apt to be rather long in proportion to their width. In colour they generally appear to be a uniform pinky grey. Actually they have a pale greenish or cream ground, but this is usually evenly covered over with freckles of a pinkish hue. In size they average 35·8 by 24·8 mm.

GARHWAL ACCENTOR

Laiscopus collaris whymperi Stuart Baker

BOOK REFERENCES. *F.B.I.*, II, 188 ; *Nidification*, II, 163.
FIELD IDENTIFICATION. (2) Only found in summer at extreme elevations on bare open ground. Is noticeably larger than the next accentor and in this species the breast is greyish-brown while the flanks and sides of the abdomen are chestnut. It also lacks the pale supercilia with prominent dark lines above them.

DISTRIBUTION. This race, which was first separated from the Turkestan Accentor (*Laiscopus collaris rufilatus*) by Stuart Baker in 1915, is a bird of very high elevations throughout its range from Garhwal to Northern Kashmir and Ladakh, while the Turkestan bird has a wide distribution in Turkestan, the Pamirs, and the higher hills of the North-West Frontier. Osmaston alone appears to have noted it breeding in our Kashmir territory, having on 13 August come upon a family party, consisting of the parents and well-fledged young, fully a month old, at 12,500 ft above the Gangabal Lake. On 16 September Meinertzhagen found them not uncommon on Aphawat and obtained one at 12,000 ft. In winter they evidently drop to considerably lower elevations as Osmaston obtained a pair at 6,000 ft near Srinagar on the Takht-i-Suleman.

HABITS AND NESTING. This race is in summer a bird of high elevations inhabiting inhospitable slopes well above the tree line. It is then to be found only on rocky precipitous ground from some 12,000 ft upwards, Osmaston stating that in Ladakh he came upon a pair on the Khadong Pass which were evidently breeding at no less than 16,500 ft. There are no actual nest records for our area—the bird has been included here on the strength of Osmaston's Haramukh fledglings—but elsewhere in its range it places its nest on steep slopes under a stone or in a cleft in the rocks. Ward, however, got a nest, taken on 8 July at a great height near Gilgit (perhaps belonging to the Turkestan race, *rufilatus*), which was placed very low down in wild-rose brambles: this nest is said to have been like that of the British Hedge-Sparrow. Whymper, who took nests in Garhwal in June and July, states that they were made exclusively of moss and were placed well under flat stones at nearly 15,000 ft. The eggs taken by him were the usual hedge-sparrow blue and average approximately 22·7 by 16·5 mm.

JERDON'S ACCENTOR

Prunella strophiata jerdoni (Brooks)

(See Plate 23 facing p. 92.)

BOOK REFERENCES. *F.B.I.*, II, 197; *Nidification*, II, 170.
FIELD IDENTIFICATION. (2) The darkest of the accentors. The streaked and mottled plumage, pale red breast and prominent stripes above the eyes are dominant characteristics. Particularly

common in juniper patches at high elevations, but also occurs in the upper portions of open forests.

DISTRIBUTION. Very numerous at high elevations throughout our area both on the Pir Panjal and main Himalayan ranges. Though most numerous in bushy tracts above tree-level, it also occurs in open forests down to as low as 8,000 ft; for example, near Koragbal in Gurais, Tanin in the East Lidar Valley, and near Inshan in the Wardwan. This bird, though not a migrant in the true sense of the word, must surely be forced down to lower levels in the winter though we cannot find to what extent this occurs, but Meinertzhagen obtained two specimens in the Vale near the Wular Lake on the 18th and 21st of March, and Waite obtained a female as far away as Sakesar, the highest point in the Punjab Salt range, on 30 January 1929 during a spell of exceptional cold.

In India this species of the widely-spread accentors is represented by two races. *Jerdoni* is found at medium and high elevations from South Waziristan and the Border hills, thence through the Himalayas to Garhwal. Here it meets the typical race which spreads on to Northern Burma, Yunnan and Western China.

HABITS AND NESTING. We have kept the trivial name of Accentor in preference to Hedge-Sparrow for this species, as in Kashmir, and no doubt elsewhere in its Indian habitat, this sweet little bird lives in a hedgeless land. To this a further objection might be added; it is not really a sparrow, but belongs to a subfamily of the thrushes and is a mountain bird at that. Jerdon's Accentor has a sweet voice, though some of its notes are a bit harsh, and its phrases have a decided resemblance to those of the Kashmir Wren. It is, however, softer, not so exuberant and boisterous. We have seen them singing 15 to 20 ft up, but normally they stick much to low cover, often perching on rocks as well as on bushes.

To come up with it in large numbers one has to ascend to the bush-dotted uplands above the trees. It does, however, frequent forests of not too close a character down to as low as 8,000 ft —the subject of the illustration was taken at such an elevation at Nildori in the Kazinag mountains—but in the juniper tracts, from some 10,000 ft upwards where the scrubby bushes form smudgy patches against the green upland slopes, this bird is in its element. Here in the breeding season, from the last weeks of June and throughout July, it is almost an impossibility to explore a juniper bed without coming upon nests of Jerdon's Accentors. Junipers are not easy bushes to search and the neat cosy nests are tucked away well surrounded by the thick protecting foliage which scratches and pricks the arms unbearably as one carefully parts the branches.

In the juniper the nests are compact thick-walled affairs of bents filled in with moss and wool with a deep felted lining of fur, wool and hairs, sometimes containing a few feathers. They are beautiful little structures strongly reminiscent of the hedge-sparrows' nests in the British Isles. When breeding amongst the trees the nests are built occasionally as high as 10 ft up near the extremities of low drooping fir branches in such a manner as to be well screened from view. Here lichen-covered twigs often form the outer framework and lichen is also made good use of in the body of the nest and in its camouflaging. Nests are at times found in leafy bushes, but wherever placed they are always well concealed. The exception proves the rule, however, for we found one at Suknes amongst straggling shoots sprouting from the top of a perrottia stump.

The sitting female often remains boldly on her eggs until almost completely exposed, slipping away quietly at the last moment. The subject of the photograph became so tame that we used periodically to raise the branch which formed a protecting umbrella over the nest, to see whether the typical immaculate hedge-sparrow blue eggs had hatched, without her even contemplating quitting her task.

The number of eggs laid seems to vary from only 2 up to 5, but 3 is the commonest number found. Buchanan took an abnormal clutch of 6 on Aphawat on 23 June. In size they average 19·6 by 13·8 mm.

KASHMIR SOOTY FLYCATCHER

Hemichelidon sibirica gulmergi Stuart Baker

(See Plate 25 facing p. 100.)

BOOK REFERENCES. *F.B.I.*, II, 205 ; *Nidification*, II, 175 ; *Handbook*, 124.

FIELD IDENTIFICATION. (1) This earthy-brown flycatcher is characterized by a habit of hawking insects from a fixed perch, to which it returns again and again following each short sally. No other Kashmir flycatcher is so persistent in this method of feeding. It is also the dullest in coloration, being entirely brown above and paler below with no relieving splash of colour anywhere about it. The large eyes are a noticeable feature.

FLYCATCHERS

DISTRIBUTION. Being widely distributed throughout the entire Western Himalayas, it is consequently found everywhere in our area, where in summer it inhabits the better wooded tracts and forests from about 7,000 ft up to the birch woods at 11,000 ft, being commonest in the upper half of its range. It is particularly numerous in the forests of Silver Fir and Blue Pine. The Sooty Flycatcher migrates to winter in the outer ranges and adjacent plains, returning to its breeding-grounds towards the end of May.

The species is divided into three races found from the borders of Afghanistan in the Sufed Koh to the Himalayas and on across Northern Burma to Yunnan and Siam. The Kashmir bird is the Western form, breeding as far east as Garhwal.

HABITS AND NESTING. The Kashmir Sooty Flycatcher is at once marked out from its fellows by its characteristic habit of feeding from a perch. To this it returns time after time from fluttering little parabolas in pursuit of its winged prey. It is certainly connected in our minds with the topmost pinnacles of the tallest fir trees, where it often appears to the observer below as a tiny speck. On the other hand this preference for an unlimited view does not deter it from making use of the most lowly vantage-points, particularly in rainy weather when a plentiful supply of gnats is only to be found close to the ground. We once saw an immature bird, judging by its streaky appearance, use a tent peg as its main stance for the whole of one rainy afternoon, while perches at any height from the ground may be utilized, particularly the tips of vertical dead branches which afford an uninterrupted outlook. It is undoubtedly amongst the commonest of the birds of the more heavily wooded portions of Kashmir, having an exceedingly wide distribution from the lower mixed forests at about 7,000 ft to the limit of the trees in the neighbourhood of 11,000 ft. Although the pine and fir forests, particularly in the neighbourhood of 8,000 to 10,000 ft, are its main stronghold, it will also be found among the birches at the highest elevations at which these trees grow.

They arrive from winter quarters in the outer foothills in May, few commencing nest-building before the last week of that month; in fact it is doubtful whether many pairs have reached their breeding territory before then. We have taken one or two nests with eggs at the beginning of June, but most full clutches are to be found at both extremes of their altitudinal range from 15 June to 15 July, while there are a few records of late nests in the early part of August. Although these little birds soon become tame and confiding, their nests are difficult to find, due partly to position and partly to camouflage. For the structure is always a beautifully built thick-walled

PLATE 25

Kashmir Sooty Flycatcher at nest

Western Slaty-blue Flycatcher, female, on nest

PLATE 26

Kashmir Red-breasted Flycatcher, the tail characteristically raised in the lower illustration

compact cup of moss about $3\frac{1}{2}$ in. across, often nearly as deep as it is wide, but covered on the outside with lichen and web to match the colour of its surroundings. It stands up as a substantial excrescence on the top of a horizontal bough of a fir tree, often just where the leafy portions commence, but at times, as in the illustration, it is up against the trunk; in fact amongst the birches this is the position in which we have most often seen it. It may be at any height from about 7 ft upwards, frequently being as much as 40 ft from the ground and at times considerably higher. The nest in the illustration was about 12 ft from the ground with three more similarly placed nests within a hundred yards of it. Whether this was mere coincidence due to the commonness of the species it is hard to say, but this pointer to the possibility of a preference for breeding in company has been remarked upon before. On the whole they are silent birds, drawing attention to themselves solely by their tameness and feeding habits. We have seen the incubating female fed at the nest by the male, but when doing so he hovers before her for but a brief instant without alighting so that we have not been able to record this thoughtful attention.

The number of eggs laid is either 3 or 4, but we see Stuart Baker records that he had a clutch of 5. The tiny eggs are so profusely marked with microscopic specks of reddish on a greenish ground that they appear an almost uniform pale grey-green or reddish-green, the markings often being densest at the larger end. They have a brittle shell, are glossless and measure $16 \cdot 0$ by $12 \cdot 1$ mm.

KASHMIR RED-BREASTED FLYCATCHER

Siphia hyperythra Cabanis

(See Plate 26 facing p. 101.)

BOOK REFERENCES. *F.B.I.*, II, 212; *Nidification*, II, 181; *Handbook*, 121.

FIELD IDENTIFICATION. (2—) There are three salient features to help distinguish this little bird from all other flycatchers of Kashmir. First, the red throat and breast, which are of a particularly deep colour in this Indian species, being in the case of the adult male a brick red with black lines on either side. The second point to note is the black-and-white tail, the median feathers of which are wholly black, the remainder being white on the basal half. Lastly, this bird has a habit of cocking the tail over the back

FLYCATCHERS

so that the tips of the wings show prominently beyond the vent and pale undertail-coverts. It is a bird of the mixed forests, in which it haunts the lower and middle foliage.

DISTRIBUTION. A fairly common bird throughout our limits in well-wooded country and mixed forests up to about 7,500 ft. We have not met with it on leaving the Vale until reaching approximately 6,000 ft. We refer, of course, to the breeding season only, as it is a migratory species, leaving Kashmir in September. It will be seen in the Jhelum and Kishenganga Valleys on the return passage in April and early May. We found many pairs of this spry little bird breeding in the diminutive Rampur-Rajpur valley above the Wular Lake, the mean elevation of which is about 6,500 ft. This is the elevation at which it is most numerous.

The Red-breasted Flycatchers are contained in a widespread genus which breeds throughout a great part of Central and Southern Europe to Central and Northern Asia, wintering farther south. The Kashmir bird can no longer be considered a race of *Siphia parva*, whose typical form, together with an Eastern race, are winter visitors in India in some numbers. *Siphia hyperythra* breeds only from Chitral to Garhwal, wintering as far south as Ceylon.

HABITS AND NESTING. An unobtrusive flycatcher, rather furtive in its quick movements, its presence being brought to notice generally by a glimpse of that contrasting black-and-white tail as it darts from a bush to some adjacent sapling. Its nimble sallies after flies are not too obvious, since it takes a great part of its food when flitting from one perch to another under the shade of leafy foliage. It lives in a zone up to say 20 ft, descending but momentarily to the ground and seldom exploiting the upper layers of the forest, while it appears to prefer the smaller trees and saplings. In fact, when searching for its nest, natural hollows at shoulder height in the trunks of small trees should always be investigated, particularly those in wild pear, cherries and perrottias, even in those with trunks barely 4 in. in diameter. At times nests are high up in tall trees, Davidson having taken four near Gund between 20 and 40 ft up, while we took one at Kitardarji in a slit about 30 ft from the ground in the trunk of a forest giant. It fancies light mixed forest with plenty of scrub, and during an evening stroll from the Rampur-Rajpur Forest Rest House, along a path through just such a patch on the forest's rim, we found three nests in quick succession each about 4 ft from the ground. One was in a cavity in a large pine ; the other two in knot-holes in very small trees, the nests being just beyond reach of the finger-tips. Holes and slits of small size, often leading vertically downwards, are usually chosen, while woodpeckers' borings may also be appropriated. We have come upon nests inside large hollows or wide depressions wherein the whole nest was open to view, but these are exceptional, and the usual indication of a nest is the appearance of the bird from a tiny hole as the result of rapping the bark, a luminous black eye giving it an air of injured innocence as it pauses before quietly leaving the cavity.

Whenever it alights, and after almost every movement, the tail is raised at an acute angle, sinking back again to the normal position fairly slowly, while as it flits from one stance to another it utters a sharp little *chack*. When momentarily at rest, a subdued but harsh purr is sometimes indulged in conjointly with a flicking of the wings and the up-and-down movement of the tail. It is this characteristic pose which we have attempted to catch in the otherwise mediocre photograph. The full song is sweet, loud and robinlike, but short.

The nest is a collection of dead leaves, grasses, and dry moss, lined with a few hairs or bark strips, and contains from 3 to 5 fragile eggs, their ground colour being a dull white or very pale green spotted with pinkish-brown at the larger end. They average 16·6 by 12·5 mm. The third and fourth weeks of May, and June, mark the breeding season, many nests with eggs being obtainable in the first half of the latter month.

WESTERN SLATY-BLUE FLYCATCHER

Muscicapula tricolor tricolor (Hodgson)

(See Coloured Plate III facing p. 136 and Plate 25 facing p. 100.)

BOOK REFERENCES. *F.B.I.*, II, 219; *Nidification*, II, 186; *Handbook*, 123.

FIELD IDENTIFICATION. (1) The flashy little male, so well depicted in the plate, is more often heard calling *ei-tick*, *ei-tick* than seen, and that not very frequently for he is a self-effacing bird very partial to the undergrowth. The female, dull brown with a slight rufescence about the tail, is very like the Rufous-tailed Flycatcher though she is somewhat smaller and lacks the inconspicuous narrow ring of buff feathers around the eye. Both sexes are prone to droop the wings and elevate the tail after the manner of the Red-breasted Flycatcher, a habit quite foreign to the Rufous-tailed species.

FLYCATCHERS

DISTRIBUTION. A common bird in the forest tracts of Kashmir up to at least 10,000 ft. It will be met with anywhere outside the rim of the Vale but is particularly numerous between 6,500 and 9,000 ft in mixed forest and well-wooded areas with plenty of undergrowth. Whether or not it entirely leaves the Kashmir valleys in winter we cannot say, but after dubbing it a resident in the Himalayas, Stuart Baker goes on to say that in winter it is found throughout the plains adjacent to the hills and at that time of the year is given to frequenting heavy reed-beds and elephant grass as well as forest. Jones records that in the Simla area it descends into the foothills at elevations between 1,000 and 3,000 ft, which seems to us more in keeping with its general habits.

The species ranges from the Western Himalayas to Yunnan and down to Siam, the Western race extending to Eastern Assam. Birds south of the Brahmaputra in the Khasi hills are intermediate, but Stuart Baker refers them to the Eastern race.

HABITS AND NESTING. This is the most secretive of Kashmir's flycatchers. In fact with its love of remaining hidden away low down in shady undergrowth, its ways are more on a par with those of the shortwings and bluechat, for it is far more often heard than seen, moving about in the thickets where it keeps up a subdued *tack-tacking* noise as it follows one about within the limits of its territory. The tiny olive-brown female, in parts of whose plumage one is able to discern touches of ferruginous, occasionally shows herself on the outskirts of a bush, perching in a robinlike manner with wings adroop and tail cocked upwards as she eyes the intruder, but it is on quite rare occasions that one catches even a glimpse of her spouse. This is indeed a pity, for he is quite handsome with slate-blue upper plumage and conspicuously black-and-white tail, but he likewise prefers to remain out of sight and is sadly lacking in devotion to duty where nest-guarding comes in. In consequence he seldom approaches his home even when it is apparently threatened by the proximity of the intruder. Certainly he does on occasion feed the female on the nest and no doubt takes his fair share in keeping the chicks supplied with food once they have emerged, but he does not consider it necessary to hang about in the vicinity of the nest so continuously as does the female.

Young males attain adult plumage only gradually, and consequently often breed in an intermediate attire in which, along with the females, they are rather strikingly like smaller editions of the Rufous-tailed Flycatcher, from which, however, they may at once be distinguished by their habit of dropping the wings and cocking up the tail after the manner of a robin, a habit not indulged in by *ruficaudus*. The male possesses a pleasant but feeble little song.

The nest is placed from near ground level to 10 or 12 ft up, but by far the most sought-after site is in a depression or notch in the trunk of a large tree about 3 or 4 ft from the ground, just sufficiently deep to hold the nest, whose mossy rim is thus flush with the surface of the bark. It may at times be most cunningly hidden in the uprooted end of some fallen giant pine or screened by bushes or weeds and tucked right in out of sight on a ledge or in a cavity on the underside of a log spanning some depression. Not uncommonly it is quite open to view, but usually it is screened in some manner from direct observation such as by the foliage of an adjacent bush or through being located within the confines of a secluded ditch. Occasionally, however, nests may be come across well away from cover of any sort. The nest itself is a small but thick cup of moss compactly and neatly built with a deep central cavity plentifully padded with hairs, wool, or fur, and an occasional feather or two.

The eggs, which number either 3 or 4, quite often the former, have a faint pink hue, often deeper round the cap, where the minute reddish freckles are closer and more numerous, but the markings are so minute and the colouring often so faint that the eggs may appear an almost even faint buff colour. They measure $15 \cdot 6$ by $12 \cdot 1$ mm. The breeding season does not commence until the latter end of May, but extends well into July. Our latest record of a nest with fresh eggs is 10 July.

WHITE-BROWED BLUE FLYCATCHER

Muscicapula superciliaris superciliaris Jerdon

(See Plate 23 facing p. 92 and Plate 24 facing p. 93.)

BOOK REFERENCES. *F.B.I.*, II, 221; *Nidification*, II, 189; *Handbook*, 123.

FIELD IDENTIFICATION. (1) An unobtrusive forest bird in which the sexes differ, the male being deep blue above, the blue continued on to the breast in a broad collar which is always interrupted to some extent in the middle. The underparts are a clear white. A white supercilium almost meets on the nape. The female is brown, paler below, with indistinct darker streaks on the head.

FLYCATCHERS

DISTRIBUTION. In the breeding season we have seen this tiny flycatcher as low as 6,000 ft in the foothills of the Kazinag and in the lower Sind. On the other hand we have seen it at Lidarwat at 8,500 ft and nearer 10,000 ft at Sonamarg. It is commonest throughout our whole area in the mixed forests round about 7,000 or 8,000 ft. In the Kishenganga we noted a pair prospecting for a nesting site in early May at Keran, which is barely 5,000 ft. In winter it migrates to the Plains, penetrating as far south as the Deccan.

The two races of this species breed throughout the outer Himalayas, and, according to the *Fauna*, in the hills south of the Brahmaputra between India and Burma and on to Yunnan, the dividing line being about Western Nepal. Wardlaw Ramsay secured a male from a pair he saw on the Peiwar Kotal at the head of the Kurram Valley and Whitehead also obtained a single bird there.

HABITS AND NESTING. Out of the six members of the family breeding commonly in Kashmir, the Paradise Flycatcher alone surpasses this little bird in the possession of fine feathers, and then only by virtue of taking into consideration the glories of both sexes. Unfortunately the blue and white male is a quietly disposed little fellow seldom flaunting his neatly-attired person before the public eye and, when disturbed near his nest, sitting quietly amongst the branches uttering a soft trill which is barely audible 30 yards away. The song, although so subdued, is unmistakable, being an oft-repeated *che-chi-purr, che-chi-purr*, the final purring notes being lower in the scale. Nor do its flycatching abilities attract attention, but they are obviously considerable judging by the short intervals between appearances at the nest.

We have found this bird commonest in fairly thick forest, both of a mixed character such as one finds at Pahlgam and Lidarwat in the 7,000 to 8,500 ft zone, and in the more exclusive pine woods. Against this, at Kitardarji, at barely 6,000 ft, we found a nest in a fruit tree in a small orchard standing well away from a tongue of the forest. More often than not the nest is at no great height from the ground in a narrow cleft or natural hole in a forest tree. The last one found was only 4 ft up, but we have seen a bird enter a slit in a birch about 25 ft from the ground. Davidson took a nest as high up as this, and one of his nests taken near Gund was in a disused woodpecker boring 20 ft from the ground. The nest is generally out of sight, but this rule does not always hold good, and at times both the site chosen and the nest are similar to those of the Slaty-blue Flycatcher. This is evident from the illustration, the nest here being wedged into a cavity in a stump pitted with holes where the wood was rotting away. It was quite open to view though the birds had chosen the deepest ledge.

It is a soft little structure hardly 3 in. across, mainly of fine moss with a lining of hair, but there are often fine shavings and strips of bark skin in the foundations. A nest taken many years ago at Lidarwat had the foundations composed of many fine strips of the skin of birch bark so that it looked as if the bird had filched the shavings out of a chocolate box. As already stated the only other nests with which the open nests are likely to be confused are those of the Slaty-blue Flycatcher, but the eggs are different, being in the main a biscuit colour as opposed to skimmed milk, while a glimpse of the parents will of course settle the matter at once.

The White-browed Blue Flycatcher lays 3 to 4 eggs in May, June, or early July. They have a pale olive-greenish or dull stone-buff ground, the latter type usually so closely stippled with minute reddish-brown specks as almost to obliterate this ground colour. The greenish eggs are less profusely marked and the freckles coalesce into a cap at the larger end. They average 16·0 by 12·2 mm. They have little or no gloss. The male, as is the case with many, if not all, of our flycatchers, feeds the female on the nest, at any rate when the young are at a tender age, in order to relieve her of the necessity to leave them.

RUFOUS-TAILED FLYCATCHER

Alseonax ruficaudus (Swainson)
(See Plate 27 facing p. 108.)

BOOK REFERENCES. *F.B.I.*, II, 250 ; *Nidification*, II, 217.

FIELD IDENTIFICATION. (2—) The most unobtrusive of the Kashmir flycatchers. Characterized by its dull chestnut tail and a chatlike habit of bobbing forward at the same time flicking the wings. Draws attention to itself in the breeding season by the monotonous repetition of its danger-call.

DISTRIBUTION. Widely distributed throughout Kashmir in the mixed as well as in the pine and fir forests up to between 9,000 and 10,000 ft. We found it to be commonest amongst the deodars which commence in the Kishenganga Valley as low as 4,500 ft, and in the Kazinag range and Lolab Valley at from about 6,000 ft.

FLYCATCHERS

It is in fact a most numerous bird in the open woods bordering the lower portions of all the side valleys around the Vale, decreasing proportionately as the elevation increases.

Outside Kashmir it is found from the Sufed Koh to Gilgit and thence throughout the Western Himalayas to Garhwal. In winter it spreads through North-West and Western India down to Travancore. On the Eastern side it has been noted in Orissa.

HABITS AND NESTING. The Rufous-tailed Flycatcher is the dullest representative of its family, not alone in habits but, except for *Hemichelidon*, in coloration as well. Were it not for the persistence of its call-notes many would escape notice, but it is a common bird in parts and draws attention to itself by its habit of at once objecting to entry of man or beast within the confines of its territory. The danger-note in question is a plaintive *peup* uttered *ad nauseam* while the bird stands forlornly on a shady perch, occasionally flicking the wings and bobbing forward in a distinctly chatlike manner. On its approach to the nest the notes often change, the same *peup* being followed by a soft *churr* in a lower key or preceded by a short *te*. Osmaston states that it also has a song of a few clear notes which reminded him of that of the Blue-headed Rock-Thrush. These notes are loud and pleasant and uttered with great persistence, particularly in the evenings and generally from a high vantage-point. Monotony is to a certain extent avoided by considerable variations in all but the first couple of notes, which are usually the same.

Its search for food is likewise unobtrusive, for it sticks largely to the upper and middle foliage and seldom indulges in fluttering little flights, but snatches its prey while in transit from one tree or perch to another. Though widely distributed in Kashmir, until we visited the Kishenganga Valley and the Kazinag range at the western end of the Vale, we had little first-hand knowledge of it, but in the deodar forests of the Kazinag in particular we found it extremely common. It is a bird of the lower rather than of the high altitudes, though it undoubtedly occurs and nests sparingly up to some 10,000 ft. In the lower limits of the deodars, in the Kazinag certainly as low as 6,000 ft, and in the middle Kishenganga considerably lower still, as also in the lower Sind Valley, it is a common bird. For instance, on one occasion in June at Sanzipur near Handowar we had three pairs at once all complaining bitterly at our presence. We must not, however, give the impression that its presence is to be missed elsewhere for it is without doubt the commonest flycatcher at the lower and median elevations, being

PLATE 27

Himalayan Paradise Flycatcher, female, on nest

Rufous-tailed Flycatcher on nest

PLATE 28

Himalayan Paradise Flycatcher, adult male

found wherever there are woods to hold it. It permeates the pine forests as well as affecting secondary growth and perrottia scrub so long as there are a few large pines about.

The nest is not unlike that of the Sooty Flycatcher, but it is less neat and apt to spread about its supports. Placed as it so often is at a junction on a horizontal branch, it is not easy to locate unless very low down, and the bird does not readily disclose its position. But it is sometimes placed up against or near to the main trunk, and from quite low elevations in saplings to some 30 or more feet up, the majority perhaps being at or below 15 ft. It will also be found low down in perrottia scrub. Both Whymper and Rattray report finding nests in banksides within or close to forest, and we too can claim one such nest which was tucked into the roots of a small bush on an exceedingly steep slope in thick forest. The material used is moss, liberally coated on the outside with lichen and lined within with hair and a few feathers. Nest-building is mainly carried out towards the end of May and in June and takes a considerable time, but eggs will also be found in July. Normal clutches number 3 or 4, most often the former. In colour they have a greenish or bluish tinge and are freckled lightly, mainly at the larger end, with faint rufous markings. The markings are very occasionally almost if not entirely wanting. They have no gloss and measure 17·2 by 12·8 mm.

HIMALAYAN PARADISE FLYCATCHER

Terpsiphone paradisi leucogaster (Swainson)
(See Plate 27 facing p. 108 and Plate 28 facing p. 109.)

KASHMIRI NAMES. Male, *Fhāmbasir* (Cotton-flake), *Literāz* (in the Lolab) ; Female and Young, *Ranga Bulbul*.
BOOK REFERENCES. *F.B.I.*, II, 268; *Nidification*, II, 235; *Handbook*, 131.
FIELD IDENTIFICATION. (2) Fully adult male, mainly white with metallic blue-black head and straight sharply defined crest ; a conspicuous rim of cobalt blue feathers round the eye. The central pair of tail-feathers is elongated into fine streamers which may be as much as 14 to 18 in. in length. In the female and young male the white is replaced by a chestnut red and the streamers are lacking. Males breeding in an intermediate red plumage with red streamers may also be met with. The striking resemblance

of the female and young male to a bulbul is denoted in the local name of *Ranga Bulbul*.

DISTRIBUTION. This race of the Paradise Flycatcher is highly migratory, but its full movements still require elucidation. In April large numbers may be seen moving up the Jhelum and lower Kishenganga Valleys, the latter possibly Turkestan birds. In the third week of that month many appeared at Pateka and we saw one or two at Tithwal, but though we halted for ten days at Keran in early May none appeared there. The Paradise Flycatcher is common in the summer months throughout the groves and gardens of the main Vale. Few pairs leave its confines and these penetrate for but a short distance into the large side valleys, although we once saw an adult male at the Surphrar camping site. We were told that they are sometimes seen in the little forest-encircled valley of Rampur-Rajpur, 1,500 ft above the Wular Lake.

This striking species, divisible into a number of races, covers Turkestan and Afghanistan, India, Burma and Ceylon, and a great part of the Oriental region. Our pale Himalayan race is the one found breeding in Turkestan and Afghanistan, and the hills of Northern India from North Baluchistan to Eastern Nepal.

HABITS AND NESTING. 'If there be a Paradise on Earth, it is this, it is this, it is this.' So spoke the Persian poet Firdausi, but not as it happens of the Vale of Kashmir; the Taj Mahal was the cause of his ecstasy, but the Mogul creators of the Shalimar and Nishat Baghs must surely have felt the same thrill when the panorama of the lake-studded Vale was first revealed beneath them; that green valley backed by the snow-capped glory of the great Himalayan range from the domes of Nun Kun and Kolahoi past Haramukh to the sharp pinnacle of Nanga Parbat. Is it not fitting, therefore, that the natural beauty of a country, which at once strikes the eye of the beholder with such force, should possess living adjuncts of surpassing loveliness in the form of liquid-voiced orioles, bejewelled kingfishers and Paradise Flycatchers? The former's brilliant hues catch the eye, but it is the ethereal quality of the form and flight of the aptly-named Paradise Flycatcher which is so entrancing. The bird ripples through the air, or rather such is the impression as the white male darts in and out of the apple-laden trees of an orchard, or flutters around the clusters of chenar leaves of the giant trees of the Nishat Bagh. For in flight the long tail-feathers wave behind like paper streamers fluttering in the breeze; at rest they droop in a graceful curve.

In the canal-intersected orchards these birds often catch flies on or close to the water's surface, at times almost completely immersing themselves in the process. We have heard many complain that this lovely bird should be possessed of such harsh notes, but in addition to the oft-repeated call-note, the male Paradise Flycatcher has a short and rather charming song which we can only describe as a tumbling cascade of bulbul-like notes, but fuller and with a flutelike quality. Unfortunately it is the scratchy call-note (not unlike the *pench* of a snipe), which is far more often heard. Its frequent utterance does, however, serve the purpose of advertising the Ribbon-bird's whereabouts, so we ought to be grateful for it. It has also an additional call which might be syllabized as *weni-wedi-weech*. The white plumage and long tail do not extend to the female and young male, in whom the white is replaced by chestnut-red. Long-tailed red males may also be seen, and at times birds in a transitional stage having the underparts white or nearly so, for it is said that the male takes from three to four years to reach his full glory.

They are not shy birds ; in fact we have had one return to its nest before we had moved away ten yards, vouchsafing us the pleasing sight of a fussy white male settling on the nest preparatory to taking his fair share of incubation in relief of his less conspicuous mate. For the nest adds to the picture, being a thin-walled cone or cup built up from a horizontal bough or frail upright fork. The neatly-coiled fine grasses, sometimes with a little moss intermixed, are cemented together with cocoons, web, and spiders' egg-bags into a smooth whole. Egg-bags, discarded cocoons, and other scraps are also at times attached as decorations. This beautiful structure is about $3\frac{1}{2}$ in. across and about the same in depth, though it may not infrequently be a flattened cup rather than a cone. It is placed on any thin horizontal or slightly sloping bough, occasionally in an upright fork, from some 6 or 7 ft up to very considerable heights. In the chenars, for instance, we have noticed nests in most inaccessible positions well over 50 ft up amongst the smaller branches. On the other hand we have seen a nest in a fruit tree at eye-level. A nest in a Baramullah garden about 20 ft up in a chenar was supported solely by a thin almost vertical twig which ran the entire depth of one side of the nest.

The eggs have a pale pink ground giving the impression that the contents are showing through the delicate shell, but sometimes the colour is a definite salmon pink. The markings are spots and small blotches of red and reddish-brown, scattered about in no great profusion mostly towards the larger end. Three or 4 are laid, the former number being quite common. Nest-building, which is said to be undertaken by both sexes, although prolonged

SHRIKE

observation by us has failed to show that the male helps, commences early in May, so eggs may be found from the second or third week in that month until July. They are rather elongated ovals averaging 20·8 by 15·4 mm.

RUFOUS-BACKED SHRIKE

Lanius schach erythronotus (Vigors)
(See Plate 8 facing p. 29.)

KASHMIRI NAME. *Harawātij*.
BOOK REFERENCES. *F.B.I.*, II, 295; *Nidification*, II, 265; *Handbook*, 141.
FIELD IDENTIFICATION. (2+) The Rufous-backed Shrike has the head and underparts grey with a conspicuously wide black stripe through the eye continuing the line of the stout slightly-hooked black bill. There is also some black in the wings and in the tail which is noticeably long and narrow. A harsh call-note is frequently uttered from a bush top or other vantage-point on a lower branch whence it planes to the ground to seize grasshoppers, beetles and other insect prey. When returning to its lookout, it often does so with a final almost vertical upward sweep on outstretched wings.

DISTRIBUTION. Numerous in the Vale and widely distributed in the more open country of the surrounding hills and side valleys, becoming progressively scarcer up to some 8,500 ft. Occurs throughout the Jhelum Valley, and in the lower Kishenganga up to about Keran. It then appears to be absent from the remainder of that valley until the milder climate of Gurais is reached. There we found it common and Meinertzhagen says it breeds along the Gilgit road up to some 10,400 ft. A few pairs are to be encountered in the Wardwan Valley. September witnesses the beginning of its migration to the Plains, but some pairs remain in the Vale until mid-November. They return towards the end of March, Meinertzhagen noting first arrivals (males) in the Vale on the 24th of the month, and stating that they were abundant by the 27th.

The genus *Lanius* is extremely widespread in Europe, Asia, Africa and parts of North America, and has consequently been split into a number of species with many races. *Lanius schach* is found from Turkestan and Afghanistan throughout India to China. The Rufous-backed race, to quote Whistler, ' breeds in

Turkestan, Gilgit, Kashmir, the outer Western Himalayas, North-West Frontier Province, Baluchistan, Sind and the Punjab, and winters in peninsular India '. It also breeds in Afghanistan.

HABITS AND NESTING. The first glimpse of this ferocious little hunter in the Vale of Kashmir will almost certainly be from the car when passing through the now sadly depleted lines of poplars between Baramullah and Srinagar, for the lower branches and twiglets of these trees not only provide many pairs with nesting sites, but also serve as vantage-points from which they glide down to seize an unsuspecting beetle or some more succulent insect. The Rufous-backed Shrike hunts from a perch, watching the ground around him for any movements to betray the presence of his legitimate prey. Legitimate prey is here a wide term, as the shrikes as a family have not earned the pseudonym of Butcher-Birds for nothing. Lizards, crickets, and any of the smaller creeping things do not come amiss, and we regret to say they are not above seizing the odd chick from an unguarded nest. As their eggs near hatching point, some species impale their luckless victims on thorns near the nest to form a larder, but we have not so far come upon evidence of this trait in the Rufous-backed Shrikes of Kashmir.

Amongst his other achievements this bird is an excellent mimic, and in a Baramullah garden we were once regaled for a quarter of an hour by a bird which poured forth unceasingly his ideas on the song of Tickell's Thrush and the Himalayan Starling, adding many drongolike notes and a subdued rather metallic, but by no means unpleasant, song which may have been his own composition ; another had obviously lived in the marshes and had picked up the Reed-Warbler's harsh utterances. A third startled us by producing the *Did-he-do-it* of a Red-wattled Lapwing from the summit of a willow. After hearing the shrike's grating call-note, which proclaims his whereabouts so unmistakably, it comes as quite a surprise to find he is no mean songster, even though his lay lacks volume and the timbre of his voice is not all that it might be. Phillips notes that an irregular flirting of the wings always accompanies his efforts at song.

In Kashmir the Rufous-backed Shrike appears to be the main fosterer of the Asiatic Cuckoo. We say 'appears' advisedly, because, although Stuart Baker's collection of 96 Kashmir-taken cuckoos' eggs contains no less than 20 from nests of this species, it must be remembered that the Rufous-backed Shrike's nest is easily traced, far more easily than those of the other known Kashmir fosterers. Being innately bold, it is at no great pains to conceal the position of its home, and when carrying building material or food for its family, a few minutes' observation generally suffices to track it down. We have found the nest in dense wild roses and other bushes, in mulberry trees, in the clusters of leaves which one finds on the lower limbs of the magnificent chenars, and supported by the smaller branches against the trunks of poplars. The bush or tree chosen may be quite isolated and is usually in fairly open country where the bird's territory provides it with good observation, but it is just as happy in orchards and gardens within the towns as in scrub jungle, cultivation, and bush- and tree-dotted waste ground outside them. The nests may be concealed, but many are not, concealment obviously not being one of the bird's main considerations.

Some nests are compact, while all are fairly massive and deep, but others are a rough cup of bents, straw, and dry grasses, the lining of thin roots often including oddments such as bits of rag, wool, tow, vegetable fibre, and the odd feather picked up in a nearby village. They are seldom more than 10 or 15 ft from the ground, but when built in bushes they may be as little as 3 or 4 ft up. The coloration of the eggs is rather insipid, a dull white, pale green or buff ground, with large and small blotches of sepia and reddish or yellowish brown with secondary grey markings. The markings often form a zone round the larger end. They average 23·0 by 17·9 mm. Breeding commences in May and continues throughout June. Clutches of 5 are common; 4 are not infrequent, while occasional 6's may also be found.

INDIAN SHORT-BILLED MINIVET

Pericrocotus brevirostris brevirostris (Vigors)

(See Coloured Plate II facing p. 40.)

KASHMIRI NAME. *Wozul Mini.*
BOOK REFERENCES. *F.B.I.*, II, 323; *Nidification*, II, 289; *Handbook*, 148.
FIELD IDENTIFICATION. (2+) The coloured plate illustrates the fully adult male. In the female the black is replaced largely by slate-grey and the scarlet by varying shades of yellow. Young birds of both sexes are olive green above with black barring, and yellow below mottled with brownish-green on the flanks and breast. These minivets are sleek rather long-tailed birds, seldom seen away

from firs and pines. They trail one another from tree to tree, in the breeding season usually in pairs or family parties, at other times in small flocks.

DISTRIBUTION. Where the pines start, at approximately 3,000 ft, there too, in the breeding season, will be seen the Short-billed Minivet. With the exception, therefore, of the central plain of the Vale of Kashmir, and the lowest reaches of the Jhelum and Kishenganga Valleys, this bird will be found throughout our area up to an elevation of approximately 10,000 ft, at which elevation we have seen it in the little valley leading to Astanmarg where the forests soon give way to scattered woods of birch and open bush-covered hillsides no longer to its liking. It is perhaps commonest from the rim of the Vale at about 6,000 ft up to 8,000 ft or so. In the winter it moves down to the foothills and plains adjacent to the Himalayas.

The Short-billed Minivets are widespread in Northern India, Burma and Eastern China. The typical, Indian, race breeds from about 3,000 ft at the head of the Kurram Valley to the North-West Himalayas, along which it spreads as far as Nepal. It winters in the northern half of India from the Punjab to Eastern Bengal and down to the Central Provinces.

HABITS AND NESTING. The Short-billed Minivet is one of our most arresting birds. The bright scarlet plumage of the male imprints itself on the vision like the flickering of a dying flame as each bird, intermittently spreading and closing the wings in series of rapid beats, launches itself in follow-my-leader fashion from tree to tree, maintaining a pleasant twittering whistle to apprise its companions of its whereabouts. Regarding these calls, Magrath (*Notes on the Birds of Thandiani*) wrote as follows : ' When travelling in search of food along the tops of the pines they are constantly using a titlike chatter and a call of *Switswitswititatit*. Possibly this latter note gave rise to the name " Minivet ". Often they use a pretty note like *Swisweet-sweet-sweet*.' This same routine of trailing each other is kept up in the breeding season, even when the bands have paired off, so where one bird is seen its pair is almost certain to appear, while couples breeding in adjacent areas have no objection to joining up into small foraging parties. As a rule they remain in the middle and upper foliage of the pines, but at times they descend to lower trees and large bushes to search their leaves for insects. They affect parkland, provided there are scattered pines, the open mixed forests, and particularly the edges of the pine woods, but do not penetrate the heavier unbroken forests to any great extent.

DRONGO

The nest is difficult to locate, as it is constructed high up in some pine tree—Osmaston says they have a predilection for the Blue Pine—where it is placed on a horizontal branch seldom as low as 30 ft from the ground and more often than not twice that height up. It is a very beautiful structure, a much flattened cup of fine twigs and roots cemented together with spiders' webs and coated on the outside with bark chips and lichens to match the bough upon which it is fabricated. Davidson reports one nest in the extreme top of a walnut, and we too have taken a nest in that tree, but his other nests, all in fir trees, he describes as ' most beautiful cups of moss lined with fine roots, a little down and hair, and covered outwardly from top to bottom with green lichen '. He goes on : ' They are, I think, the most beautiful nests I have ever seen. The number of eggs in all full clutches was four.'

Four eggs is the maximum laid, but as few as 2 in a clutch are by no means uncommon. They are broad ovals, white, pale cream, or pale green in ground colour, rather heavily marked with spots and blotches of brownish-red with underlying grey or lavender markings. They measure 19·8 by 15·1 mm., and are to be taken in May, June, and possibly July at the higher elevations. Although both sexes build, and feed the young, prolonged observation indicates that the male takes no part in incubating the eggs.

INDIAN GREY DRONGO

Dicrurus leucophæus longicaudatus Jerdon

(See Coloured Plate IV facing p. 232.)

KASHMIRI NAMES. *Gunkots, Telakots.*
BOOK REFERENCES. *F.B.I.*, II, 362; *Nidification*, II, 329; *Handbook*, 158.
FIELD IDENTIFICATION. (2+) Upper plumage a deep blue-black, beneath greyer. Young birds are browner and often show mottling on the abdomen owing to the white bases of the feathers showing through. Tail proportionately long and deeply forked, the outer rectrices turning outwards with the tips bent up. Possesses many harsh calls as well as some musical whistles, and, in addition to flycatching from high perches in the tree-tops, may also be seen hawking insects disturbed by the passage of sheep or cattle, using the animals' backs as perches.

PLATE 29

Indian Great Reed-Warbler

Witherby's Paddy-field Warbler : nest among weeds on sloping ground

PLATE 30

Hume's Lesser Whitethroat

Hume's Yellow-browed Willow-Warbler near nest

DISTRIBUTION. Common in the Jhelum and lower Kishenganga Valleys. It occurs in and around the Vale, but is not particularly numerous. In the wider side valleys it may be found up to about 6,500 ft. We have seen it at Gund in the Sind Valley, but have not recorded it in the Lidar beyond Aishmakam. A summer visitor to the Vale and the higher elevations.

This species has a number of races in India, Burma, Siam, and Malaya, *longicaudatus* occurring throughout a great part of India west and south of Nepal, being absent from Sind, Gujarat, much of Rajputana and the Punjab. Except in the Himalayas it is mainly a winter visitor and its breeding in other parts of India appears doubtful.

HABITS AND NESTING. This race differs from the well-known Black Drongo or Kingcrow of the Plains mainly in its lighter underparts and in a slight matter of size; it is a degree smaller and more slender. In habits it is almost identical. Bold and courageous, the little black fury will sally forth from its favourite lookout with many shrieks and shrill whistles to drive away from its territory even the largest interlopers, be they birds of prey bent on violence or just innocent trespassers. We have seen one dart down on a Black-eared Kite to such good effect as to remove a couple of tail-coverts. Timid spirits, such as orioles and doves, are not molested, and are said often to place their nests near those of the Kingcrows for the security thereby vouchsafed them; we have ourselves found orioles' and Kingcrows' nests in the same trees. Most of their notes are far from musical, consisting of these harsh shrieks and grating whistles, but they have also some quite pleasant bell-like notes and are reputed to have considerable powers of mimicry as well as a short tinkling song of their own.

The Grey Drongo is particularly agile in its dipping flight, and twists hither and thither with considerable dexterity to seize fleeing insects. This is one of the few species which we have noticed persistently hunting butterflies, but the dodging abilities of butterflies are excellent and many attacks upon them prove abortive. These Drongos, too, love to follow flocks and herds of all kinds, alighting on the animals' backs whence they make short sallies after flies or descend momentarily to the ground to pick up small game flushed in the flock's progress. When so engaged they show little fear of man. They prefer for their habitat fairly well-wooded territory and are particularly fond of the open fir and deodar forests on the more abrupt slopes which offer such scope for their steep swoops and ascents after winged prey.

In Kashmir the nest, which is to be found in May and June, is generally in a large deciduous tree at some considerable height from the ground, often in one of the tall trees surrounding a village, such as the chenar or walnut. We have also seen nests in the taller fruit trees. The nest differs in no way from that of *macrocercus* except that it may be neater, being the same flat saucer, sometimes thin enough to see through and four or five inches in diameter, built into a more or less horizontal fork of a slender outer branch. It is composed mainly of thin roots, but fine twigs may be incorporated, while the lining is of finer roots, grasses and a few hairs. It is generously plastered on the outside with lichen, bark scraps, and web. It is almost invisible from below and the sitting bird is said to place her tail along the bough so that she too remains well hidden.

The eggs, 3 or 4 in number, sometimes only 2, are very variable in colour, the ground ranging from white (rare) to a rich salmon or terra-cotta. Stuart Baker says many have a rosy tint, which is most exceptional in the *macrocercus* group. They are usually handsomely marked with small blotches of dark red-brown, but the markings, too, are variable both in extent and colour. Secondary markings of lavender or grey are generally present. The markings are always more numerous at the large end and often run into each other to form a zone round the egg. They average 23·2 by 17·9 mm.

INDIAN GREAT REED-WARBLER

Acrocephalus stentoreus brunnescens (Jerdon)

(See Plate 29 facing p. 116.)

KASHMIRI NAMES. *Korkuch, Karkat.*
BOOK REFERENCES. *F.B.I.*, II, 389; *Nidification*, II, 352; *Handbook*, 161.
FIELD IDENTIFICATION. (2+) Creeps about the interiors of reed beds giving vent to loud strident calls whence it derives its onomatopaeic Kashmiri name of *Korkuch*. In coloration it matches the rushes amongst which it lives. It can hardly be confused with any other bird, except perhaps the next which is much smaller and possesses a sweet little song in place of the harsh call-notes.
DISTRIBUTION. Confined to the Vale, where from April to September it infests every reed bed capable of supporting it. Occurs in very large numbers in those of the Dal and Anchar

lakes and indeed throughout all the marshes. In winter it migrates to India, penetrating to the south of the peninsula. Colonies are also resident in permanent jheels in Northern India and the United Provinces, and probably in Khandesh and the mangrove swamps of the West Coast at least as far south as Bombay.

The resident birds in India are greatly augmented by a host of migrants from more northerly breeding areas as this race of the Great Reed-Warbler nests from Transcaspia through Persia, Turkestan and Northern Afghanistan, moving farther south in the winter.

HABITS AND NESTING. We feel quite sure that every bird-lover who has traversed the waterways of the Kashmir Valley must have the strident calls of these reed-warblers indelibly recorded upon his brain. Certain it is that we can never think of our first trip upon the lovely waters of the Dal lakes and their canals without calling to mind the tall flanking reeds and the many varied voices which issued from their depths; the clucks of the Waterhens, the croaks of the Little Bitterns, now and then the plaintive bubbling voice of a Dabchick, but above all the raucous vibrant *kori-kuch kori-kuch kriti-kriti*, with many variations, of the Indian Great Reed-Warblers. Wherever a reedy patch occurs, therefrom will this bird proclaim its noisy presence at all hours of the day, calling a great deal even after sundown. Occasionally one appears on the outskirts of its patch, clambers expertly to the summit of the tallest wand and gives tongue to quite a passable song consisting of a most varied repertoire of harsh outpourings based upon the notes given above. On occasion, it even ascends to the added vantage-point of a willow tree from which to exercise its vocal powers, but nine-tenths of its time is spent creeping unseen in the fastnesses of its reedy forest.

It is about the beginning of June that the reed beds of Kashmir attain a sufficient growth for nesting to begin. A few eggs may be found before then—23 May marks our earliest full clutch—but June and July are the main breeding months of the reed-warblers. Numbers of eggs may still be had in August, but most of these belong in all probability to birds whose first nests have been swamped by sudden rises in the water level. Three or four, occasionally only two, close-growing reeds are harnessed to their purpose, and around the stems, from two to about four feet above the water, the nest of coarse rush grass is woven. The materials are those at hand with no lining other than the finer blades. The cup is deep, as the soughing wind rippling across the bed sways the reeds in unison so that a deep enclosure is advisable to prevent the eggs from being tipped into the water. Very occasionally a nest may be found built in a young willow growing in close proximity to, or even amongst, the rushes. In fact, in the extensive willow nurseries between the Anchar Lake and Shadipur these willow tree sites are common. The owners of a nest are bold even before the young appear, and often approach close to the intruder scolding loudly until the danger is past. On one occasion an excited bird pitched on the rim of its nest while our fingers were actually resting on the opposite side.

The eggs are usually 4 in number, quite often 3 only, averaging 22·6 by 15·8 mm. They vary considerably in coloration, but a normal egg is a glossless rather blunt oval, white, greyish, or greenish-white in ground colour, and has large irregular markings of brown and blackish-brown generally with underlying secondary markings of lavender or stone colour. The shell is rather coarse and brittle.

The Asiatic Cuckoo, whose familiar voice is to be heard around the jheels of the Vale as well as in the mountains, sometimes lays its eggs in the nests of this species, perhaps more frequently than might be supposed. On 6 June a nest was taken containing a cuckoo's egg and one of the rightful owner's. On the following day a second reed-warbler's nest was discovered about 300 yards away which also contained a cuckoo's egg along with two of the warbler's. These nests were in inundated willow plantations near Shalabug. In this case it is likely that both eggs were the product of the same cuckoo. J. E. Scott also took cuckoos' eggs in early June from two nests in the Anchar area.

WITHERBY'S PADDY-FIELD WARBLER

Acrocephalus concinens haringtoni Witherby

(See Plate 29 facing p. 116 and Plate 80 facing p. 317.)

BOOK REFERENCES. *F.B.I.*, II, 396 ; *Nidification*, II, 361.

FIELD IDENTIFICATION. (2—) We have heard this bird described as a miniature of the last species, but in point of fact it differs from it very considerably, especially in habits. For it avoids the coarser reed beds and is a shyer bird. It rather affects the softer grasses and thin reeds around the edges of the more secluded waters, and as we have shown below, even patches of weeds on open hillsides. It lacks the harsh notes of the Indian Great Reed-Warbler, the male possessing a rather sweet song which he proclaims in the vicinity of his nest from the summit of a reed or plant overtopping its neighbours. In coloration it is a sober olive-brown above with a well-defined eye-stripe, which extends farther back and is more conspicuous than that of the much larger Great Reed-Warbler. The lower plumage is buff.

DISTRIBUTION. Occurs in considerable numbers around the lakes and marshes of the Vale, while Brooks recorded a nest bordering the extensive rice-fields near Shupyion. We also found it to be not uncommon near Suknes (9,000 ft) in the Wardwan Valley, but have reason to believe that it nests in that Valley from at least the vicinity of Inshan.

This is a bird which has given a lot of trouble in the past and there is still something to learn regarding its distribution and status. It belongs to the *concinens* group and not *agricola*. The late Hugh Whistler separated it from the Khagan bird, *haringtoni*, under the name *Acrocephalus concinens hokræ*, partly on colour and partly due to the fact that it had hitherto been found in Kashmir breeding only in the jheels in the Vale. A series collected by Meinertzhagen in Afghanistan, and birds found by us breeding at some distance from water in weeds on hillsides in the Wardwan, caused Whistler to review the available specimens in the light of the dates on which they were collected and satisfied him that *hokrae* is not separable from *haringtoni*. This then is the form breeding in Afghanistan, the Khagan Valley of Hazara, and Kashmir.

HABITS AND NESTING. Generally speaking this little reed-warbler is to be found on the same jheels and marshes as the far more conspicuous Indian Great Reed-Warbler. It, however, prefers the softer grasses around the margins rather than the coarse reed beds. It is, too, of a retiring nature. For these reasons it is scarce or wanting on more open frequented waters such as the Dal lakes, although we have seen it in the vicinity of the entrance to the waterway to the Shalimar Gardens. We found it very common at Hokra Jheel, but it doubtless occurs all over the Vale where conditions are favourable. For instance, mention might well be made here of the nest found by Brooks and described by him in the following terms : ' Near Shupyion I found a finished nest of this truly aquatic warbler in a rose-bush which was intergrown with rank nettles. This was in the roadside where there was a shallow stream of beautifully clear water. On either side of the road were vast tracts of paddy swamp.' It possesses a sweet song and males may often be seen on the summit of a reed or tall weed, probably not far from their sitting spouses, puffing out their throats while singing a sweet refrain.

In the marshes the nest is generally well hidden where soft grasses grow amongst the thinner reeds, although we have seen nests anchored to three or four thin rushes unconcealed in any way. We also came upon this bird breeding in fair numbers in tall rank weeds around Suknes up to an altitude of 9,500 ft. It may breed along the Wardwan Valley at considerably lower elevations than this, as we stumbled upon a nest in tamarisk scrub on an island opposite Inshan, 16 miles lower down the river. This nest had been swamped and filled with silt, but it contained one broken egg. The nests in the Wardwan Valley were built in beds of tall weeds mostly well removed from water. The first one to be found was on a steep hillside, the remainder on almost level ground. Nidification was then, late July, nearing completion as we saw numbers of young birds and only in the first nest were there eggs. In 1932 we saw birds, which we now consider probably belonged to this species, at the mouth of the Nichnai Nullah not far from the Thajwas bridge over the Sind River. This valley junction is a definite storm-centre where the rainfall would appear to be greater than elsewhere in the Sind Valley so that the vegetation approximates to that around Suknes. The elevation, too, is similar, so this little warbler may well breed there.

They appear to construct two different types of nest which, however, grade one into the other. One is a small replica of that of the Great Reed-Warbler of coarse grasses with no real lining, finer material of the same kind being used as the inside nears completion. This type of nest is at times to be met within the swamps as well as being the type we found in Kishtwar. To the other nest moss is added in varying quantities. A nest found at Hokra Jheel on 18 June was made almost entirely of moss and was anchored to three thin reeds. It was barely a foot off the wet ground and quite unconcealed. Wilson, writing in the *Journal* in 1899 (XII, 637), says of five nests found by him at Sumbal on 11 July : 'All were beautiful cups of moss lined with grass and a few feathers.' Osmaston also notes that his nests invariably contained some moss, saying of thirteen nests taken in the Krahom swamp that they were composed of dry grass and rush leaves and occasionally some moss, neatly woven together with animal wool or vegetable cotton and lined with fine grass, moss, fruiting stems, etc. As a rule the nest is rather close to the ground, whether in the swamps or on the hillsides, at times being scarcely a foot from it and seldom as much as three feet up. Its degree of concealment varies considerably, but when placed in a tangle of marsh grasses it is sometimes very difficult to find.

Three eggs generally constitute the full clutch although 4 may at times be found, but one of our Wardwan nests contained 2 incubated eggs. Both in the hills and in the Vale, June and July would appear to be the breeding season. The eggs are replicas of those of the Great Reed-Warbler, but they are usually less profusely marked and at times in fact the markings amount to only a few bold spots. They are, of course, smaller, averaging 17·1 by 12·7 mm.

LARGE-BILLED BUSH-WARBLER

Tribura major (Brooks)

BOOK REFERENCES. *F.B.I.*, II, 403; *Nidification*, II, 364.

FIELD IDENTIFICATION. (1) A small brown warbler paler on the underparts, with a whitish, often indistinct, eye-stripe. A skulker in thick grass, weeds and bracken at elevations of from 8,000 to 12,000 ft. Has a monotonous but distinctive ticking note.

DISTRIBUTION. A common enough bird wherever it occurs throughout our whole area, but rather erratic in its distribution owing to choice of a particular type of habitat. In the Sind Valley Davidson found it only around Sonamarg at approximately 9,000 ft, but we noted it as quite common on the open slopes of side nullahs beyond Baltal at about 11,000 ft. It is probably commonest between those limits, but Buchanan, who also took its nest at Sonamarg, states that it does not appear to breed below 7,500 ft nor above 10,000 ft. The latter figure is definitely wrong and we think few pairs will be found below at least 8,000 ft.

The Large-billed Bush-Warbler breeds in Eastern Turkestan, Ladakh, and in the Western Himalayas from the Khagan Valley of Hazara to Tehri-Garhwal. In winter it is said by Stuart Baker to descend the hills to about 4,000 ft.

HABITS AND NESTING. Osmaston describes this bird's notes as persistent and monotonous, resembling *chipi-chipi-chipi*, repeated indefinitely at the rate of about three to the second generally from the topmost twig of some bush or small tree. To anyone unconversant with these ticking notes, this arrant skulker is easily overlooked, for it possesses an extremely retiring nature, moving furtively through the herbage more like a timid mouse than a bird.

In fact in the vicinity of its nest it is almost an impossibility to flush it. On one occasion we were within a few feet of a nest, but, search as we would, all that we achieved was to hear the bird's rustling movements in the thick grass and to catch a fleeting glimpse of it sneaking off between the stems. We went through that patch of grass and thick weed twice, but failed to locate the nest, nor did the bird let itself be seen a second time.

The Large-billed Bush-Warbler is a bird of open waste lands. It appears to prefer slopes dotted with bushes growing amongst patches of rank grass and weeds. According to Osmaston it also occurs in irrigated grass lands and cultivated fields, but Davidson, at Sonamarg, failed to find it anywhere except amongst the long grass and weeds fringing the forests and for no great distance out from the forest edge. It never, of course, enters the woods.

It seems to be a late arrival, for Davidson neither heard nor saw it at Sonamarg until 8 June. Like most of the later migrants to move in, it gets down to nesting immediately and by the 15th he was able to find completed but empty nests. Its nesting time is undoubtedly the latter half of June and July, in which month Buchanan also took nests in Sonamarg.

The nest is a small but deep cup of dry grass placed on or near the ground in thick grass or weeds. It is often amongst the smothered roots of small bushes and is always well concealed. The normal clutch is 3 or 4, although 2, and even 1, incubated eggs have been recorded. These are broad ovals, large for the size of the bird, of a decided pink hue, and are marked with bright terra-cotta spots and a few underlying greyish streaks. These markings are well distributed all over the egg, but are at times more numerous in a zone around the large end. They average 18·9 by 14·3 mm.

HUME'S LESSER WHITETHROAT

Sylvia althæa Hume

(See Plate 30 facing p. 117.)

KASHMIRI NAME. *Chet Hyot.*
BOOK REFERENCES. *F.B.I.*, II, 450; *Nidification*, II, 407; *Handbook,* 174.
FIELD IDENTIFICATION. (2) A neatly attired warbler which appears mainly ashy above with the sides of the face slightly darker.

PLATE 31

Tickell's Willow-Warbler building in juniper

Large Crowned Willow-Warbler with food for young

PLATE 32

Indian Oriole at nest. Male and female

Sullied white below, greying on the flanks. The throat, being a purer white than the rest of the lower plumage, is rather conspicuous, contrasting well with the dark cheeks. It inhabits open rocky ground with low scattered bushes and scrub.

DISTRIBUTION. Common on the barer hillsides around the Vale in patches of low bushes, but found in smaller numbers in like country up to at least 8,000 ft. In the side valleys we have taken its nest at Pahlgam, Aru, and Praslun in the Lidar Valleys, in the Rampur-Rajpur Valley above the Wular Lake, and below Tragbal, and have seen it in the lower Kishenganga Valley in early May where at Keran, elevation 4,900 ft, it appeared about to breed. It is a summer visitor, wintering, according to Whistler, in Southern India. It is to be seen on the autumn passage in the Vale in some numbers. It is a breeding bird from Transcaspia to Turkestan, and North-East Baluchistan to Kashmir.

HABITS AND NESTING. This is a common breeding bird in late April, May and June on the scrubby promontories and stony slopes around the Wular Lake, such as enclose the bays at Kyunus and Watlab. The lower slopes are composed of rather dried-up stony ground with here and there tufts of coarse grass, scattered boulders and little cliff-like declivities. Everywhere are scattered low prickly bushes of cotoneaster, berberis and other thorny kinds, while in places the brushwood is more varied and thicker with briars and larger-leafed plants appearing in the more sheltered areas. This in the early nesting season is ideal whitethroat country and wherever its type appears, up to some 7,000 or even 8,000 ft, there, too, is one likely to find this bird.

Often the nest is well concealed in the midst of a thorny bush, but the bird cannot be termed shy and at times proceeds with its home-making in conspicuous situations, for we have found it in a scraggy briar within reach of one's hand without moving from a much frequented path; in a village hedge; and quite open to view, nearly 8 ft up in a young fir. The latter site is of considerable interest for it was in the month of July, the clutch was incomplete, and the locality between Pahlgam and Aru. It has been averred that after nesting at lower levels in May some birds at least proceed to higher elevations to raise a second brood, and it is then that these nests in young firs and saplings are to be found. Personally we have seen no evidence of a later upward movement and Davidson's remarks (*Ibis*, 1898, 16) point rather to birds making use of tree sites for second broods in the same areas in which they had built their bush nests. 'This bird', he writes, 'we found in great abundance on the bare hills around Srinagar in the end of April, and among the scrub jungles along the Sind River as far as Koolan (6,800 ft) four or five miles farther up than Gund. It was breeding from the end of April to the end of May in low scrub, generally along the nullahs ... On our return to Srinagar in the end of June the hills had got very much burnt up, and we were very surprised to find this bird again breeding, but instead of being among the scrub the nests were on the outer branches of pine trees, 15 and 20 ft from the ground. We found four or five nests in this situation on the Takht-i-Suleman, all with fresh eggs, and the birds seemed at that time to be restricted to the small scattered pine wood.' It must not be inferred from the above that the Lesser Whitethroat proceeds to raise its second brood exclusively in nests in tree sites, for at the end of June and in early July we have seen nests both in trees and in scrub and in briars.

The nest is rather a loose cup of dry grass, thin bents and rootlets, lined with fine roots and a little hair; Stuart Baker says a few leaves may be incorporated, but we have not noticed this. It is thin-walled and frail-looking, but tough enough to stand all and more than is required of it. The eggs number 3 to 5, usually 4, the ground colour being a greyish-white with a number of rather bold dark brown or blackish spots with secondary markings of a neutral or lavender tint. They average 19·5 by 13·0 mm.

As incubation proceeds the birds become very confiding, coming to the nest quite readily when one is still in its vicinity. The breeding season is prolonged, lasting from late April until at least the first half of July, rendering it probable that to bring up two broods in the year is general. Of the song Osmaston writes that it is bright with occasional harsh notes, and on the whole the effect is pleasing. We would add that on being disturbed at the nest it is apt to utter a very characteristic *churr*. An interesting and fuller note is to be found in Magrath's *Notes on the Birds of Thandiani*, in which he writes : ' One bird was singing most exquisitely and its notes were different to those of the others; it was, I think, Hume's Lesser Whitethroat. The song may be rendered in words something as follows : " *Karu-karu-karu, kari-kari-kari, chirri-chirri-chirri, chup-chup-chup-chup, chawai-chawai-chawai, ih* . . .".'

TICKELL'S WILLOW-WARBLER

Phylloscopus affinis (Tickell)
(See Plate 31 facing p. 124.)

KASHMIRI NAME. *Viri Tiriv*.
BOOK REFERENCES. *F.B.I.*, II, 454 ; *Nidification*, II, 410.
FIELD IDENTIFICATION. (1) This minute warbler may be recognized from others of that large fellowship of willow-wrens by being the only one to be noted in the breeding season at such high elevations, that is on the open juniper- and bush-dotted hillsides well above the tree line, where it will be found feeding in low bushes and even on the ground around them. Secondly, the bright yellow underparts are diagnostic. The upper plumage is brown with a tinge of olive about it, which is relieved only by a conspicuous supercilium. No other small greenish warbler with these characteristics is found in the treeless highlands.
DISTRIBUTION. Distributed generally in our area throughout the uplands above forest level, that is between about 10,500 and 15,000 ft, arriving on its breeding grounds in late May. In April we saw numbers in the lower Kishenganga Valley on passage. They also use the Jhelum Valley route and many pass through the Vale on their way to the surrounding heights.

The summer distribution of this species is a wide one, and includes the frontiers of India from North Baluchistan to the Himalayas, in which they breed throughout the entire length, and onwards through South-East Tibet and Western China to Kansu. Ludlow states that they are common in summer in Lhasa. In winter they migrate south into India and Burma excepting the drier north-west areas of the former.
HABITS AND NESTING. Early in June this hardy little warbler betakes itself to the uplands well above the highest limits of the forests to those open hillsides and valleys where there are patches of juniper, dwarf rhododendrons or other small bushes in which to build its untidy but cosy little globular nest. This it constructs from a half to two feet from the ground, often in conspicuous positions on the outer edges of the bushes. Tickell's Willow-Warbler is a trusting little bird and seldom makes any real attempt at concealing its home. It shows little fear of mankind and visits its nest freely during one's near presence. Stuart Baker in the *Nidification* says that in Kashmir it breeds from 9,000 to 12,000 ft, rarely as low as 7,000 ft. On what authority he gives it this low range we cannot say, and our experience, which appears to tally

with that of others, is that 12,000 ft is nearer the lower limit rather than the upper. So far we have never taken a nest below 11,000 ft, and in the breeding season we have seldom even seen the bird much below that elevation. For instance, it is comparatively common in the juniper patches between Zojpal and the top of that back-breaking climb from Tanin in the East Lidar Valley commonly known as the Pisu. Again we have met with it at Sekwas at about 12,500 ft, but not down at Lidarwat. Lastly, the bird in the illustration was photographed on the slopes above Killenmarg, a good forty minutes' walk above the ski-hut. We have, in fact, looked for its nest on a number of occasions at lower elevations but in vain. In the winter it spreads itself over the greater part of the plains of India, commencing its return migration to the hills about April.

In construction the nest is a flimsy sphere of dry grass very roughly put together with bits and pieces sticking out untidily in all directions. Its lining, however, is a compact warm one of feathers, often with a few hairs intermingled, to form a very necessary and cosy blanket as it were, to guard the occupants against the bitter cold of the nights at such lofty altitudes. The first nest we found was heavily lined with soft black feathers, undoubtedly those of choughs which inhabited the ledges and caves of an adjacent cliff. All accounts seem to agree that feathers form the main item in the lining, but hair and fur are also used to a lesser extent. These warblers appear to have only one brood and to crowd building, laying and incubation into the period between the middle of June and the middle of July, presumably due to the shortness of the warm weather in an elevated habitat. At any rate by the end of July many young are strong on the wing, and in the first week of August we have quite failed to locate occupied nests.

If it is possible to judge by one nest alone, building is rapid and performed by both birds. At midday on 16 June the subject of the photograph was just laying the foundations of its nest about one foot from the ground in a low juniper bush; elevation about 11,000 ft just above Killenmarg. On the 17th nest-building proceeded furiously, and by 10 a.m. on the 18th, when we left the hiding-tent, the grass envelope was ready to receive its lining. On that day we remarked that both birds were building, for, although we never had the two of them actually at the nest at the same moment, one was exceedingly bold whilst the other was very nervous and jumped violently at every clap of the shutter. Building was carried out in frenzied bursts during which grass was often collected from the ground close to the nest and around the hide. Other evidence that the male takes a full share in all operations is provided by Whitehead, who caught the male incubating.

As to its vocal powers we quote A. E. Osmaston : ' The bird has a tack-tack note similar to that of many other Warblers. It also possesses a song, if such it may be called, composed of a single note uttered some four to six times in rapid succession and preceded by a single rather high-pitched note.'

Four eggs is undoubtedly the normal clutch, though the first nest we ever took contained, on 1 July, only two showing the first signs of incubation. These eggs were white with a few very minute red spots here and there. The spots are always few and small, and one or two eggs of a clutch may be immaculate. Stuart Baker, who says that 5 eggs are laid in about one nest in four, remarks that as a rule in each clutch one egg is quite distinctly spotted, one or two faintly spotted, and the others pure white. They are fragile eggs with a fine gloss and average 16·2 by 12·3 mm.

TYTLER'S WILLOW-WARBLER

Phylloscopus tytleri Brooks

KASHMIRI NAME. *Viri Tiriv.*
BOOK REFERENCES. *F.B.I.*, II, 455 ; *Nidification*, II, 411.
FIELD IDENTIFICATION. (1) One of the plainer willow-wrens with little about it to aid identification. Like the last bird its underparts are yellow but the tint is pale, and it also has no wing-bars. It is essentially a bird of the forests, nesting at the extremities of fir branches at considerable heights from the ground, but often feeding low down in the viburnum bushes and other undergrowth.
DISTRIBUTION. Found in summer throughout all the pine and fir forests from moderate elevations to the trees' highest limits. It is not particularly common, but is perhaps more numerous on the Pir Panjal side of the Vale. It is said to winter mainly on the western side of India as far south as the Nilgiris, but does not appear to occur in Gujarat, and is, of course, absent from Sind and apparently also from the Punjab plains. Some merely drop into the outer hills and plains adjacent thereto.

The *Nidification* gives the full breeding range of this bird as the Western Himalayas from the Afghan boundary to East Garhwal, but its occurrence is not mentioned by A. E. Osmaston in his *Birds of British Garhwal*.

HABITS AND NESTING. This little willow-warbler is one of the least obtrusive of its kind, for it has nothing particularly distinctive about it to attract attention either in habits or in its coloration. It is on the whole a detached bird avoiding the limelight, though its notes, as Osmaston has it, resembling the words ' Let's kiss him ', are uttered at frequent intervals. It makes no lamentations on the discovery of its nest, like those indulged in by the fussy Crowned Willow-Warbler, and in fact to trace it to its home is no easy matter. Although it may often be seen in the undergrowth, especially in bushes bordering sunny blanks in the forests, it prefers to nest in the upper layers of the fir trees, the majority of homes being 30 to 40 or more feet from the ground. Nevertheless, on 28 June we found one, in the vicinity of Tanin, so low that it could be reached from an adjacent boulder. This nest was domed with a large side-entrance, not unlike a tiny dipper's nest, and constructed of moss, lichen, short lengths of fine grass, and other oddments. It was lined with a considerable quantity of feathers. It was, in fact, very like that of Ticehurst's Willow-Warbler. Davidson's and Osmaston's descriptions of the nest also tally with this, for the former describes two nests taken near Sonamarg on 9 and 11 June as neat and globular with the entrance at the side. They were composed of feathers, grass, birch-bark and hair ; the last described as either horse or mouse : surely rather queer that doubt should arise on that score ! Far more peculiar, however, is Major Cock's description of a nest about which he wrote to Hume as follows : ' Of all the birds' nests I know of, this is one of the most difficult to find. One day in the forest at Sonamarg, Cashmere, I noticed a Warbler fly into a high pine with a feather in its bill ... so allowing a reasonable time to elapse (nine days or so) I went and took the nest. It was placed on the outer end of a bough, about 40 feet up a high pine, and I had to take the nest by means of a spar lashed at right angles to the tree, the outer extremity of which was supported by a rope fastened to the top of the pine. The nest was *a very solid deep cup* [the italics are ours] of grass, fibres, and lichens externally and lined with feathers. It contained 4 white eggs ... I forgot to add that this nest, the only one I ever found, was taken early in June.' Did Major Cock remove this nest himself or was it brought down to him with its large side-entrance turned upwards ? Whether domed or cup-shaped, it seems to have been more substantial than any nest of *proregulus*. Cock's nest was high up, but both of Davidson's were in ' forks of small pollarded trees some 12 to 15 feet from the ground'. Cock's nest contained 4 eggs, but one of Davidson's held 5, and one 4 small young. The eggs are plain dull white and average about 16·0 by 12·2 mm. An occasional egg may, however, be faintly spotted with red.

TICEHURST'S WILLOW-WARBLER

Phylloscopus proregulus simlaensis Ticehurst
(See Coloured Plate II facing p. 40.)

KASHMIRI NAME. *Viri Tiriv.*
BOOK REFERENCES. *F.B.I.*, II, 467 ; *Nidification*, II, 420.
FIELD IDENTIFICATION. (1—) This tiny warbler is readily distinguishable from other willow-wrens by its yellow rump. This bright patch is very conspicuous when it flies up at an angle into the fir trees in the outer branches of which it places its fragile domed nest. It also possesses a well-defined yellow coronal band and two indistinct pale wingbars.
DISTRIBUTION. Ticehurst's Willow-Warbler is a bird of the fir forests, particularly those containing Scotch Firs and Blue Pines, but it will also be seen hunting about in the lower cover at the forest's edge. It is fairly common from about 7,000 to 10,000 ft, and occurs throughout our whole area including Gurais, the Wardwan Valley and the forests of the Pir Panjal running down to the main Vale.

Besides the three races nesting or wintering within the limits of India and Burma, the species as a whole is widespread in Asia and penetrates into Europe ; an example of the typical race was taken in England in 1896. Ticehurst's race nests on the North-West Frontier in the Kurram Valley (Sufed Koh) and thence along the Western Himalayas as far as Garhwal. In winter it drops into the foothills, to low levels in the coldest spells, but is not found in the Plains. It appears to carry out its upward movement in April.
HABITS AND NESTING. This is one of the smallest and most delightful of that great band of willow-warblers in which Kashmir is so rich, both in regard to numbers of individuals and to species. It is not particularly common although it is well distributed, but its nest is so flimsy, well-concealed and small, and the bird itself not given to advertising its presence, that but for the tell-tale rump it might easily escape detection. When flying up at an

WILLOW-WARBLERS

angle, however, to reach the branches 20 or so feet from the ground, this broad yellow band across the rump shows up well and, as no other willow-wren in Kashmir has this distinctive mark, its identification becomes simple. It is an unobtrusive quiet bird, often feeding in the undergrowth as well as to be seen fluttering about the ends of the fir branches. It does not stick always to the interiors of the forests, but may on occasion be found where there are more or less isolated pines. For instance, we noticed a pair with a nest tucked into a mass of sticks and pine-needles wedged into the branches of a tree standing quite by itself on the marg at Bagtor in Gurais.

Nests may be found from half a dozen up to 50 ft or more, but many of them, as in this case, are about 15 to 20 ft from the ground. We became used to marking them down by listening for the sharp, monosyllabic *tsip* of a bird flying over our heads and then, having followed it up, watching it fly direct to its well-concealed home. Where there are thick tufts of needles formed by the end of one branch drooping on to another are the favoured points, so that the snug domed structure, which takes up little room, is hidden from above as well as from below. In fact, it is generally quite invisible until the branches are parted, and almost the only way to discover it is to note the prolonged disappearance of the bird from view. Stuart Baker, however, writes in the *Nidification* : ' Some nests are very carefully concealed and can only be discovered, especially when they are high up, by watching the birds on to them ; others are easily spotted and can quickly be detected once the breeding tree is marked down, not a difficult thing if the actions of the bird are understood.'

The nest is a domed or semi-domed affair with a comparatively large entrance in the side and matches its surroundings well. It is composed of scraps of moss, particularly on the outside, strands of lichen, skin of birch-bark, and then a little grass, and is lined with soft feathers. In spite of this long list of materials the whole thing can at times be compressed in the hand to the size of a walnut, but, with its protecting needles and foliage around it, it is sufficiently strong and snug. The eggs, 3 or 4, and rarely 5, are white profusely speckled with reddish-pink, brick red or reddish-brown, the markings often forming a ring round the larger end. Nest-building begins about halfway through May and fresh eggs may be found until the end of June. They average 14·1 by 10·9 mm.

PLATE 33

Kashmir House Sparrow, male, leaving nest

Himalayan Starling at nesting hole.
Note the absence of spots in the worn breeding plumage

PLATE 34

Blackheaded Myna at nesting hole

Common Myna with food for young

HUME'S YELLOW-BROWED WILLOW-WARBLER

Phylloscopus inornatus humei (Brooks)
(See Plate 30 facing p. 117.)

KASHMIRI NAME. *Viri Tiriv.*
BOOK REFERENCES. *F.B.I.*, II, 469; *Nidification*, II, 422; *Handbook*, 176.

FIELD IDENTIFICATION. (1) A ground-nesting willow-wren of the lighter forests, laying spotted eggs in a domed nest usually placed in the grass on steep banks and hillsides. Olive-green above and whitish below with pale eye-stripes, an indistinct coronal band and double wingbars, the forward one of which is also rather indistinct.

DISTRIBUTION. A common bird throughout Kashmir, particularly in the Silver Fir forests from about 7,500 ft up to more or less the top limits of these forests. It is equally numerous in the forests of the Pir Panjal range as in those on the northern side of the Vale. It is particularly common in woods on steep hillsides which are well provided with sunny glades. We were surprised to find a nest at Astanmarg at over 12,000 ft in the shattered birch wood. In late summer these birds leave the Himalayas, spreading throughout nearly the whole of India with the exception of Sind and the extreme South-East. They pass through the Vale in late September and October, returning towards the end of March and in April.

Of the three races of Yellow-browed Willow-Warblers breeding over a large part of Central Asia, Hume's nests between some 7,000 and 12,000 ft in Turkestan, the Tien Shan, Afghanistan and the Western Himalayas to Garhwal. Whitehead stated it to be common in the Sufed Koh between Afghanistan and the North-West Frontier.

HABITS AND NESTING. This bird is almost as common in Kashmir as the Large Crowned Willow-Warbler. In the open forests around Sonamarg, for instance, the loud call, which may be syllabized as *tissip tissip*, or as Whistler has it *te-twee-up*, is continuously to be heard from all directions throughout the hours of daylight. It is in this area that we have taken more nests than in any other part of Kashmir, but from all accounts it is equally common in the vicinity of Gulmarg and no doubt around any of the forest-encircled upland meadows where these delightful little birds help by their cheerful and lively presence to add to the happiness of those who visit these pleasant haunts. Like nearly all the willow-warblers they are restless energetic birds, forever searching the trees and even the lowliest undergrowth for insects. They may be seen fluttering against the bark of a tree, threading their way through bushes and brambles or even descending to the ground, occasionally breaking into a short attempt at a song and uttering endless *tissips* if one ventures near the inconspicuous nest, a small globe of grass concealed in the herbage on a steepish bank and usually protected by a bramble or plant of some sort as are so many willow-wrens' nests in England. The nest at Astanmarg, found on 2 July with fresh eggs in it, was in the roots of some straggly juniper growing amongst the gaunt birch wood about the centre of the marg. Above all, Hume's Willow-Warbler loves the forest glades, particularly those on steep hillsides which catch the sun and which Osmaston so aptly dubs sunny blanks in the forest. It is around these that the nests may be found by watching the birds, as the small entrance-hole, hardly an inch in diameter, is to say the least of it not easy to spot. One nest we took at Sonamarg in the month of June was made almost entirely of fresh pine-needles with a foundation of the older fallen needles. It was lined with a number of horse-hairs as well as finer hairs of some other kind. The lining is usually of grass, finer than the outer ball, but a few hairs are generally incorporated. Within this little sphere the bird lays 3, 4, or occasionally 5 small white eggs spotted sparingly with tolerably conspicuous spots of red or reddish-brown. These average 14·0 by 10·9 mm. The birds commence to arrive in Kashmir in late April or in May and at once begin nesting, but we have taken fresh eggs well on into July. Late May, June, and early July will be the main months, the greatest number of eggs being laid in late June.

LARGE-BILLED WILLOW-WARBLER

Phylloscopus magnirostris Blyth

KASHMIRI NAME. *Viri Tiriv.*
BOOK REFERENCES. *F.B.I.*, II, 476 ; *Nidification*, II, 426.
FIELD IDENTIFICATION. (1) The largest and darkest of our willow-warblers, being generally dark olive-green above and sullied below. It possesses a well-defined yellowish-white supercilium and two wingbars, only one of which, however, is at all noticeable. Its presence is far more likely to be remarked by its

distinctive five-noted call, the second and third, and the fourth and fifth, of which are lower in the scale than the first note and the preceding pair respectively. Generally heard near fast running water.

DISTRIBUTION. Widely distributed between about 7,000 and 10,000 ft, generally along the banks of rivers and streams running through forested areas. Ward, however, records a clutch taken at 6,000 ft in the Dachhgam Nullah. A migratory species into the Indian Plains as far south as Central India to Bengal, and to Northern and Central Burma.

Beyond our limits it breeds at the head of the Kurram Valley on the slopes of the Sufed Koh, and in the inner Himalayas generally from the Khagan Valley in Hazara to at least Garhwal, possibly to Sikkim. The *Fauna* also adds Ladakh and Tibet to Kansu.

HABITS AND NESTING. This, the most elusive of the willow-warblers, arrives in Kashmir in May. Until it reaches its breeding ground it is an unobtrusive, rather secretive bird, moving about the lower branches of the fir trees, in the undergrowth and amongst the weeds on grassy banks along the streams, for its unmistakable clear notes are not added to the avian chorus until it is in occupation of its chosen territory. In 1935 we were in Lower Gurais when numbers were moving up the valley, but until we reached Koragbal on the last day of May not once did we hear its call. It is not in our opinion particularly common, but most streams worthy of the name, with open banks not too overshadowed by the encompassing forest, and every stretch of the larger rivers from 7,000 ft up to the thinning off of the pines and firs, will bear evidence of the presence of usually one pair of these warblers within hearing. The song, however, carries a considerable distance so that this does not imply a numerous population. Besides the five-noted song this bird has a quieter double call-note, the second and higher note pitched in the same key as the first note of the song.

Both the bird and its nest are hard to track down for the bird itself is not the familiar fussy little being that so many of the willow-warblers are, and the nest is well concealed, being a somewhat loose ball of grass, dead leaves or moss, etc., tucked well into a hole in the roots of a fallen tree, in a cavity in an overhanging bank, in a crevice in the rocks, or out of sight in a tangle of growing roots ; in almost every case on or near ground-level overhanging or close to water. Sometimes the nest may be composed almost exclusively of moss, or grass, bits of fern or leaves may predominate, while the inner lining will consist of finer grasses and possibly a few hairs.

WILLOW-WARBLERS

Although the birds commence to call at the beginning of June, they do not appear to start building until towards the end of that month, whilst their eggs are taken mainly in July, August, and even in early September. A completed nest with the bird seated in it was, however, noted on 30 June at Sonamarg. The eggs are pure white, up to 5 in number, usually 4, and average 15·3 by 11·9 mm.

LARGE CROWNED WILLOW-WARBLER

Phylloscopus occipitalis occipitalis (Blyth)

(See Plate 31 facing p. 124.)

KASHMIRI NAME. *Viri Tiriv*.
BOOK REFERENCES. *F.B.I.*, II, 479; *Nidification*, II, 429; *Handbook*, 178.
FIELD IDENTIFICATION. (1) Upper plumage olive-green; lower plumage white with a yellowish tinge. Has a coronal band and pale yellow supercilia and indistinct wingbars. Nests in holes in or near the ground and is very fussy in its vicinity. It is the commonest of the willow-warblers to be found breeding in Kashmir. Has a peculiar habit, when searching the foliage, of restlessly waving one wing.
DISTRIBUTION. Surprisingly common throughout Kashmir in all types of forest and well-wooded country. It is most numerous between, say, 6,000 and 8,000 ft, but occurs and breeds up to the topmost limits of the forest. Down to what elevation it breeds in the Jhelum and Kishenganga Valleys we cannot say, and it is not generally met with in the lower-lying parts of the Vale. It is a migratory species, arriving in our area in April and May and leaving by September, but the dispersal seems to commence at the end of July. It spends the winter throughout the plains of Northern India excepting Sind and Gujarat, and spreading down south as far as Travancore on the west and Orissa and Bengal to the east.

Two races of this species breed from Turkestan and Afghanistan across to Japan, our western race being found in summer as far east as Garhwal. So far as India is concerned it breeds on the outer face of the Himalayas to their northern extremity and thence down the North-West Frontier to the Kurram Valley.
HABITS AND NESTING. The Large Crowned Willow-Warbler is one of the most widespread birds in Kashmir, for it permeates

PLATE III

1. Northern Indian Pied Bushchat (adult male). 2. Crested Bunting (adult male).
3. Western Collared Pigmy Owlet. 4. Orange Bullfinch (adult male). 5. Himalayan Goldfinch. 6. Indian Grey-headed Bunting. 7. Eastern Swift. 8. Western Slaty-blue Flycatcher (adult male). 9. Japanese Wryneck. 10. Pink-browed Rosefinch (adult male).
11. Himalayan Greenfinch (adult male). 12. Stoliczka's Mountain-Finch

woods of all types and well-wooded country from almost valley level to the tops of the fir forests, being excessively common round about 6,000 to 8,000 ft. At the beginning of the breeding season males are very combative and quarrelsome, Magrath recording that on one occasion a pair fell at his feet locked in mortal combat. They are indeed excitable birds and from May to July it is impossible not to find numbers of their nests, for they nest on or near the ground, though occasionally making use of holes in trees up to 20 ft, and by their very habits and fussiness draw the attention of the searcher to the vicinity of the home. The intensity of the plaintive *tee*-ing increases in proportion to the nearness of one's approach thereto. To quote Magrath : ' a characteristic habit of this little bird, as it creeps about branches and shrubs, is its restless waving of one wing ; at the same time it is continually uttering its monotonous note of *chip-chip-chip-chip*.' According to Osmaston they stop calling in August, preparatory to disappearing altogether in September.

The nest is a loose conglomeration of dried moss and leaves, sometimes intermixed with a little wool. Often the material partially lines the roof of the cavity, for it is placed out of sight in all sorts of queer places. At times it is well under a stout root of a tree ; frequently in old rat-holes ; pushed into a cavity in a stone wall, or yet again in a hole in a tree. Magrath records one built in the rafters of a cowshed and another high up in the gable of a bungalow's roof, while Osmaston writes of one at Gulmarg between the double boarding of the hut he was occupying. We have on a number of occasions found them down vertical holes in the floor of the forest, and the dogs, digging feverishly, as we have thought for a mouse, have instead unearthed a nestful of young warblers.

This willow-wren is much victimized by the Himalayan Cuckoo which lays for its benefit a white egg, of course considerably larger than that of the willow-wren. By what method the cuckoo succeeds in introducing its egg into some of the nests is somewhat of a mystery. Davidson and many others found numerous nests of this warbler in May and June, recording from 4 to 6 eggs. He and Rattray also claim to have taken eggs of *Cuculus poliocephalus* (the Small Cuckoo) from their nests. Ward notes that they breed in June and July and we have also taken their eggs in the latter month. The Large Crowned Willow-Warbler normally lays 3 or 4 frail white eggs measuring on an average 16·0 by 12·3 mm. They are devoid of markings and have little or no gloss.

PALE BUSH-WARBLER

Homochlamys pallidus pallidus (Brooks)

(See Plate 75 facing p. 300.)

KASHMIRI NAME. *Dofa Pich.*
BOOK REFERENCES. *F.B.I.*, II, 507; *Nidification*, II, 460; *Handbook*, 188.

FIELD IDENTIFICATION. (2—) Far more often heard than seen, for its distinctive long-drawn-out whistle, followed by a tinkle of short notes, cannot escape notice. The bird itself is a small brown skulker amongst the bushes and rough herbage of the hillsides, and, at lower levels, in the lighter forests.

DISTRIBUTION. This bush-warbler has a wide altitudinal range, being found in the breeding season from below 3,000 ft in the Jhelum and Kishenganga Valleys, around and on slopes within the Vale from about 5,500 ft, and in the side valleys up to approximately 8,000 ft. In the Kishenganga Valley we heard none beyond Keran (4,990 ft) although much country around Sharda appeared ideal. We left Sharda on 10 May, so it is possible the bird had not then returned and was moving up the valley slower than we were, for ' in winter it descends to the outer foothills and is even found commonly at that season on the bush-covered slopes along the Soan and Leh rivers near Rawalpindi' (Osmaston, *Journal*, XXXI, 992).

Its full range includes the North-West Himalayas as far east as British Garhwal.

HABITS AND NESTING. The very name Pale Bush-Warbler savours of mystery and indeed the ways of this little bird are mysterious in the extreme, for the drabness of his uninteresting plumage is compensated for in full by the intriguing qualities of his voice and his ability to shun observation. In twenty-five years we must confess to having seen the Pale Bush-Warbler on but few occasions ; yet a frequent and most arresting sound in any of the side valleys, and indeed within the Vale itself at such places as Achabal and Verinag, is this warbler's song, if song it can be termed. The notes are not unduly loud yet they carry an amazing distance and the long-drawn-out whistle in each of the phrases catches one's attention immediately. Osmaston syllabizes the song as,

                You . . . mixed-it-so-quick
                He'll . . . beat-you.

Anyone memorizing those phrases, unduly prolonging the first word until out of breath, cannot fail to recognize the Pale Bush-Warbler's presence.

It must be an arrant little skulker for to follow up its voice is of little avail; its whistling ceases on approach and it conceals itself in the bushes and rank herbage. We used to consider it in Kashmir a bird of bushy hillsides, also entering cultivated areas which possess rank grasses and waste patches, such for instance as are found around Koolan and Gund in the Sind Valley or Pahlgam in the Lidar. But later we came across it well within mixed forest at Achabal, of course where low cover abounded, and also in the forests of the many small valleys running into the Kazinag mountains. It would appear, therefore, that at any rate in its lower range it is a bird of tree forest provided blanks occur, as well as of open bush-dotted hillsides and waste land. Brooks, in fact, wrote of it, ' It is found in dense jungle at lower elevations '.

The nest is difficult to find, being a ball or purse of coarse grasses lined with finer grass and a few feathers ; an untidy affair showing little architectural merit. It is placed from six inches to 2 ft up, well hidden in rank herbage or the roots of a thick bush.

The eggs are nearly as remarkable as the bird, being uniform dark chocolate-brown, sometimes almost purple in tone. Occasionally they show faint signs of mottling. They are broad ovals, slightly glossy, very fragile, and measure approximately 17·5 by 13·2 mm. Four is almost invariably the number of eggs laid. The breeding season appears to extend from late May to well into July as Osmaston took three nests with fresh eggs on the 19th of that month.

HIMALAYAN GOLDCREST

Regulus regulus himalayensis Jerdon
(See Coloured Plate II facing p. 40.)

BOOK REFERENCES. *F.B.I.*, II, 539 ; *Nidification*, II, 492.
FIELD IDENTIFICATION. (1 —) This minute little bird is usually seen moving about the ends of pine and fir branches. It is somewhat rare and is not easy to distinguish from members of the host of willow-warblers in which Kashmir is so rich, but its markings are well defined so that at close quarters or with binoculars its recognition is not difficult. A study of the coloured plate

GOLDCREST

therefore merits the attention of those who are not well versed in its appearance.

DISTRIBUTION. Found in small numbers in the summer-time well distributed in our area in the Silver Firs and also in the deodar forests. We have met with it in Gurais, west of the Wular Lake, on the Lolab slopes, in the Sind and Lidar Valleys, and in the Wardwan Valley, and Osmaston says he has seen it at Gulmarg. The lowest elevation at which we have found it when breeding was near Bagtor in Gurais at 7,600 ft. In winter it is reputed to drop to 5,000 ft or lower, and Ward obtained a specimen at 5,500 ft passing through the Vale on 2 March.

This bright little bird, found from Europe to Western China, has been divided into a number of races. The Himalayan form has been noted breeding from the Sufed Koh on the North-West Frontier, and in the Himalayas in the Khagan Valley in Hazara and in Kashmir. It occurs as far east as Nepal, the Sikkim bird having been separated as a different race.

HABITS AND NESTING. This exquisite little bird is unfortunately rare, though one can usually count upon coming across one or two in the course of any trek which takes one through the upper fir forests. It is not an obtrusive bird and is usually come upon hunting industriously about the tufts at the outer ends of the lower and medium branches of the firs and deodars, generally along the fringes of the forests or around the blanks. It has been pointed out by more than one observer that the Goldcrest has a peculiar habit of alternating between two favourite feeding-grounds, never staying long at one or the other, so that through this restless habit it seldom gives one a chance of protracted observation.

Although the nidification of this bird is well known—since the genus has a wide distribution throughout Europe and Asia in the palearctic region—comparatively few of its nests have been recorded from the Western Himalayas. In the Lidar Valley Buchanan took its nest in May and Ward's collectors took others on 19 June, 2 and 6 July, in each case with one or both of the parent birds. The 19th of June nest was watched being built. Stuart Baker in the *Nidification* alludes to later nests taken by Ward on the Ladakh border of Kashmir between 10,000 and 12,000 ft in June and July 1908. It seems probable that these were taken on the southern face of the main Himalayan range and so within our area. On 22 May we shot a male on the edge of the Silver Fir forest at Bagtor. Its organs showed that it was breeding, but unfortunately we were unable to locate the nest. This is invariably a felted

PLATE 35

Nest and eggs of Hodgson's Rosefinch

Hodgson's Rosefinch, female, on nest

PLATE 36

Kashmir Cinnamon Sparrow at nesting hole. Male on left

hammock or pouch suspended from twigs near the end of a branch and so placed that it is well screened by other branches growing closely above and below it. It is therefore difficult to locate except by flushing the sitting bird. It is deep and entered from above and consists of interwoven lichens, moss, and moss roots, sometimes only the former, knitted into a flexible whole with the wall up to about half an inch thick. It is often irregular in shape according to the placing of the attachments from which it is slung. It is a beautiful little structure usually three to four inches overall in both depth and diameter. It is said to be built at heights of from 10 to 40 ft from the ground, but we lately discovered a disused nest at shoulder height.

The eggs are of two varieties, but most if not all the Kashmir-taken eggs have been white spotted sparingly, but fairly boldly with red. Stuart Baker, however, writes in the *Nidification* : ' When Ward's eggs came into my possession there were none of the red-spotted eggs among them and they had probably been given away, but there were three clutches and one single egg all of quite a different type.' These presumably were like the eggs of the British Goldcrest or of the Firecrest, which vary from pale dull creamy-buff blurred—one cannot say blotched—at the larger end with dull grey, to a warm buff exactly like the eggs of the Firecrest. Thirteen eggs average 14·3 by 10·7 mm. Ward says the clutch consists of 4 eggs, but a nest taken by Sir E. C. Buck in the Sutlej Valley contained no fewer than 7 fully-fledged young birds.

FIRE-CAPPED TIT-WARBLER

Cephalopyrus flammiceps flammiceps (Burton)

BOOK REFERENCES. *F.B.I.*, II, 545 ; *Nidification*, II, 495.
FIELD IDENTIFICATION. (1—) In ways, song, size, but only partly in appearance, this little bird is close to the Goldcrest. In one important respect it differs widely from that bird, for it makes a cupped nest of grass, which it places in a hole of a tree, and it lays blue-green eggs of a deeper shade than those of the Hedge-Sparrow. In the male the forehead and forecrown are a flaming scarlet, the chin and throat orange-scarlet. The sides of the head, breast and flanks are golden yellow, paling off on the abdomen and vent. The rump and upper tail-coverts are also yellow. The female is a much duller bird, lacking all traces of scarlet. Found

mainly in mixed forest as opposed to the Goldcrest's preference for the firs.

DISTRIBUTION. This is not a common bird, but seems to be generally distributed in our area, in the mixed rather than in the fir forests, from near Valley level up to about 10,000 ft. It has, however, been seen at 11,500 ft in dwarf willows near Astanmarg. We have found it spaced around the Vale on the Pir Panjal, in the Lidar Valleys near Astanmarg and Pahlgam, in the Sind Valley from Gund upwards, in Gurais, and above the Lolab. In winter it descends to the better wooded parts of the plains of Northern and Central India, returning to the hills in March and early April.

It breeds from the Afghan border to Garhwal where it meets the Eastern race *saturatus* which extends the range of the species to Bhutan.

HABITS AND NESTING. This bright little bird is called by some writers the Fire-Cap, whilst others dub it the Fire-capped Tit-Warbler, for it is one of those creatures which puts the systematist in rather a quandary. Stuart Baker retains it with some misgivings amongst the *Regulidæ*, that is linked with the Goldcrest, on account of its apparent affinities with that bird in size, structure, voice, and the fact that both are entirely insectivorous in their diet. On the other hand, certain of its habits, and its eggs, are so utterly different from those of the Goldcrest and approximating in some respects to those of a couple more families, that it may have eventually to constitute a family of its own.

The nest-habits are the cause of this confusion, for it nests in holes in trees, as far as we can see having a preference for those holes whose rims protrude somewhat from the trunk, so beloved of nuthatches. Although the hole is usually of small size, wider cavities in rotten trunks or branches are at times made use of. Opinions seem to differ in regard to the positions of the nests. Stuart Baker says the Fire-capped Tit-Warblers nest most often at a great height from the ground, but from the records at our disposal height does not enter unduly into their calculations. We have twice taken nests at about 20 ft up : Davidson took four between 20 and 40 ft, while other accounts speak of nests at 13 to 15 ft from the ground. A. E. Jones of Simla states that they nest at heights of from 4 to 40 ft.

The nest is a cup of dry grass and rootlets woven into the bottom of the hole, generally well built but at times rather scrappy. Sometimes the lining may be partially or even wholly composed of feathers. The eggs are 3 to 5 in number and are broad ovals of blue-green, somewhat darker in tone than those of the Hedge-Sparrow. They average 14·5 by 11·0 mm.

This bird returns early from the Plains, apparently moving up in small flocks, and at once gets down to nesting, with the result that many clutches must be complete by the end of April. Nests with eggs have also been taken in May and even in June. It has already been stated that its habits approximate to those of the Goldcrest and it has the same way of flying backwards and forwards between two favourite feeding-grounds. It picks its insect food from the leaves of both small and large trees in the mixed woods, often in their upper foliage, and does not particularly favour the pines and firs. It is credited with a pleasant twittering song, and Magrath likens its call-notes to the *tsit-tsit* of the White-Eye. It is shy in its approach to the nest, but a habit which is quite foreign to the Goldcrest, and indeed to any warbler, is its titlike way of puffing out its feathers and hissing at the intruder when disturbed on its nest. In spite, therefore, of its length we proclaim ourselves definitely in accord with those who advocate the name of Fire-capped Tit-Warbler.

INDIAN ORIOLE

Oriolus oriolus kundoo Sykes
(See Plate 32 facing p. 125.)

KASHMIRI NAMES. *Poshnūl, Poshinūl.*
BOOK REFERENCES. *F.B.I.,* III, 6 ; *Nidification,* II, 498 ; *Handbook,* 191.
FIELD IDENTIFICATION. (3) A bright golden-yellow thrushlike bird with black wings and tail, an undulating flight as it flashes swiftly from one tree to another, and loud flutelike notes, which might be rendered in words as ' Phew-to-you ', and an emphatic ' Champignon '. Frequents orchards, groves, and gardens well provided with trees. In the female the yellow of the male is replaced by greenish-yellow and the black by brown.
DISTRIBUTION. Common in all well-wooded areas in the Vale, especially in the groves around villages and in the orchards and gardens. Indian Orioles are not as a rule seen far up the side valleys. They seldom, in fact, breed above 6,000 ft, though Ward says he has seen them at 7,500 ft. We saw one at about 7,000 ft in the Surphrar Nullah. Osmaston noted a straggler at Gulmarg

ORIOLE

in June, and Stuart Baker records that he received a nest from Sonamarg, which is quite exceptional and raises serious doubts in our minds. Might this nest by any chance have come from Sonarwain, where the bird is common, and not from Sonamarg? They, of course, frequent the Jhelum Valley from Baramullah downwards. We also saw one in the lower Kishenganga in April near Dhani when they would still be arriving from their winter quarters, for Orioles are only summer visitors to Kashmir, putting in their appearance towards the end of April. They leave in September and Meinertzhagen says that in 1923 all had left the Vale by the 20th of that month.

The Indian race of this well-known oriole of Europe, Asia, and Africa breeds in Turkestan, East and South Afghanistan, and from the hills of Northern Baluchistan to the Western and Central Himalayas; also in the Indian plains and peninsula as far south as Mysore. It shuns the drier north-western areas and is absent from all but Western Bengal. In the winter it deserts the northern plains as well as the hills, at which season it extends its range south to Cape Comorin.

HABITS AND NESTING. Of Kashmir's many beautiful birds the Indian Oriole is undoubtedly one of the most attractive. Being on the whole a shy bird and entirely arboreal in its habits, its brilliant plumage becomes inconspicuous when it is taking shelter amongst the green foliage of the larger trees of the groves and gardens in Srinagar and around the villages, or in the extensive orchards of apple and pear trees which it loves to frequent. It is, however, quite impossible for such a gem of a bird to escape detection for but the briefest of periods, for its shyness is offset both by its unmistakable liquid notes, which are far-reaching though soft and flutelike in tone, and by the brilliant flash of golden-yellow, greenish in the female, as the bird passes swiftly from one refuge to another. The flight, by the way, is undulating but strong.

In its nesting this bird still retains its cautiousness, although their well-woven hammocklike structures are not infrequently rather conspicuously placed within a few feet of the ground towards the end of a drooping branch of a fruit tree in the middle of an orchard, or even overhanging some secluded waterway along which passing shikaras glide within a couple of yards of them. On the other hand we have spotted nests near the summits of gigantic chenars and tall poplars, while the majority are perhaps built at heights between 15 and 30 ft from the ground. No matter what its position, it is seldom that an Oriole will permit itself to be observed at the nest, at any rate until the young are hatched, whilst the discovery of nests still under construction is apt to lead to immediate desertion.

The nest is a hanging-basket affair suspended between a fork towards the extremity of a horizontal branch. Strips of bark, coarse fibre or grass are wound round the arms of the fork to form a cradle into which other materials, such as more bark fibres, grass, cotton, rags, etc., are woven. The nest varies considerably both in size and in the thickness of its walls, but is always a deep cup from which the eggs could scarcely be tipped, however much the branch should sway. The eggs rest on a final lining of thinner grass and number 2 to 4, but we once came upon a nest between Ganderbal and Shalabug which contained 5. Three, however, is the number most commonly found. They are white, spotted somewhat sparingly with black, but at times clutches are met with in which the markings are a deep red. They average 29·3 by 20·3 mm. The nesting season is late May, June, and July.

HIMALAYAN STARLING

Sturnus vulgaris humii Brooks

(See Plate 33 facing p. 132.)

KASHMIRI NAME. *Tsinihangūr*.
BOOK REFERENCES. *F.B.I.*, III, 31; *Nidification*, II, 515; *Handbook*, 199.
FIELD IDENTIFICATION. (3—) The general impression of a starling is of a highly iridescent black bird closely spotted with buffy-white over the greater part of the body. By the time, however, that the Himalayan Starling is nesting, the light tips to the feathers are so worn that the spots are scarcely noticeable even at close range, particularly so in the case of the male which often retains but a few spots on the vent and undertail-coverts. The bill is sharp and pale yellow, the legs and feet of a reddish-brown colour. Starlings feed on the ground in pairs or flocks according as to whether it is the nesting season or the winter months. They may often be seen in the fields busily following the grazing herds or the plough to snap up insects and grubs disturbed in their passage.
DISTRIBUTION. Common in spring and summer in the Vale, becoming progressively scarcer in the lower reaches of the wide side-valleys up to 6,500 or 7,000 ft at the most. A migratory species from the adjacent plains of Northern India, arriving in

packs, some as early as February, and leaving again in late August and September, though Meinertzhagen states that he saw no downward movement in 1925 up to the end of the latter month. According to Ward it is rare in the Vale in winter, but Unwin writes that a number remain throughout that season, congregating to roost in trees near swamps. The majority of these winter birds may, however, belong to the race *porphyronotus* which breeds in Turkestan and the Tien Shan. Starlings of other races traverse Kashmir on passage, but none of these more highly migratory forms remains to breed.

This widespread species, common from the British Isles to Eastern Asia and down into Africa, consists of numerous races. Two of them nest in different parts of India while others visit it in winter. Hume's race breeds in a restricted area in the North-West Himalayas and Stuart Baker, both in the *Fauna* and in his *Nidification*, appears to give it too wide a distribution. It does not appear to occur in Baluchistan at any time of the year; Whistler states that he can trace no evidence of its nesting in Afghan territory, although he looked out for it during six consecutive seasons; Briggs was unable to confirm Hume's note of its breeding in the Peshawar Vale; lastly we do not know upon what evidence Stuart Baker includes the Simla Hills, and A. E. Jones does not list any of the starlings as nesting round Simla.

HABITS AND NESTING. Starlings are important, or rather that seems to be the impression they wish to convey. They are always purposeful in their actions and strut energetically here, there and everywhere on the lawns or in the fields, in pairs or parties, with erect carriage and pointed actions. In fact, everything about them, both habits and build, is pointed; the rather narrow bill, the streamlined body and the sharply pointed wings. The speckled plumage is not conspicuous in the breeding season, the general tone except at close quarters appearing an almost uniform glossy black owing to the wearing down of the spotted tips to the feathers of the body and wing-coverts. During its sojourn in Kashmir, the Himalayan Starling is an abundant and fairly tame bird only in the main Vale. They penetrate the larger side-valleys for a comparatively short distance, ceasing to be found as soon as they narrow down and the ground commences a serious upward trend; for instance, in the Sind Valley Gund, or at most Koolan, seems to be their limit, and in the Lidar Aishmakam, while we recall seeing none beyond Handowar west of the Pohru River. In fact, we can find nothing to support Stuart Baker's contention that they breed in Kashmir up to 8,000 ft.

They are numerous within the orchards, groves and gardens of the Vale of Kashmir and along both sides of the Jhelum, where they nest in holes in trees of all kinds, seeming to have a particular predilection for cavities in gnarled mulberries. They also nest in holes in vertical banks and to a certain extent in the walls of buildings. Betham gives an account of nests in holes in the old matted reeds at the edges of the jheels. To quote from the *Nidification* : ' My shikari told me that these Starlings nested in holes in the reed beds and, on going round these in a boat, I certainly took eggs out of the passages made in the dense reeds and saw the birds coming out. Regular holes had been made at the ends of the tunnels and in these they bred.' To this Stuart Baker adds : ' Later other collectors confirmed Betham's observations and proved that he had not been gulled by his *Shikari* placing eggs in the tunnels and then showing them to him.' Many years ago in Ireland we took two untidy but normal cup-shaped nests of the Common Starling in a dense reed bed.

Often the nest is no great distance from the ground ; in fact it may just as well be 2 ft up as 20. It consists of a very loose conglomeration of straw, dry grass and feathers, sometimes in considerable quantities, sometimes scanty. In it there are laid from 4 to 6, occasionally 3 only, immaculate pale blue eggs. Like the bird the egg is rather broad and often tapers to a well-defined point. The gloss is considerable and they average 29·7 by 20·5 mm.

Starlings are early arrivals in Kashmir, many forming into flocks and leaving the Plains in February. They are not present in any great numbers until the second week of March, but they are already nesting by halfway through April. Numbers of dull-greyish young ones are to be seen greedily following their parents even in the latter half of June, and by August the packs are forming preparatory to their return to the Plains. In this connexion Meinertzhagen notes that the adults collect in flocks outside the limits of the towns, the birds of the year remaining in the vicinity of towns and villages—in other words the adults and juveniles, as is so often the case, separate.

Starlings are beneficial to agriculture, for their food consists of a percentage of noxious insects out of all proportion to the limited amount of grain consumed by them. When breeding they are naturally well spread out over the countryside in their pairs, but on arrival, and as soon as nesting is finished, they collect into flocks often of a couple of hundred birds. These flocks may be seen quartering the fields, the birds walking quickly this way and that to pick up whatever catches their quick eye, sometimes running or fluttering to pick up a grasshopper or fly endeavouring to make its getaway. When a flock takes to the air, its members fly swiftly and closely packed on pointed wings, and are capable of mass manœuvres of great precision, wheeling, dipping or rising swiftly upwards in complete unison while the swiftly moving throng equally rapidly changes its formation. Now it is a long wavy line, now a compact ball. As it nears the ground, the rapid wing-beats cease and the birds come down in a swinging glide to resume once more their feeding. When settling down for the night, many flocks choose a communal roosting place in some sheltered trees in which parties continue arriving till sometime after dark, the shrill squeaking and chattering which goes on as they search for vacant perches being audible at a considerable distance.

The starling possesses a pleasing subdued song largely imitative of other birds, as well as many of the screaky harsh noises reminiscent of the Myna.

BLACK-HEADED MYNA

Temenuchus pagodarum (Gmelin)

(See Plate 34 facing p. 133.)

BOOK REFERENCES. *F.B.I.*, III, 47; *Nidification*, II, 521; *Handbook*, 201.

FIELD IDENTIFICATION. (3—) A warm buff bird, grey on the back, with black flight-feathers and a long black crest lying flat on the head and neck. The legs are bright yellow and the bill tipped with that colour. A typical myna in its ways, but more solitary and not so noisy as the Common Myna, than which it is slightly smaller.

DISTRIBUTION. Only to be met with in small numbers in the Jhelum Valley up to Uri. We have no information as to winter movements, but it probably retreats to the Plains as it is said to be only a summer visitor elsewhere in the outer Himalayas.

This species is resident over the greater part of India and Ceylon, shunning only the barren portions of the north and north-west and the more humid parts of Bengal. In the Himalayas it occurs normally up to about 4,500 ft, but round Gilgit and Chitral it is found much higher.

HABITS AND NESTING. This sleek little myna, so common in the Plains, has not so far penetrated to the Vale, but it undoubtedly

PLATE 37

White-capped Bunting at nesting site. Male on right

PLATE 38

Nest and eggs of Eastern Meadow-Bunting

Eastern Meadow-Bunting

breeds in Uri where we have seen a pair carrying straws to a hole in a tree on the roadside just by the dak bungalow. We have noted it at other halts along the road and its numbers increase as one drops towards Kohala. Observed closely, its colour scheme and appearance are rather pleasing ; it is so well-groomed and keeps its glossy black crest well brushed down. It is not as boisterous as the Common Myna, although capable of producing many raucous noises, sinking its bill to the breast in their production in the same manner. It also possesses quite a pleasing song and is a good mimic. Although it enters villages and towns freely, it is on the whole more of a jungle bird, and, we are thankful to say, much less vulgar and quarrelsome.

The nest is generally in a natural hollow of a tree at any height from, say, 10 ft up, but holes in the walls of buildings, whether inhabited or not, may also be used. We once watched an unequal struggle, which lasted for nearly three weeks, between a pair of Common Mynas and a couple of these birds for the possession of a ledge on the top of a pillar. After stealing each others' building materials constantly, two separate nests side by side were almost completed, but the rougher Common Mynas eventually destroyed the smaller birds' nest.

In the Plains the Black-headed Myna's nesting season is prolonged ; in fact they lay throughout the greater part of the hot weather, but the period is probably curtailed in the Hills by the greater length of the severe weather. At Domel, we have seen birds carrying nesting materials as far apart as May and early July. The nest is an untidy collection of straw and feathers, often considerable quantities of the latter in the lining. In it 3 to 5 spotless blue eggs are laid. In depth of colour they are rather pale, but have a high gloss. They measure 24·6 by 19·0 mm.

COMMON MYNA

Acridotheres tristis tristis (Linnæus)

(See Plate 34 facing p. 133.)

KASHMIRI NAME. *Hōr.*
BOOK REFERENCES. *F.B.I.*, III, 53; *Nidification*, II, 525; *Handbook*, 203.
FIELD IDENTIFICATION. (3—) Numerous in the Vale, mainly in and about habitations along the main river. The rich brown

plumage, with white patch in the wings so conspicuous in flight, the white tips to the darker tail-feathers, yellow bill and legs, and above all the bare yellow skin beneath and behind the eye, are sufficient pointers to this bird's identity. Its noisy fratching, varied harsh utterances and extreme tameness as it struts about the ground are also diagnostic.

DISTRIBUTION. Common along the Jhelum Valley road and in the Vale. It does not penetrate the side valleys as a resident for any distance, certainly not above 7,000 ft. In the Kishenganga Valley it was noted solely at Tithwal. In Kashmir it is not nearly so common as down in the Plains and is not such a bazaar bird. In addition it is only locally distributed, chiefly along the main river, so that it is by no means always with us. Away from the Vale it is positively scarce and not to be seen in any numbers above 6,000 ft. About halfway through June, however, a spirit of adventure seems to possess it and small parties make temporary excursions into the hills. We have seen such parties in the Ahlan Nullah at over 7,000 ft where they are perhaps resident, up the Marbal Glen at a good 9,000 ft, at Pahlgam and at Sonamarg.

This Indian species ranges from the Eastern boundaries of Baluchistan including the Sibi plain, and Eastern Afghanistan through India and Burma as far south as Mergui. It is also found in parts of Siam. It does not cross the main Himalayan range into the northern provinces of Kashmir. Birds from Ceylon belong to a darker race, *melanosternus*. It has been introduced into other countries, where its aggressive habits are proving deleterious to more interesting indigenous species.

HABITS AND NESTING. This must surely be the best-known bird in India. The kite and the house crow run it close, but the Myna is so fearless and so unconcerned at our presence that it will hardly get out of our way in the street and will even enter the house to snatch scraps of food or material for its nest. In addition the quaint manner of depressing the bill on to its breast in producing all manner of harsh noises is bound to attract our attention. The Myna, however, has a subdued and rather thin song which it often proclaims from the shade of the trees during the heat of the day. It feeds almost entirely on the ground, strutting backwards and forwards with fixed purpose, the yellow-rimmed eye singling out a grain of rice here, an insect there. In the fields it follows the bullocks to seize with the utmost dexterity grasshoppers and other prey flushed in their progress, a sudden fluttering flash of the white wing-patches presaging the demise of an insect fleeter than the rest. Injurious insects provide the great bulk of its diet, though it is practically omnivorous. It is, therefore, a species of benefit to agriculture.

It is a most aggressive bully, for is it not quite common to come upon a luckless Myna on its back in the dust with four or five of its fellows seemingly pecking the life out of it? Even the spectators are not content just to watch the outcome of the fight, but needs must join in the mêlée against the luckless underdog, although their participation may only amount to adding their raucous screams to the general hubhub. In spite of this propensity for squabbling with its own kind, the Myna roosts communally, parties coming in to the agreed roosting trees from a wide area. Until all are accommodated pandemonium reigns, often carried on till well after nightfall, and at intervals thereafter should anything disturb the gathering. One evening, when moored to the bund at Shaltin, we could hear a great uproar emanating from the bagh beyond the wood depot where a great many Mynas must have congregated in the tall trees.

The same aggressiveness is carried into the search for a nesting site, and if a desired cavity already has a smaller or less virile occupant it will soon change hands. For the Myna nests in a hole of some kind, often in a tree, but also in a river bank or the face of a quarry, the sides of a well, in a wall, or under the eaves of a house. The sites are usually within the confines of a town or village, but to a lesser extent they will be found in the trees around the fields or even in light forest. In this cavity an untidy conglomeration, often of considerable bulk, of straw, feathers, rags, leaves, and other oddments, is collected, the egg-cavity being a rough depression to hold the 3 to 5, rarely 6, uniform blue eggs. These are typical of the starlings, slightly glossy when fresh and somewhat deeper in colour than those of the Himalayan Starling. They average 32·3 by 21·9 mm. The Mynas are resident birds, the pairs obviously remaining together throughout the year, being very affectionately disposed towards one another, but nesting does not commence until May. In the Plains they are said frequently to bring up two and even three broods in the year. The season is more curtailed in Kashmir so two broods is probably the limit, but we have no information on this point.

BLACK AND YELLOW GROSBEAK

Perissospiza icteroides (Vigors)
(See Coloured Plate IV facing p. 232.)

KASHMIRI NAME. *Wyet Tōnt*.
BOOK REFERENCES. *F.B.I.*, III, 102; *Nidification*, III, 34; *Handbook*, 218.

FIELD IDENTIFICATION. (3—) A bird of tall fir and mixed forests, often seen in small parties. The brilliant yellow body-plumage of the male with the usual pattern of black wings and tail, in conjunction with the massive apple-green bill, are quite unmistakable characters. In the female and young birds the yellow is replaced by a dull grey except that the underparts are a bright tawny-fulvous. Has a loud but pleasing double or triple call-note which carries a considerable distance.

DISTRIBUTION. The Black and Yellow Grosbeak is evenly distributed in the breeding season from about 6,000 to 10,000 ft in all the fir and heavier mixed forests throughout our whole area. It is a resident bird in the main, having a small vertical movement according with the seasons and coming down to at least the level of the Vale in the winter and early spring.

Whistler gives the distribution of this form as throughout the Western Himalayas from Kumaon to Hazara and Chitral. It is also abundant in the Sufed Koh.

HABITS AND NESTING. This grosbeak with his large size, brilliant plumage and massive bill is the most striking finch in the Western Himalayas. It is also widespread and far from timid, so the chance of studying this fine bird at close quarters is seldom long deferred. True, they often remain but voices in the pine woods, for the loud liquid whistle, *tit-te-tew, tit-te-tew,* or as Magrath has it, *trekatree-trekatree-trekup-trekup,* carries afar over the tree-covered slopes and is frequently uttered as the members of a flock work their way through the tree-tops. The song-note, uttered at rest, and also used by both birds calling to each other from the nest, is a clear musical *tookiyu, tookiyu*.

The Grosbeak is essentially a bird of the fir and mixed forests found only where the trees are tall and stately, but it does not spend all its time aloft and there is no surer way of attracting them than to pitch camp within the pines' fringe in woods where they are numerous. Time and again we have had them cracking the fallen pine-seeds disturbed in our peregrinations around the tent, being apprised of the birds' presence close to the tent's walls by the clicks of the breaking seeds being crushed in the strong mandibles. On one occasion a bold male, accompanied by a female and two ashen-grey youngsters, hopped right into the mouth of our tent in face of a growling terrier. In addition to seeds they feed on the fresh green shoots of the pine trees and on the fruits of shrubs and bushes in the undergrowth.

Grosbeaks appear to have a habit of roosting together but whether this trait is sufficiently ingrained to persist throughout the breeding season, or whether it only appertains to the males at that time, we cannot really say. Above Tanin, however, late in June, more than fifty birds were turned out of a patch of tall thick bushes which were well separated from the adjacent forest. Unfortunately it was too dark to be certain that all were of the male sex, but at this period we have noticed that the small parties feeding by day around the tent have frequently consisted entirely of males. Once breeding is over, they collect into flocks, not usually of any great size, but certainly consisting of more than just family parties.

The nesting season is said to be rather prolonged, commencing in May and continuing well into the summer, but June must normally end the period of egg-laying, as fledglings still being fed by their parents are seldom seen after the end of July. The nest, composed of an outer foundation of lichen-covered twigs, is flattish and contains a certain amount of moss and weed stems, being lined with dry grass and rootlets. It is generally placed against the main trunk of a deodar or fir tree, occasionally out on a horizontal branch. It is often built at considerable heights from the ground, 60 ft or more, but there is a record of a nest at a height of only 18 ft, while Captain Cock found a nest 'upon a sapling lime'. Between the 19th and the end of June we observed one under construction. This, at an elevation of 8,200 ft, was being built 50 ft from the ground on two close-growing boughs springing from the trunk of a deodar. It had a fine canopy protecting it from above. Both birds took part in its construction and spent a considerable time at the nest. They were most cautious in their approach, calling as they came. By 29 June the female was sitting, and throughout the period of incubation she alone was noted on the eggs.

The eggs are 2 or 3 in number, broad and rather finely pointed. They have a pale grey slightly glossed ground and are marked mainly in a narrow zone towards the large end with irregular intertwined lines of dark umber-brown with underlying fainter clouds of inky-purple. Amongst the scrolls and lines are usually a few bold spots. They measure $28 \cdot 3$ by $19 \cdot 9$ mm.

BULLFINCH

ORANGE BULLFINCH

Pyrrhula aurantiaca Gould
(See Coloured Plate III facing p. 136.)

KASHMIRI NAME. *Sāma Sonatsar.*
BOOK REFERENCES. *F.B.I.*, III, 109; *Nidification*, III, 38.
FIELD IDENTIFICATION. (2) In the field the coloration of this handsome finch often appears a clear yellow. This may in part be due to tricks of the bright sunlight at highish altitudes, for the fact remains that the skin from which the illustration was painted is, if anything, deeper and duller in tone than shown on the plate. This particular skin, however, was collected early in March. Two males, obtained when breeding in August in the Sind and Lidar Valleys, are far brighter in tone, having the underparts a definite orange yellow, on one the abdomen being yellow with no orange tinge in it at all. La Personne obtained a further skin at Lidarwat on 17 August which is definitely greenish in tone but still far brighter than the March specimen. The females are much duller beneath and have the upper parts grey. This bird may often be seen alone or dotted here and there on the ground feeding on fallen weed and grass seeds on cropped lawns and blanks within the forest or at the edges of the margs. It is a tame, quiet little bird blessed only with a pleasant soft whistle and a couple of twittering notes. It is rather partial to the vicinity of young firs and birches.
DISTRIBUTION. No one appears to have seen the Orange Bullfinch in the Pir Panjal mountains, although it occurs in the Galis. On the northern side of the Valley it has been recorded as not uncommon over a wide area from the Wardwan, through Astanmarg and Tanin and the West Lidar Valley to Kolahoi, around Gund, Sonamarg and Baltal in the Sind Valley, and over to Gurais. In the winter it descends almost to the level of the Vale, living between 5,500 and 7,000 ft, but in May it moves upwards and is then found from 9,000 to 11,500 ft or more. Ward states that his collectors took two nests in August 1906 on the slopes of Kolahoi at between 12,000 and 13,000 ft. More were taken in 1908 in the Rewil Nullah nearby.

The bullfinches contain a number of species widespread in Europe and Asia, *aurantiaca* being a West Himalayan form found at comparatively high elevations in summer from Chitral to Garhwal. It is resident except for a limited vertical movement.
HABITS AND NESTING. Our first meeting with this typical

little bullfinch took place at Tanin in the month of June, where our tents were pitched on a small lawnlike patch of green turf below the woods enclosing the Astanmarg path. Here in the early mornings anything from four to a dozen birds were to be seen feeding quietly on the sward, like sparrows where fallen grain is to be found. So tame and confiding were they that we could stand in the mouth of the tent and watch them at the closest range. Quite often they would be feeding amongst the tent ropes, and when disturbed would fly off with subdued whistles to the adjacent firs, only to resume their feeding shortly afterwards. Colonel Unwin records that, ' on one occasion while awaiting " chota hazri " before sunrise, two or three quite entered my tent and hopped about close to my feet '. One might add that it is usually in the earlier hours of the morning that they are to be found feeding thus in the open. We particularly like Colonel Meinertzhagen's account of his meeting with these birds in March in the fir forests opposite Kangan. ' I was sitting quite still,' he writes, ' watching some Tits and Tree-Creepers, when a cock Orange Bullfinch flew down from a neighbouring tree and commenced feeding but a few feet from me. He was soon joined by another and yet others until a party of nine had assembled on the mossy, leaf-strewn carpet of pine-needles. They scarcely moved as they fed, each bird picking up and crushing what it could reach from where it sat. The brilliant orange of their plumage showed up like some precious stone on the drab dull green of the ground on which they fed. After a moment they would fly up into a tree, soon to return one by one, and sit quite motionless munching small seeds. When in a tree they also sit quite motionless and hardly visible.'

In the breeding season they seem rather partial to the small firs and copses and to the birches in forests where there are herbaceous weeds for their food, and it is in such localities apparently that they breed. We say ' apparently ' as in spite of their unobtrusive tameness there are few records of authentic nests. Osmaston describes one, found on 4 August at 9,000 ft in the Lidar Valley from which he says the young had just flown, as exactly resembling that of the English Bullfinch, being composed of thin twigs and sticks lined with fine roots. The nest was placed on a horizontal branch of a Silver Fir sapling, sheltered above by a second branch, and 4 ft from the ground. It was in a thick patch of similar young trees. We found an empty nest agreeing essentially with this description in a like situation at Baltal on the last day of June. Ward's nests, in the Rewil Nullah and each containing 3 eggs, were obtained by his collectors. They are described as having been found at 12,000 ft on 19 and 21 August, each in a bush, and the descriptions are virtually the same : ' a round cup, outside coarse grass and stems of flowering plants ; inside roots and lining of musk deer and goats' hair. Outside open and coarse, inside very neat.' His 1906 nests are described as small cups of dry grasses lined with the hair of the musk deer. The eggs, one a clutch of 4, ' are of a dull white marked with reddish-brown in small streaks and spots chiefly at the thicker end '. They average 21·3 by 15·0 mm. Lastly Marshall in a list entitled *Birds Nesting in India*, records under July, ' Orange Bullfinch, 11th, eggs '.

KASHMIR WHITE-BROWED ROSEFINCH

Propasser thura blythi Biddulph

BOOK REFERENCES. *F.B.I.*, III, 124 ; *Nidification*, III, 44.

FIELD IDENTIFICATION. (2+) In Kashmir this rosefinch is found only at very high elevations except when on passage in the spring and autumn. It is noticeably larger than the other rosefinches which nest within our area. Although the name implies that the supercilium is white, it is a glistening pale pink with white shaft streaks and a white end to it. The general tone of the plumage is brown well suffused with rosy-pink, the whole lower plumage being rosy-pink. The female lacks the rosy tints and is much streaked with black both above and below, while her throat, breast and flanks have a rufous tinge.

DISTRIBUTION. In addition to the extreme North-Western Himalayas this bird is reputed to occur in the North-West Frontier hills, Whitehead recording it as not uncommon in the Sufed Koh in summer from as low as 8,000 ft. Other races extend the range of the species eastwards from the Simla Hills through the Himalayas and Tibet to Western China. Ward says considerable numbers pass through Kashmir in spring and may be seen in small flocks pecking about under the bushes. They appear to start the return journey in September, but their winter quarters do not seem to have been defined. As regards the summer distribution in Kashmir, the White-browed Rosefinch is only to be found at high elevations amongst the scattered juniper beds, and the few records all point to a restricted area comprising the glacier-dominated highlands about the sources of the Sind and Lidar rivers and at the head of the Wardwan. In addition Magrath says he came across a small

PLATE 39

Nest and eggs of Hodgson's Yellow-headed Wagtail

Nest and eggs of Himalayan Goldfinch

PLATE 40

party of Rosefinches at 11,500 ft at Sonamuss at the head of the Surphrar Nullah (about halfway up the Sind Valley) which contained members of this species and the Dark Rosefinch. As this was in June or July, they may well have been still on their way to the above breeding grounds.

HABITS AND NESTING. The Kashmir White-browed Rosefinch nests in our area at high elevations in wild country above the tree line where beds of stunted juniper or the dwarf rhododendrons alone provide cover. Our present knowledge points to a preference for the glacier-dominated plateaux and hillsides in that much broken-up area where the sources of the Sind and Lidar rivers have their being and massive ridges divide them from an adjacent patronized area of like country at the head of the Wardwan Valley. Ward is the only oologist who appears to have obtained its eggs. Of these he wrote in the *Journal* that on the 6th, 9th, and 10th of August, three clutches of eggs, two of 4 and one of 3, were found high up in the hills where the Sind and Lidar rivers rise. These nests were large saucer-shaped structures in bushes and the creeping juniper, and were composed of soft grasses and seed-pods. The eggs were greenish-blue, sparingly spotted at the thick end with black. They averaged—converted into millimetres—22·2 by 15·2 mm. Osmaston, returning from Suru, came upon these birds early in August on the Kanital slopes below the Bhotkol glacier at the head of the Wardwan Valley. Here they were living at from about 10,500 to at least 12,000 ft in the dense patches of juniper so common at those altitudes. They were then in pairs. He found no nests but considered they were about to breed in the juniper. He describes their call as consisting of a loud harsh whistle repeated six to eight times, being reminiscent of the call-note of the White-cheeked Nuthatch.

PINK-BROWED ROSEFINCH

Propasser rhodochrous Vigors
(See Coloured Plate III facing p. 136.)

BOOK REFERENCES. *F.B.I.*, III, 129; *Nidification*, III, 48.

FIELD IDENTIFICATION. (2) These little rosefinches may possibly be come upon in the same areas as the Common Rosefinch, though they are at times to be found in open forest, but both sexes are readily distinguished from it, not only by their somewhat larger size but by the possession of well-defined supercilia and a generally lighter coloration. The Common Rosefinch has no eye-stripes, and the male is a considerably darker bird, crimson rather than pinkish.

DISTRIBUTION. Generally distributed in our area in summer between 9,000 and 12,000 ft both on the Himalayan range and in the Pir Panjal mountains. We have found them in the breeding season as low as 9,000 ft in the undergrowth in open fir forest, but they are perhaps more often to be come upon above the trees in juniper and other bushes. Though more numerous than the last bird, they are not common anywhere and are usually found only in pairs. They appear to remain on the southern side of the main Himalayan range, not crossing into Ladakh. In winter they drop down to lower levels, but are said never to be found below 4,000 ft. At this time they pack to some small extent and may be noted in small flocks. The downward movement is not completed until November. By March they are on the way back.

This is a purely Himalayan species, being found from Kashmir to Western Sikkim.

HABITS AND NESTING. The nests of this pretty little finch have been taken by Buchanan, Whymper, Osmaston and others, many of their records appertaining to the Sonamarg area, whence Buchanan first described the nest and eggs which he took there in 1902 at an altitude of about 10,500 ft. There are skins in the Prince of Wales Museum, Bombay, collected by La Personne at between 9,000 and 12,000 ft in the Lidar Valley, and we came upon a pair with young nesting in some thick bushes below Tanin in the East Lidar at somewhat below 9,000 ft. Osmaston says they nest at times in small fir trees, but their nests are usually in bushes of some kind, often thorny ones, including rose bushes, at heights of from a few inches to four or five feet from the ground. The majority are perhaps at and below two feet up. Although these birds frequent open forest, this will only be of a type where the trees are thinning out and there are consequently plentiful patches of bushes amongst the trees or, at higher altitudes, juniper scrub starting hard by.

The nest varies somewhat, always comprising more than just grasses, and is consequently distinguishable from that of the Common Rosefinch. It often has a foundation of twiglets and on this a compact nest of grass is imposed with a lining of finer grasses and hair. Moss and the skin of birch-bark too is often incorporated. They are bold confiding birds at the nest, the female sitting very close. The clutch consists of 4 to 5 eggs—Osmaston mentions one clutch of 6—which are a deep blue with a few bold black or dark reddish-black spots. Unspotted eggs are not uncommon, this applying at times to entire clutches. They average 18·8 by 14·2 mm. Ward's men got eggs in August, on the 7th and 9th, while Buchanan's eggs were also taken in the first week of that month, but they are said to be procurable from the middle of July to the end of August.

HODGSON'S ROSEFINCH

Carpodacus erythrinus roseatus (Blyth)

(See Plate 35 facing p. 140 and Plate 80 facing p. 317.)

KASHMIRI NAME. *Gulāb Tsar.*
BOOK REFERENCES. *F.B.I.*, III, 137; *Nidification*, III, 52; *Handbook*, 220.
FIELD IDENTIFICATION. (2) In the breeding season the general plumage of the male is brown with a ruddy tinge; the head, throat, and rump are rosy crimson. The female, as can be seen from the photograph, is in no way distinctive and might almost be taken for a female House Sparrow by the unobservant, except that she lacks all trace of an eye-stripe. The stout bill is a noticeable feature.
DISTRIBUTION. This is not really an uncommon bird and is to be found in summer in bush-covered tracts above 8,500 or 9,000 ft throughout our area excepting the Pir Panjal range, for which we can trace no records. It is, however, very patchy in its distribution, some areas though apparently suitable holding none, while in others they are excessively numerous, such an area being the upper Wardwan Valley between Basman and Wanhoi. This little rosefinch is highly migratory, after completion of the autumn moult in September, spreading in packs over the greater part of India to as far south as the hills of the Madras province.

The Common Rosefinch—or let us raise this rather beautiful finch above the ordinary, as some British bird books do, by calling it the Scarlet Grosbeak—has a wide breeding range from East Prussia to Central Asia and Siberia, wintering farther south from India to China. The species comprises a number of races, and *roseatus*, though in the main leaving India, breeds in some numbers throughout the Himalayas and in the higher hills of the North-West Frontier from North Baluchistan to Chitral.
HABITS AND NESTING. Hugh Whistler in his *Note on the Weavers*

ROSEFINCHES

and Finches of the Punjab (*J.B.N.H.S.*, XXX, 185), writes: 'The breeding range of this Eastern form of the Scarlet Grosbeak is somewhat difficult to define; it appears to breed at high elevations in most of the mountains that bound the Punjab both to the North-West and the North-East . . . but this bird is a late migrant from the Plains and nests in portions of its range as late as August . . . The published records for Kashmir and Eastern Turkestan are so unsatisfactory and indefinite that there is no point in discussing them in detail; the species, however, appears to breed through a large portion of this area.' We find these last remarks only too true and can elucidate the position only to a limited extent. For one thing it is probably not until June that the majority reach their breeding areas. It also seems likely that the migratory flocks retain a certain amount of cohesion and breed almost colonially in a limited area. We also think it possible that the area chosen is apt to vary, so that country which one year contains plenty of nests, may the next prove void or virtually so. Osmaston points out that the majority pass over the main Himalayan range, only a small fraction of the vast numbers which visit India remaining to breed at the heads of the wider valleys such as the Sind, Lidar, and Wardwan. Gurais needs checking up, for that district must also have its quota. In 1937 we came upon a great concentration in the Wardwan Valley around Basman at 8,500 ft where in the valley bottom are many acres of dwarf willows, bushes of various kinds, and wild briars. We soon found a score of nests between $2\frac{1}{2}$ and 7 ft from the ground. The majority were fairly well concealed in leafy bushes, but a couple in willows and one in a briar were quite open to view. Later, amongst others, two nests were found at Suknes in thick bracken, and we have reason to believe that numbers were breeding in the extensive patches of bracken around Rikinwas. The nests were all normal, that is, rather untidy cups of dry grass and weed-stems lined with finer grass and roots, sometimes with a few hairs. It was noticeable that the eggs, 3 to 5 in number, seldom the latter, were all well incubated and one or two nests held newly-hatched young. We estimated that the eggs had been laid about the same time, probably in the last week of June. Later, however, at Suknes we found nests still being built in late July.

The eggs, rather elongated in shape, are a clear blue, sparingly spotted with black or dark reddish spots which at times are almost absent. They average 20·8 by 14·9 mm.

Another factor which tends to confirm the cohesion of packs throughout breeding is the song. In this connexion we cannot do better than again to quote Osmaston. 'The song during the breeding season', he writes, ' is a bright cheery refrain of from 5 to 8 notes repeated at intervals by the cock bird, usually from a tree or bush not far from the nest. Each individual has its own particular combination of notes which is invariable, and this is often shared by the other birds in his immediate vicinity, but in different localities there is considerable variation. These different songs do not, however, differ in character and are always easily recognizable as belonging to this Finch.' Rikinwas is but a few miles beyond Basman, but the songs of the rosefinches at these two places bore striking differences.

When nesting they are rather tame and confiding. On our arrival at Suknes a pair were found to have commenced building in the shoots growing from a small stump within ten yards of the cook's tent. Not only did they persist in their efforts, but by the time we left on 27 July the nest contained three eggs. In this connexion Meinertzhagen's remarks are worth quoting. ' There is no skulking about this Rose-Finch. He seems to delight in displaying his plumage from the tops of bushes, and to see several cocks sitting in the sun on a rose bush, a solid mass of pink and scarlet flowers, is a sight not to be forgotten.'

HIMALAYAN GOLDFINCH

Carduelis caniceps caniceps Vigors

(See Coloured Plate III facing p. 136 and Plate 39 facing p. 156.)

KASHMIRI NAMES. *Sehāra, Sēra.*
BOOK REFERENCES. *F.B.I.*, III, 150; *Nidification*, III, 61; *Handbook*, 222.
FIELD IDENTIFICATION. (2—) These little finches with their bright colours, melodious call-notes and sweet song, may often be seen feeding their way from tree to tree in follow-my-leader style, the yellow-barred wings and indented tail showing up well. As can be seen from the illustration they differ from the European bird mainly in the pattern and coloration of the head markings.
DISTRIBUTION. Breeds commonly in spring and early summer in and around the Vale and in the lower reaches of the side valleys up to about 6,000 ft. After raising a first brood many repair to higher elevations where they may be found nesting even in the birch zone at 11,000 ft, though the majority remain a couple of

GOLDFINCH

thousand feet lower. They are then numerous at such places as Gulmarg, the Marbal Glen, Pahlgam, Sonamarg and Baltal, in Gurais, and in the Wardwan Valley. Some few probably breed at lower elevations than that of the Vale : Davidson recorded seeing a pair at Uri on 26 June. This bird is resident in the Vale, collecting in winter into flocks of at times considerable size. At this season, however, numbers move down into the lower hills and a few into the adjacent plains, appearing not infrequently as far out as Rawalpindi and Taxila.

We appear to have two races of Indian Goldfinches which are found nesting in the hills of Northern India from North Baluchistan to the western Himalayas as far as Kumaon, the typical race being the west Himalayan form.

HABITS AND NESTING. This cheerful little finch can be recognized as well from its pleasant notes as from its distinctive colours. According to Osmaston, the birds begin to sing by the end of February, often in unison, and in May the parties break up into pairs to breed. These may then be seen flying from one pine or willow tree to another, uttering mellow call-notes, which once heard are easy to recognize, but which are difficult to express adequately in words. The song is very pleasing and like that of the European bird. Though they seem often to stick to the upper foliage, Goldfinches are most partial to thistles and other weeds, fluttering round their heads in search of the seeds for food and the thistledown with which to pad their nests. Consequently, where the valuable *kuth* plantations are, there too will the Goldfinches be found. We well remember missing a grand opportunity to obtain Goldfinches' photographs when staying at the Wurjwan Forest Rest House in the Wardwan Valley, in front of which there is a small *kuth* nursery. In addition to Goldfinches these thistles were visited by many species of willow-warbler and a pair of Redbrowed Finches.

Goldfinches' nests, which they begin building towards the end of May, are amongst the most beautiful structures imaginable, and that in the illustration was probably unique. It was composed largely of interlaced fresh-cut edelweiss and forget-me-not stems, a number of the tiny blue flowers of the latter forming a little wreath around the softly-felted inner cup. The nests may be built of moss or other fine vegetable matter compactly put together with an inner lining of willow or thistledown and sometimes a little hair. It is generally placed on a horizontal branch of a Blue Pine or willow tree from about 10 to some 60 ft up. Sometimes it may be in a fork of a large bush. The nest mentioned above was approximately 20 ft up one of the first of the young pine trees bordering the track from Baltal to the famed Zoji La. The elevation would be about 10,400 ft and the date was the 26th of June. When found the nest contained only one egg, but the owner was so tame that while a man was climbing the tree, she approached to within three feet of his head.

The end of June is a little early for the second nests, the majority of them being found in July and August. At the lower levels, late May and June are the months. The straggling pine wood on the slopes of the Takht at Srinagar usually holds a nest or two at this time, but we must confess to never having found large congregations nesting together there or elsewhere, and the pine wood referred to by Rattray (*Nidification*, III, 61) has largely ceased to exist. In addition to pines we have found nests in willow trees lining irrigation channels and in perrottia scrub and other bushes. To these Osmaston adds the birches at the higher elevations.

The eggs, 3 to 5 in number, are thin-shelled, a pale skimmed-milk blue in colour, sparingly and faintly marked with spots of reddish-brown and greyish. Eggs are not infrequently unmarked. They average 18·0 by 13·2 mm.

RED-BROWED FINCH

Callacanthis burtoni (Gould)

BOOK REFERENCES. *F.B.I.*, III, 152; *Nidification*, III, 63.
FIELD IDENTIFICATION. (2+) A quiet bird often come upon in the woods feeding on the ground in small bands, the members of which fly up one by one to a low perch, emitting soft *tews*, only to resume their feeding soon after one has passed. The yellow marks in the dark wings and the wide band of the same colour round the eye are distinctive features.
DISTRIBUTION. A denizen of the fir forests and well-wooded country from about 7,500 to 10,000 ft. Probably, as Osmaston says, it prefers the deep forests of Silver Firs, but we have come across it in mixed forests at Pahlgam, Lidarwat, Tanin, Sonamarg, and near Gurais, and at times just outside the fringe of the main forest in what one might term very well wooded country. It is also not uncommon amongst the Blue Pines in the Wardwan Valley. Around Gulmarg is another of its recorded strongholds, and we met with it in the Marbal Glen, so it almost certainly

RED-BROWED FINCH

occurs throughout the length of the Pir Panjal range. It is probably a resident with but a slight vertical movement in winter. Unwin writes of it : 'The Kashmiris say that it comes lower in winter, but I have never seen it out of the upper forests.'

Outside Kashmir this interesting bird is found in Chitral and to the south-east as far as Sikkim.

HABITS AND NESTING. Although widely distributed and far from uncommon, we have still much to learn regarding the nesting of this intriguing finch. In spite of its unruffled and seemingly confiding ways, its nest has been taken on but few occasions. Colonel Ward's description of a nest as consisting almost wholly of the spines of the Blue Pine is at variance with the accounts of Rattray and Skinner, while unfortunately none of these three workers mentions dates. In a *Miscellaneous Note* (*J.B.N.H.S.*, XLVI, 721) Major Waters now gives an excellent and full description of a nest taken by him in 1946 at about 9,500 ft in the vicinity of Gulmarg. This confirms our suspicions that Ward's nest, probably brought in by his collectors, was in fact only the lining of the structure. Taking Major Waters' account in conjunction with those of Rattray and Skinner, we have at last got a picture sufficiently complete except as to the extent of the breeding season. To complicate matters, the birds are at all times liable to be come upon feeding quietly on the ground in small parties whence they fly up to nearby perches only on being closely approached, uttering soft *tews* as they go. Nor should it be taken as definite that when seen singly or in pairs they are indeed nesting. It would appear, in fact, that the season is extended, some nesting early, probably commencing in May or even late in April, others not until July. Osmaston says that young ones, strong on the wing, were seen near Gulmarg at the beginning of June. On the 12th of June at Sonamarg we encountered a newly-fledged young one, while at the end of that month a pair we had had under prolonged but unsuccessful observation on the Thajwas Marg were seen for the first time feeding two fledglings on the ground near some isolated pines. At the Wurjwan Forest Rest House in the Wardwan Valley, on 4 July, we were entertained by a male displaying himself to a very inattentive lady on a post by the veranda. He was vibrating his drooping wings after the manner of a sparrow, at the same time uttering a little trill, the only time we have ever heard any but the *tew* call. This perhaps is the monotonous song, more or less on one note, referred to by Magrath (*J.B.N.H.S.*, XXVIII, 277).

Major Waters' eggs were taken from the nest on 10 July. A

PLATE 41

Nest of Scully's Red-rumped Swallow

Eastern Grey Wagtail, female, leaving nest

PLATE 42

Hodgson's Yellow-headed Wagtail, female, leaving nest

Hodgson's Pied Wagtail

précis of his account amounts to the following : On 25 June the birds were seen in semi-open forest taking small dry twigs and lichen to near the top of a fir-tree which was thinly foliaged compared with many of the giants around it. The nest appeared to be under construction by both birds, who never approached it direct, always alighting some distance below the site which was some 70 ft up and 4 ft out on a horizontal branch. It was well screened from above by the next higher branch. Construction took about fourteen days and the nest appeared to be complete by 5 July. It was a massive shallow cup, 5 in. across and $2\frac{3}{4}$ in. high. The cup was $2\frac{1}{2}$ in. in diameter and half that in depth. The outer nest consisted of a mass of wiry lichen, pale greenish-white in colour, held together by about a dozen stoutish twigs. The inner nest was a shallow cup of fine dry herb-stalks and rootlets. It had a thick lining of goat or cattle hair with a combing of long black human hair, evidently taken from a nearby Gujar settlement. A few feathers were also incorporated. The two eggs it contained, when removed on 10 July, showed slight signs of incubation. In colour they were a clear greenish-blue of medium depth marked with spots and specks, the former blackish-brown with a brownish nimbus and the latter almost jet black. One egg had only one small spot and a couple of specks, the other being quite well marked at the broad end with a larger spot and a patch about 3 mm. square of closely set specks. The smooth shell showed but a slight tendency to gloss.

Unfortunately Major Waters omits to give the measurements of the eggs. The description, however, agrees well with Ward's and that of the three eggs taken by Rattray, which averaged 23·4 by 16·4 mm. Ward's pine-needle nest was found at 9,000 ft near Kolahoi. Rattray's nest, to which Skinner's also approximated, had the twig framework filled with an admixture of fern and weed stems, and both green and dry moss, lined finally with roots and grasses. These latter nests were only 10 and 15 ft from the ground on large boughs of pine trees, so the bird seems to be rather versatile both as regards the materials it employs and in the siting in relation to the ground.

HIMALAYAN GREENFINCH

Hypacanthis spinoides spinoides (Vigors)
(See Coloured Plate III facing p. 136.)

KASHMIRI NAME. *Sabōz Tsar.*
BOOK REFERENCES. *F.B.I.*, III, 160; *Nidification*, III, 68; *Handbook*, 222.

FIELD IDENTIFICATION. (2—) The trivial name of this striking finch is rather misleading, for its predominant hues are brightest yellow with blackish patches about the head, wings and tail. The female is only slightly less brilliant than the male. The bill is short and stout. They are birds of the pine forests but favour those bordering fallow lands and cultivation, where they feed on the ground off fallen seeds, remaining even in the breeding season in twittering little bands rather than in couples.

DISTRIBUTION. This Greenfinch is capricious in its choice of habitat, seeming to have a preference for certain localized areas, yet shunning others which would appear equally suited to its ways. It is said to breed commonly in the Sind and Lidar Valleys, although we happen to have met with it only in the former in the vicinity of Gagangair, and earlier on when still moving up, near the Woyil bridge; and in the latter in small parties between Pahlgam and Aru. Osmaston records them as common in and around Gulmarg, while we found them most numerous in the Wardwan Valley from Inshan up to Suknes, that is between 8,000 and 9,000 ft. We did, however, see a few as high as Rangmarg, which is over 10,000 ft.

This Greenfinch, which occurs in the Kurram Valley and throughout the entire length of the Himalayas, is a partial migrant, in the winter appearing in the foothills and for some distance into the bordering plains. It may then be seen around Kohat and in the Peshawar plain as well as in those districts of the Punjab and western United Provinces adjacent to the hills. A darker race extends through parts of Burma to Yunnan.

HABITS AND NESTING. We doubt if the Himalayan Greenfinch reaches Kashmir from its winter quarters in the lower foothills and adjacent plains much before the end of May. We have noticed them around Pahlgam towards the end of that month, and when we visited the Wardwan Valley in the latter half of June we found them in very large numbers but still in bands of considerable size which were not then contemplating building. Prior to nesting they spend the greater part of their time out in the meadows and ploughed fields where trees are dotted about into which they can retire when disturbed, flying off in pairs or small undulating flocks. Like most gregarious birds they call continuously to one another in soft twittering notes, while the ordinary call-note in the nesting season is a drawn-out musical *ts-weer* just like that of the European Greenfinch. Even when breeding they still remain together in small parties, and half a dozen nests have been found in close proximity to one another. Dodsworth, in fact, records finding two nests near Simla in the same tree. Blue Pines are their favourites, in which the nests are usually placed in a tuft of foliage towards the end of a more or less horizontal branch. They are also said to be partial to deodars, but in Kashmir these cedars are largely confined to the Lolab and Kishenganga Valleys where we have so far not heard of these birds being recorded.

The nests are placed at almost any height from the ground, but mostly round about 15 to 30 ft up. They are always exceptionally well hidden, but the males are said to feed the sitting females on the nest which should prove helpful in tracking them down. The nests are small and neatly made of fine grass-roots and weed-stalks and are lined with fibres and hair, sometimes a few feathers, and contain 3 to 5 eggs, but usually 4. These are a very pale blue with sparse markings of black or sometimes dark reddish. An odd egg here and there may be quite unmarked. They average 18·7 by 13·7 mm. They breed late in the season, in July, August and even September.

KASHMIR HOUSE SPARROW

Passer domesticus griseigularis Sharpe

(See Plate 33 facing p. 132.)

KASHMIRI NAME. Male, *Kantūr*; Female, *Tsar*.
BOOK REFERENCES. *F.B.I.*, III, 173; *Nidification*, III, 77; *Handbook*, 226.
FIELD IDENTIFICATION. (2) The cock House Sparrow in fresh plumage is quite handsome at close quarters. The top of his head and upper back are grey, with the greater part of the remainder of the plumage chestnut streaked with blackish. He has conspicuous white cheeks and a black chin and throat patch. Underneath he is paler. The female is a still paler edition with a well-defined eye-stripe and no black bib.

HOUSE SPARROW

DISTRIBUTION. Excessively common in the Vale in towns and villages and particularly so along the main river from Islamabad right down to Kohala. Great numbers breed in the slow-moving stretch in the Vale wherever there are vertical sandy banks riddled with holes. Elsewhere they occur in places up to 8,000 ft. They appear to be absent from Gulmarg, while in the Kishenganga Valley they are plentiful in Gurais, but absent elsewhere except for a small pocket at Tithwal. Numbers remain in the Vale in winter, but there is a large influx towards the end of March, for on the whole the Kashmir House Sparrow is a migratory species wintering from the Persian Gulf to Northern India and breeding over a wide range from Turkestan, the Frontier Hills, through Afghanistan and a large part of the north-west Himalayas into Tibet.

The ubiquitous House Sparrow is almost world-wide in its distribution, though so far as America and Australia are concerned it is an introduction. Like most *Passeres* with an excessive range, it is divisible into a number of geographical races. The late Hugh Whistler first noticed, some years ago now, that the sparrow of Kashmir was a bird with a difference. This northern race is bigger, with a heavier bill, than those resident in the Indian plains, and is also largely, though not wholly, migratory. Osmaston states that many sparrows remain in the Vale of Kashmir throughout the year and Whistler was of the opinion that these residents are vastly augmented in the spring and summer by a migratory horde from the North-Western plains. This migratory race spreads itself beyond the Himalayas in places to as high as 13,000 ft. In Kashmir proper, however, it is not as a rule seen much above 8,000 ft.

HABITS AND NESTING. The habits of this race are in no whit different from those of any other House Sparrow, for it is the same familiar hanger-on of man, feeding in numbers in his fields on fallen grain and insects and crowding the threshing-floors, or yet again grubbing industriously in twos and threes or small parties in the back yards and streets for food scraps of all kinds. It is a bird with a herd instinct, yet quarrelsome and independent, two or three birds often fighting bitterly to an accompaniment of much chattering from their excited neighbours. Particularly virile and forceful, wherever it occurs the more gentle Cinnamon Sparrow is swamped and scarce if not altogether absent.

We are not aware of the full extent of the nesting season, but Osmaston states that they start in late April, while Whistler talks of the emigration to India taking place in August and September.

With great numbers remaining resident it is probable that the season is prolonged into the early autumn, for Ward took fresh eggs as late as 15 September. Man provides hundreds of pairs of town and village dwellers with nesting sites under the eaves and in holes in the walls of houses, though in Kashmir many will be found in well-wooded country and on forest fringes breeding more or less in colonies in natural hollows in trees. The nests as elsewhere in the world are loose, often domed conglomerations of straw and feathers wedged under gables, on ledges, and amongst creepers and ivy on walls and trees, or yet again in holes in trees. We have even found them wedged in amongst the sticks of a tenanted Black-eared Kite's nest. High sandy river and canal banks in the Vale also provide cavities in which their nests are freely built. Vast numbers, in fact, breed in the banks of the Jhelum down to beyond Baramullah, and from April onwards it is no exaggeration to say that there is hardly a hundred-yard stretch free from their activities.

The eggs number from 3 to 6, and occasionally even 7, and as usual with this species, they vary considerably in coloration, often in the same clutch. The ground colour is a dull white or pale grey heavily blotched with light and dark shades of grey and even blackish, usually covering the greater part of the shell in splashes of some size. Often in the case of one egg in the clutch the marks are streaky and reveal considerable patches of the ground colour. They measure 21·7 by 15·4 mm.

KASHMIR CINNAMON SPARROW

Passer rutilans debilis Hartert

(See Plate 36 facing p. 141.)

BOOK REFERENCES. *F.B.I.*, III, 181; *Nidification*, III, 84; *Handbook*, 228.

FIELD IDENTIFICATION. (2) The cinnamon cap and upper parts at once differentiate the male Cinnamon Sparrow from the male House Sparrow. The female is more difficult to identify, but the broad fulvous stripe above the eye is a good guide. This bird is more of a jungle dweller than the House Sparrow, although it is to be seen within village limits where not ousted by its more virile cousin.

DISTRIBUTION. Not found in the Vale, but is met with all round it from about 6,000 up to 9,000 ft or even higher. It does not cross the main Himalayan range. It is the common sparrow of the countryside, found everywhere except in thick forest, and is numerous around such places as Pahlgam, Aru, Sonamarg, in Gurais, at Gulmarg, and in the many small valleys at the head of the main Vale. Wherever the House Sparrow occurs, the Cinnamon Sparrow appears to be scarcer and then shows up in its true colours as a jungle rather than a town bird. We have not noted it in the Jhelum Valley except at Rampur, though it is certain to occur up the slopes. We met with it in the Kishenganga in late April at Dhani at 3,200 ft. It was common from there upwards although it seemed scarcer in upper Gurais where the House Sparrow was once more in evidence. At Tithwal, in and around the village, only House Sparrows were noted. In the winter it descends to the foothills, often forming into large flocks.

The Cinnamon Sparrow is represented by three races which extend from the extreme Western Himalayas to Yunnan. The Western race, *debilis*, meets the Eastern Himalayan form about the Kumaon-Nepal border.

HABITS AND NESTING. This quiet little bird might well be dubbed the Jungle Sparrow as opposed to the more ubiquitous House Sparrow. Where that universal thruster occurs, there the Cinnamon Sparrow will be scarce or absent and confined instead to the fields, open country and the lighter woods. It is a confiding little bird with typical sparrow ways without being aggressive. It feeds largely on seeds and sometimes a number will collect together to make the most of some spilled grain, but it is on the whole a solitary bird sticking to its pair and leading an unobtrusive life in its own territory. Hedgerows, cultivation, parkland, and the margins of light mixed forest are all frequented by it. For instance, it is common in the middle Sind Valley, in the more open parts of the Lidar Valley around Pahlgam and Aru and in the open woods on the left bank of the East Lidar up to Tanin. It does not appear the moment one leaves the Vale; it is hard to say just why not and what constitutes its lowest limits in the summer, but while staying in a forest hut at Sanzipur in the Kazinag foothills, where we saw no Cinnamon Sparrows, it was only necessary to walk half a mile up the valley to find them any day around Vihom village. A species must of course start and end somewhere, but their appearance was so well defined with apparently nothing in particular in the type of country to account for it.

Nest-building commences about halfway through May and eggs may be taken towards the end of that month, throughout June and early July. The nest is built in any small natural hollow in the branch or trunk of a tree from low down to some 20 to 30 ft up and to some small extent under the eaves of houses. It consists of an untidy collection of dry grass or straw lined with hair and feathers. Osmaston says 4 or 5 eggs are laid, smaller and more darkly marked than the eggs of the House Sparrow, but as with that bird, one egg is usually much lighter in colour than the rest. They average 19·1 by 13·9 mm.

STOLICZKA'S MOUNTAIN-FINCH

Fringilauda nemoricola altaica (Eversmann)
(See Coloured Plate III facing p. 136.)

BOOK REFERENCES. *F.B.I.*, III, 192; *Nidification*, III, 90; *Handbook*, 228.

FIELD IDENTIFICATION. (2+) There is nothing particularly distinctive about its plumage in the field, but this ground finch is an exceedingly common bird on open stone-bestrewn and wettish hillsides and meadows above the tree line. No other member of this genus nests within our area, Brandt's and Adam's Mountain-Finches crossing the main Himalayan range into Ladakh and beyond for such purposes.

DISTRIBUTION. In summer Stoliczka's Mountain-Finch is an abnormally common bird above tree-level from 11,000 to nearly 14,000 ft, both on the Pir Panjal and the main Himalayan ranges. It is to some extent patchy in its distribution and may be found lower than this in suitable areas. Take, for instance, Glacier Valley above Thajwas where considerable numbers are to be seen in the valley bottom at barely 10,000 ft. In July incredible numbers were noted at Astanmarg at 12,000 ft, a level which appears most to their liking, and where they were just showing signs of commencing to breed. 'In the cold season', to quote Hume, 'innumerable flocks swarm on the lower ranges of the Himalayas in winter from 4,000 ft to 7,000 ft.' We have no information as to whether or not any remain in the Vale in winter, nor regarding the lower levels along the motor road from Baramullah downwards, though Whistler considered it inconceivable that they should not be found in the Murree hills. Meinertzhagen recorded huge flocks in the Vale of Kashmir (between 5,000 and 7,000 ft) throughout March. He said they had reached about 7,500 ft in the Sind Valley by the end of the month and by 13 April they were feeding on small patches of exposed ground in a continuous snowfield at 11,700 ft. The majority of the high-altitude specimens were adults, those at lower elevations being usually immature.

To put it shortly, Stuart Baker gives the distribution of Stoliczka's Mountain-Finch as Afghanistan and the North-West Frontier, Kashmir and Ladakh, north to Turkestan and the Altai, and south-eastwards to Kumaon and Eastern Tibet. It does not apparently breed in Afghanistan, and on the North-West Frontier reaches only as far west as the Samana on migration.

HABITS AND NESTING. Stoliczka's Mountain-Finch may not be a bird to catch the eye, either through beautiful plumage or pleasing habits, but wherever met it is invariably so abundant that it cannot escape notice. Buchanan speaks of it as being one of the commonest birds in the higher valleys near Sonamarg, but in fact one need not climb to any great elevation for at the end of June the Valley of Glaciers teems with them from the moment of leaving the Thajwas marg. Glacier Valley is typical of this bird's requirements. The level patches of sward are dotted with rocks and stones; some smallish and rounded, some large and flat; some lying jumbled together providing deep nooks and crannies. Others lie flat and isolated upon the ground with rat-holes drilled tortuously under their lower surfaces. The turf between is often soft and squelchy, patterned in blue and white by an endless profusion of frail anemones which surround the delicate pink splashes of rose primulas rimming the marshy depressions. On the valley's slopes, and farther up the stream where one approaches the snow-beds, tiny rivulets of icy water lose themselves in the short grass. Such is the ground beloved of these finches which rise in numbers before one, fluttering into the air or landing and sometimes running a few paces after the manner of a lark or pipit. Unfortunately they are not blessed with a song, but only with a single chirp.

Damp ground is not a necessity for we have found them on dry plateaux and hillsides, but always those of a rough nature well sprinkled with boulders. For it is in the holes beneath these in which their nests are almost inconceivably well hidden. As Whistler truly says, although this bird is so numerous, large flocks evidently breeding more or less in company in the same area, comparatively little is known about their nesting. Buchanan writes of his finds around Sonamarg: 'they did not begin to breed till the middle of July and I fortunately secured several nests at an elevation of 12,000 ft to 13,000 ft. The nest is made of grass, lined with horse-hair and a few feathers; it is placed in a hole under a rock, and the tunnel leading to the nest is sometimes

PLATE 43

Brown Rock-Pipit removing fæces from nest

Hodgson's Pipit leaving nest

PLATE 44

Kashmir Skylark

Young Asiatic Cuckoo in nest of Blue-fronted Redstart

as much as two feet long ; in two instances the holes were like those made by a rat in open flat ground. The eggs, generally 4 and sometimes 3 in number, are pure white with considerable gloss, and are somewhat pointed at the smaller end.' To holes in the ground Whymper adds as nesting sites crevices in rocks, on a sheltered ledge of rock, or under the shelter of a bank, but we consider these open nests must be very few and far between. Osmaston found these birds ' very common on the upper rocky slopes of Aphawat from 12,000 to 13,000 ft where they breed. Nests contained young in July and August '. In his *Notes on the Birds of Kashmir* he includes weed-stems amongst the nest materials and says that the eggs when fresh and unblown are a beautiful pink. He gives their measurements as about 20·9 by 15·4 mm. When he writes of nests containing young in July, he must surely have been referring to the end of the month, for the organs of specimens shot by us in the first week of July both at Sonamarg and at 12,000 ft at Astanmarg showed but faint signs of the approach of breeding. Incidentally, when in winter quarters and on their way to the uplands they rest freely in trees, often being found in forest clearings and in cultivation.

INDIAN GREY-HEADED BUNTING

Emberiza fucata arcuata Sharpe
(See Coloured Plate III facing p. 136.)

BOOK REFERENCES. *F.B.I.*, III, 199 ; *Nidification*, III, 92.

FIELD IDENTIFICATION. (2) Male sufficiently like that of Stewart's Bunting to cause confusion on a cursory glimpse in the field. But crown of head streaked black and grey instead of pure grey ; ear-coverts chestnut instead of greyish-white ; chin and throat white instead of black. Black moustachial stripes running down from the gape broaden and unite on the upper breast to form a more or less complete black gorget which encircles the white chin and throat. A broad chestnut pectoral band under this gorget extends down the flanks making practically the entire underparts chestnut except the centre of the lower breast and abdomen which are white.

DISTRIBUTION. Not common but undoubtedly well dispersed round the main Vale in the same rather dry stony ground which Stewart's Bunting affects. Also reported from the Sind Valley as far up as Koolan. We would not be at all surprised to find specimens amongst the buntings distributed along the motor road between Uri and Baramullah, but Davidson who traversed the route by tonga in more leisurely times did not report it, mentioning only Stewart's Bunting as being common from Kohala to Baramullah. We have no information about its status in winter in Kashmir, but at that season it has been recorded as far south as Central India, the Deccan (Jalna), and Nagpur.

The Indian race of the Grey-headed Bunting is found from the Kurram Valley to Hazara, and along the outer Himalayas to Nepal, the typical race, which penetrates to Eastern India in the winter, being a highly migratory bird nesting in Siberia and in North China to Japan.

HABITS AND NESTING. The Grey-headed Bunting appears to be a bird more of the outer hills from Hazara to Garhwal, penetrating to no great extent beyond the first snowy range. In Kashmir it is to be found on the same ground as Stewart's Bunting, ceasing to be met with above the 7,000-foot level, just as that bird does. Although rather scarce in our area, numbers are probably overlooked, their nests being taken for those of Stewart's Bunting. Colonel Phillips, *in epistola*, wrote regarding a nest found on the Takht-i-Suleman at Srinagar that he thought he was dealing with a Stewart's Bunting until he had actually photographed the bird. The Takht is undoubtedly a stronghold of this species and it has been seen on similar bare stony slopes near Ganderbal, north of Baramullah, and elsewhere round the Vale's rim, but Osmaston states that it is found on the hillsides on both the main and side valleys feeding on the ground amongst bushes and dense scrub of berberis, wild rose, cotoneaster, indigofera, etc. ' The cock', he says, ' has a bright short song, the best of all the Buntings with which I am acquainted. It may be expressed in words as follows: " Chick-chick-he'll get used to you, Chick ".' We heard this song on a hillside not far from Baramullah, where it was noticed that the birds were in the habit of holding forth from the bare lower branches of the scattered pines.

The nest is always well concealed in the grass and roots at the foot of a bush or tucked well into some unevenness in the hillside, and is made of grass and stems lined with either hair or finer grass. The birds appear to nest mainly in May; Osmaston says from about the middle of the month, around which time he took nests containing 4 and 3 eggs. Wilson states that his brother took a nest of 3 eggs on 30 May. Rattray took 2 nests, one of which held four and the other 3 eggs. Both these nests, however, were outside our area.

The eggs have a grey or pale greenish ground which always shows to some extent though it is well covered, more thickly at the larger end, with spots and small blotches of reddish-brown or ashy brown, sometimes a blackish-brown. These markings are in general always lighter than those of Stewart's Bunting's eggs and there are never any of the characteristic bunting streaks and lines. Osmaston's ten Kashmir eggs average 19·9 by 15·6 mm.

WHITE-CAPPED BUNTING

Emberiza stewarti Blyth
(See Plate 37 facing p. 148.)

BOOK REFERENCES. *F.B.I.*, III, 203; *Nidification*, III, 93; *Handbook*, 229.

FIELD IDENTIFICATION. (2) The pale grey head and breast offset by wide black lines through the eyes and the black stripes from the chin and throat to the sides of the upper breast are very conspicuous factors in the recognition of the male. The female is much less distinctive, the upper plumage consisting mainly of browns streaked with blackish and the underparts fulvous streaked with brown. This bird prefers rather dry and barren stony hillsides such as are found immediately around the northern and western fringes of the Vale.

DISTRIBUTION. This is the bunting of low elevations in Kashmir, for it breeds in a zone up to about 6,000 ft, after which its place is taken by the Meadow-Bunting. It is common on the barer hills around the Vale's rim, those with scant vegetation in the form of coarse grass, here and there prickly bushes, and a plentiful sprinkling of rocks and boulders. It is also the bunting which is often disturbed on the roadside from Uri to near Baramullah. In April we found it common in the lower Kishenganga Valley from Pateka (2,690 ft) to Tithwal, and in lesser numbers at Salkalla at 4,500 ft. The Pateka birds may not have reached their breeding area, as we saw numbers on unsuitable ground near the river in company with other migrants. On the other hand they are early breeders, commencing in the month of April, and it will probably be found that they nest but a short way up the barer hillsides all the way from Kohala. Large numbers winter in the foothills and in the plains of the North-East Punjab.

The White-capped or Stewart's Bunting breeds in North

BUNTINGS

Baluchistan at heights from 7,000 to 9,000 ft, also in Afghanistan and Turkestan and along the Himalayas as far as Kumaon. In winter it drops into the foothills, entering the plains of the North-West Frontier, Northern Punjab and the West United Provinces. Some reach Rajputana and Central India.

HABITS AND NESTING. This spry little bunting is by no means uncommon on the northern side of the Vale where the hills, rising directly from the plain, are rather hot and bare. A few nests can always be found on the slopes of the Takht-i-Suleman, which is typical of its favourite habitat, while we found it really common on all the promontories around the Wular Lake. Here its monotonous notes, *zwig-zwig-zwig*, may be heard from all directions, being uttered from a low bush, the top of a boulder or the lower branch of a pine tree if any be present. It of course spends most of its time moving about the ground amongst the dry grass and stones searching for seeds. It is noticeable that the habitats of this and the Meadow-Bunting do not overlap to any extent, although we once came across both on the same ground near Baramullah. Usually Meadow-Buntings are found where Stewart's Buntings cease, at about 6,000 ft, and extend thence up to at least 10,000 ft. The reason for this is mainly one of terrain, Stewart's Bunting preferring the more open arid ground which is found only at the lower levels in Kashmir, whereas the Meadow-Bunting lives up to its name, having a leaning towards cool grassy meadows and verdant bush-dotted and bracken-covered slopes, even entering and breeding in light forest. Further, while the Meadow-Bunting's nest is to be found in green surroundings concealed in lush grass, in the roots of leafy bushes, and even in the lower branches of fir trees, that of Stewart's Bunting will be found under a stone, on a ledge of rock on precipitous ground, or wedged into a cleft between two boulders. It may too be in a cavity in a bank or thrust into the roots of a prickly bush, but it is the drier ground which the bird definitely prefers for all purposes.

The nest is a cup of dry grasses and roots with a lining of the thinner rootlets, fine grass and some hairs, and the eggs are marked to tone in well with a dry and stony background. They have a dull white ground well covered with large spots and small blotches of inky-purple and dark reddish-brown, very different indeed from the light-toned eggs of *cia stracheyi* and always much darker than the rather similarly marked eggs of the Grey-headed Bunting which is sparingly found in the same areas. Four eggs is the normal clutch, with 3 and 5 as not unusual variants. On an average they measure 19·5 by 14·6 mm. This bird arrives on its breeding grounds in April and nest-building is already under way before the month is out. We have taken eggs throughout the latter part of May and early June, while Davidson, after finding many empty nests at the end of April and in the first few days of May around Srinagar and Ganderbal, took his first clutch of 3 eggs early in May. Osmaston likens the song of the White-capped Bunting to the 'Little-bit-of-bread-and-no-cheese' of the Yellowhammer, but lacking the last long-drawn-out note.

EASTERN MEADOW-BUNTING

Emberiza cia stracheyi Moore

(See Plate 38 facing p. 149.)

KASHMIRI NAMES. *Wān Tsar, Sarah.*
BOOK REFERENCES. *F.B.I.*, III, 205; *Nidification*, III, 94; *Handbook*, 230.
FIELD IDENTIFICATION. (2) The striking facial markings separate this bunting from every other species in Kashmir. The body plumage except for a grey breast is mainly chestnut-brown streaked with black. Besides being found in rough grass and in bush-dotted country, it enters the lighter parts of the forests and quite often flies up from the ground into the lower parts of the trees.
DISTRIBUTION. A bird of very wide distribution throughout the valleys and uplands of Kashmir. Once met with at approximately 6,000 ft it is numerous everywhere, other than in thick forest, up to 10,000 ft and at times even higher.

The Meadow-Buntings have a wide distribution in Southern Europe and North Africa, and from Transcaspia across a great part of Asia. On the borders of India one race, *cia par*, is found sparingly in winter in North Baluchistan and in summer in South Waziristan and more commonly on the Sufed Koh. Our race appears to be confined to the Western Himalayas from Gilgit and Hazara to about Kumaon. It is considered a resident with but a small vertical movement, but we have no information about its movements in Kashmir. The bird which visits the Punjab plains in winter is *cia par* mentioned above, which breeds from Transcaspia to Chitral.
HABITS AND NESTING. Meadowland, cultivation, scrub jungle, broken ground, and even the more open grassy hillsides, all are patronized alike provided they are not devoid of bushes or patches

of bracken and a few rocks. It may even be found well within the forests wherever a blank occurs and somewhat beyond the fringes of such blanks. It is, of course, primarily a ground bird, where it is to be come upon moving about in the grass or amongst the bushes and stones, often advertising its presence in concealing cover by subdued mouselike chirps, perhaps by way of apprising its mate of its whereabouts or of giving warning. It is, however, by no means averse to ascending twenty or more feet into a tree. Its nest, too, is generally on the ground, but at times it is to be found well up in a bush. On one occasion we found a nest near Baltal about 20 ft up, while another was 15 ft from the ground concealed in a Blue Pine where the extremity of one branch drooped over another.

When placed thus above ground-level the nest is a compact deep cup of grass lined with finer grass and hair. In the much more normal situation, that is concealed in the grass on a bank or at the foot of a bush, the grass foundation is often somewhat scanty. The hair in the lining varies greatly in quantity, sometimes being almost absent, and we have seen nests which to all intents and purposes consisted of nothing but a hollow in the bank lined with pine-needles.

The eggs are rather striking, being pale grey with a varying number of thin hair lines scrawled about them, usually around the large end. A few spots and blotches of blackish or red-brown are also often present. At times the background has a mottled appearance. Altogether they are handsome eggs, reminding one forcibly of those of the Yellowhammer. The normal clutch appears to be 3, but 4 are not infrequently laid. We have so far failed to find a clutch of 5. They measure 21·5 by 15·7 mm. The breeding season is protracted, eggs being obtainable from the end of May until the beginning of August, though May and June constitute the main period. They probably breed right up to the limit of their range, but the highest nest we have taken was at approximately 10,000 ft near Baltal.

They are said to be much victimized by the Asiatic Cuckoo. We have little doubt that this is quite true, but such is fate that out of the scores of nests we have seen belonging to this bird, we have so far been fortunate enough only to find one with a cuckoo's egg and that was rejected by the irate bunting almost as soon as it was deposited.

CRESTED BUNTING

Melophus lathami subcristatus (Sykes)
(See Coloured Plate III facing p. 136.)

BOOK REFERENCES. *F.B.I.*, III, 221; *Nidification*, III, 98; *Handbook*, 233.

FIELD IDENTIFICATION. (2) A dark bird of upright carriage with a most prominent crest. The black body-plumage is offset by chestnut wings and tail. Found mainly at low and medium elevations in broken rocky ground with plenty of rough grass and scattered bushes. In the female the plumage is browner and streaked and the crest less marked.

DISTRIBUTION. Occurs in any numbers only in the middle reaches of the Jhelum Valley up to Chenari or Uri. Davidson reported it as common all the way to Baramullah, but so far we have never seen it in the immediate vicinity of that place. It also occurs in the lower Kishenganga Valley. Although usually described as a local migrant its movements are often extensive, so if our Crested Buntings do move in the winter, it will probably be down the river to suitable ground in the immediate plains of the Jhelum District.

This bunting nests in the Himalayas from Hazara to Bhutan, and in hilly country across to Western India south to Belgaum. On the East it is found in Bihar and parts of Bengal. It extends through much of Burma and Tenasserim to China.

HABITS AND NESTING. It was Magrath apparently who first described this pert little bird as having a peacock-like stance. His actual words in an article on the Birds of Thandiani in the *Journal* read as follows : ' On the ground and walking, the attitude of this handsome Bunting is very Peacock-like. The head and breast are held very erect, while the tail, which seems to trail behind, is rather expanded.' These characteristics, so very foreign to the majority of the buntings, which move with the breast rather close to the ground, have naturally been seized upon by later workers so that Magrath's description has been perpetuated in many subsequent publications.

Being a local migrant the Crested Bunting often appears in the spring most unexpectedly in waste land suited to its habits, on which occasions it is usually in small parties of both sexes. From April onwards, however, when motoring to and from Kashmir by the Jhelum Valley route, we have often seen this bird in ones and twos in the rough ground between the road and the river.

BUNTINGS

It cannot be uncommon between Domel and Chenari and possibly farther up still. Unfortunately we can trace no published accounts of its nesting in this area, but not far from Garhi on 1 June we noticed a male just below road-level flying parallel to the car with a fine twig or piece of grass in its bill. Both birds assist in the building, and, of course, in the feeding of the young, so it can be taken as certain that this cock was on the way to a nest under construction. So far as the Himalayas are concerned the Crested Bunting is said to nest from Nepal to Murree in April to June at all elevations between 2,000 and 5,000 ft, but Stuart Baker says he can trace no April records whatsoever.

The nest, though variable, is usually a cup of grass and grass roots lined with finer stems and possibly some horse-hairs. Whistler says moss may also be incorporated, but in that part of the comparatively dry Jhelum Valley where these birds are mainly seen, it is not so likely to be used. Like those of Stewart's Bunting, whose habitat in its lower levels overlaps that of this bird, the nest is placed in crevices in rocks, in walls, on sheltered ledges on boulders, in banks or on the ground in a tussock of grass or in the roots of a bush. It is at times a loose straggling affair; at others a compact well-built cup.

The eggs, 3, sometimes 4 in number, vary a good deal, having a ground colour from a pale greenish-white to pale brownish with markings which sometimes cover the greater part of the egg, but sometimes leave considerable patches of the ground colour bare through being congregated at the larger end. These markings are spots and freckles, sometimes larger blotches or mottlings, of reddish- or purplish-brown and grey. The eggs measure 20·1 by 15·6 mm.

These buntings feed largely upon the ground, occasionally mounting to the top of a rock or bush or perching at the extremities of tall grass-stems, from which the male is apt to proclaim at frequent intervals his peculiarly soft melodious note. When the female is sitting he constantly remains in the vicinity whiling away his time in the above manner.

PLATE 45

Kashmir Pied Woodpecker at nesting hole

Scaly-bellied Green Woodpecker at nesting hole

PLATE 46

KASHMIR HOUSE MARTIN

Delichon urbica cashmeriensis (Gould)
(See Plate 40 facing p. 157.)

BOOK REFERENCES. *F.B.I.*, III, 228 ; *Nidification*, III, 101.
FIELD IDENTIFICATION. (2) A plump little swallow with a short tail and long wings which close to the sides of the body as it flits gracefully here and there across the sky in search of flies. The upper half of the plumage is black except for a broad contrasting bar of white across the rump, the lower plumage being white somewhat sullied with smoky brown. Martins nest in colonies, and even when feeding are seldom seen alone but are generally come upon in upland valleys flitting backwards and forwards in parties, sometimes of considerable strength, at varying heights above the ground from the level of the tree-tops to such heights that they are mere specks in the heavens.
DISTRIBUTION. The Kashmir Martin is distributed in our area in well dispersed colonies mainly between 9,000 and 13,000 ft. In the middle reaches of the Kishenganga we saw a party at about 7,000 ft, but cannot say whether they were breeding at that elevation, for in bad weather or when continuous rain and cloud have forced the flies to lower levels, their foraging parties will descend to elevations of little above 6,000 ft. A small colony nests in the rough-hewn part of the roof at the entrance to the Banihal road tunnel whose elevation is 9,200 ft, and we have seen no nests lower than this. Osmaston notes that he did not see this bird in the immediate vicinity of Gulmarg, but that he saw several pairs hawking insects at 13,000 ft above Pandon Pathar (Ferozpur Nullah), where they probably breed under the cliffs in that neighbourhood. There is a large colony nesting at approximately this elevation at Astanmarg and another uses a precipice at Tanin at the foot of which runs the path to Amarnath. The elevation of this congregation will be little above 9,000 ft. The colony whose nesting is described below inhabits a cliff face situated near Mojipal at 11,000 ft in a small side nullah off the Marbal Glen.

The House Martin breeds over a great part of Europe and Asia from the British Isles and Scandinavia to Japan. Our Kashmir form appears to overlap typical *urbica* of Europe in Gilgit and Ladakh and breeds in the Sufed Koh and the Himalayas from Kashmir to Sikkim, wandering farther east in the cold season, but records of its winter movements are most vague.

MARTIN

HABITS AND NESTING. We lately had the good fortune to come across a small colony of these birds nesting near the base of an overhanging cliff where some sixteen nests were so low as to be easily reached with the hand. The base of this cliff is scooped out so that when it rains a strip of the ground beneath, about six feet wide, remains quite dry. It is the roof of this overhang, sloping at about 75°, which provides lodgements for the nests. Most of these nests were separated one from another, but we noted one cluster of three and another of two. One is so apt to think of martins' nests as they usually are in England, beautifully symmetrical structures built into the top corners of windows and under eaves, that it came quite as a shock to find that these examples of the martin's art varied greatly in size and still more so in design. They were adapted perfectly to conform to the actual requirements of each particular site which their builders had chosen for them, so that those in clefts in the rock had but a narrow front wall up to a foot or so long closing in the slit; a few were replicas of swallows' nests, somewhat close up against their ceiling and open all round the top—these might well have been taken for Crag Martins' nests; two were in water-worn cups in the rock face whose circular holes had been reduced in diameter with mud pellets, somewhat after the manner adopted by the nuthatches, while the builders of a couple more in similar situations with smaller entrances had not troubled to use any earthwork whatsoever, but had merely lined the bottom of the cavity with the usual untidy scraps of grass and feathers. Nests like that in the illustration were in the minority.

On 10 July those that were not already empty, and few were, contained up to 3 large young ones—we believe one nest contained 4—and we noticed that one pair of birds was already building a new nest, adding a mud pellet every few minutes to the structure which was being raised on the remnants of a discarded one. Incidentally the pellets are small, which gives to the nest wall a neater appearance than that possessed by nests of Striated and Common Swallows. We possess no direct evidence that these birds are double-brooded, but from the foregoing it appears probable, and Colonel Magrath in a most interesting note on the breeding of this bird around Thandiani writes this: 'the first broods leave the nests from the beginning to the middle of July.' This period agrees well with our own experience of colonies within Kashmir limits. He ends up: 'I was unable to observe the second broods.'

We did not meet with the owners of the nests described above until within a mile of the cliff, but it is evident that in bad weather these martins will drop in search of food to much lower levels than their wont and will also range at any time some miles from their breeding haunts. For instance, at Bagtor in Gurais, during a ten days' stay we were visited twice by birds which appeared to come from high up the Shalput Nullah.

Martins' eggs are pure white ovals measuring 17·9 by 12·7 mm., and 3 or 4 are laid about the end of May or beginning of June, but it is usual to see only 2 or 3 young in a nest. To quote again from Magrath's note : ' Seldom more than 3 eggs are hatched and indeed three young pretty well fill the nest when half fledged. As a rule there is never more than room for two heads at a time at the aperture of the nest to receive food, consequently one young bird is generally in the background and unable to obtain sustenance till one or other of those at the entrance retires satisfied. The parents appear to use no discrimination in their feeding and it is a case of the survival of the fittest, the most vigorous securing the most food.'

Under the weather conditions prevailing we unfortunately were unable to obtain satisfactory photographs of the parent birds feeding their young ones, for the food is often transferred rapidly on the wing into a wide-opened gape thrust eagerly out of the nest, and even when the bird does remain to cling to the side of the nest, which is not often, the visit is usually of short duration. The birds often fly backwards and forwards past the nest uttering sharp squeaks, and it is most amusing to watch a row of neat little black-topped heads turning from side to side in unison as they eagerly follow the flight of their parents along the cliff face. Whenever a nest is visited much twittering at once bursts forth from all adjacent homes.

The nests seem always to be sheltered in some way from direct exposure to inclement weather either by overhanging or by being placed under a protecting ledge, but more often than not they are placed so far up the cliff faces, a couple of hundred feet or considerably more if the nature of the cliff permits it, that observation is difficult or even impossible. We have never heard of the nests being built under the eaves of houses within our limits. We have seen such nests in Sikkim, and Rattray evidently found their nests in the Galis under the eaves of houses and in verandas, but generally speaking only Gujars' huts are to be found at the elevations at which this bird breeds with us and this type of dwelling provides no suitable sites.

The flight of the Martin is similar to that of the Swallow and the birds are generally seen in parties of varying strength at any height according as to where the best feed is to be procured.

SWALLOWS

COMMON SWALLOW

Hirundo rustica rustica Linnæus
(See Plate 40 facing p. 157.)

KASHMIRI NAME. *Katīj*.
BOOK REFERENCES. *F.B.I.*, III, 240; *Nidification*, III, 111; *Handbook*, 238.

FIELD IDENTIFICATION. (2) The Swallow, with its deep blue upper plumage, chestnut throat, and the pale rufous underparts below a broad black pectoral band, is familiar to all bird-lovers. The graceful flight as it hawks flies over land or water, twisting and turning and making graceful sweeps on pointed wings with the needlelike outer tail-feathers continually separating and closing, is a feature of the Jhelum Valley and of all the Vale's waterways. It also freely enters the dwelling-houses and even the shops on Srinagar's Bund, where it places its saucer-shaped mud nest close up to the ceiling against the rafters.

DISTRIBUTION. The Common Swallow is a spring and summer visitor to Kashmir, the first birds arriving from the plains of Northern India as early as the last week of February. The movement is complete by the end of March, so that in April they are well established and already nesting. They spread throughout the Jhelum Valley and the main Vale of Kashmir, to which they are almost entirely confined. They are absent from the Kishenganga Valley and enter only the lower widest portions of the side valleys such as the Sind and Lidar Valleys.

The Swallow occurs either in summer or winter throughout almost the whole of the Old World, the Western Himalayas constituting the south-eastern breeding limit of the typical race. We cannot follow the distribution of the typical form and *gutturalis* as given by Stuart Baker in the *Fauna* and repeated in the *Nidification*. In this connexion Whistler states that he can find no evidence of *gutturalis* breeding in Afghanistan (*Journal*, XLV, 282). Nesting birds from North Baluchistan and the North-West Frontier must also be typical *rustica*, to which Ticehurst refers them in his *Birds of British Baluchistan*. In winter *rustica* is found right down to Cape Comorin and even in Ceylon, mingling in Eastern India with the Eastern Swallow, *gutturalis*.

HABITS AND NESTING. To remove the Swallows from Kashmir would be to deprive the Vale of one of its most pleasing features, for without the jewel-like flashes of the blue and chestnut kingfishers and the graceful flittings of the deep purple Swallows, the waterways and lakes would lose a great deal of their charm. As one reclines in a shikara, watching the life of the Vale glide past, the Swallows flit in and out of the willows and under the prows of the passing dungas, now skimming the water's surface to pick up a bedraggled fly, scarcely disturbing the surface in doing so, now wheeling sharply upwards to flit in and out of the casements of the chalet-like houses. As they swoop hither and thither the wing-tips constantly close almost to the sides, while the tail expands and contracts so that sometimes the attenuated outer feathers are spread wide apart, sometimes they are so close together as to be well-nigh parallel.

As the season advances, family parties may be seen in rows on the willow branches, the duller youngsters, which lack the streamers, quivering their wings with excitement and twittering loudly on the approach of their parents. But the Swallows do not confine themselves to the waterways : they are to be seen anywhere where water is not far off, which of course means all over the Vale, and few of the wooden houses are without their pair of Swallows nesting within. Within Srinagar the shops along the Bund are all invaded ; in fact, here Swallows might almost be said to nest colonially in some of the larger buildings. Even in the most crowded bazaars they are still to be found. The Kashmiris do not object to sharing their living-rooms with these trusting birds, and nail pieces of board to the rafters beneath the nests to catch the droppings. Under the prows of the larger square-ended boats runs a stout cross-piece and here too, a foot or so above the water-line, the saucer-shaped mud nests are often to be seen.

The Common Swallow does not penetrate beyond the mouths of the side valleys. Numbers of course populate the Jhelum Valley from Kohala upwards, but such valleys as the Sind and Lidar are only tenanted in their wide lowest reaches, the former to about the Wangat Nullah junction, the latter to near Aishmakam.

The swallows as a family make use of mud pellets in the construction of their homes, employing various forms of architecture, the Common Swallow fabricating a shallow half-saucer cemented to the side of a beam fairly close to the protecting roof. Meinertzhagen witnessed the arrival of a male at Baramullah (the first bird of the year) on 2 March which immediately started taking wet mud up to a nest of the previous season, but it is in April that we have most often seen birds descending to the water's edge to scrape up a pellet of mud with the bill or to collect straws, for the depression is loosely lined for the reception of the eggs with short lengths of straw and downy feathers.

SWALLOWS

Three to 6 eggs are laid. These have a tinge of pink about them but when blown they are shown to be white, the shell being so thin that the yolk tints it. Specks and bold dots of red are scattered here and there mainly at the larger end. They average 19·6 by 13·7 mm. They may be taken up to July as numbers of birds start their second broods in June. In August, with the monsoon in full blast in the Plains, some birds are already leaving, but it is not until the second week of October that the Vale once more lacks the easy grace of their ceaseless flying.

SCULLY'S RED-RUMPED SWALLOW

Hirundo rufula scullii Seebohm

(See Coloured Plate II facing p. 40 and Plate 41 facing p. 164.)

BOOK REFERENCES. *F.B.I.*, III, 252; *Nidification*, III, 122; *Handbook*, 241.

FIELD IDENTIFICATION. (2) A more slenderly built bird than the Common Swallow. In flight the outer tail-feathers are not spread so far apart, and generally the angle between them is so small that they are almost parallel. As regards the plumage, the white upper tail-coverts suffused with rufous, and the streaked lower parts tinged with the same, again differentiate it from the Common Swallow.

DISTRIBUTION. The distribution of the Red-rumped or Striated Swallow in Kashmir is peculiar. Its status, in fact, is to us still obscure. One race, the most westerly form, *Hirundo rufula scullii*, occurs and breeds in numbers all the way up the Jhelum Valley from the limits of Kashmir territory at Kohala to about Chenari, though we have seen odd birds near Uri. This race also penetrates the Kishenganga Valley as far up as the Tithwal gorge. In both cases the birds usually fade out between the 3,000 and 4,000 ft marks. Meinertzhagen obtained the female from a single pair he saw near the Bandipur Nullah (Wular Lake) at about 6,000 ft, and in May 1920 we saw many Striated Swallows in the Vale flying over Manasbal Lake (5,200 ft). After that, however, none was seen within the confines of the Vale until, in June 1937 and again on 17 May 1940, we noted a pair flying about some rocks below the Kashmir side of the Banihal road tunnel, elevation a little under 9,000 ft. Meinertzhagen's bird belonged to the race *nipalensis* and probably all those recorded by us from the Vale and above also belonged to it, this form appearing to breed throughout the Western Himalayas in a somewhat higher zone from approximately 4,000 to 9,000 ft. In the cold-weather months both these races go to swell the large numbers of Red-rumped Swallows of the Plains, *erythropygia*, and pass the winter in the districts bordering the foothills. At times really vast concourses may be seen closely packed on the telegraph wires.

Many races of Striated Swallows are found from Southern Europe to Africa and across to Japan. *Hirundo rufula scullii*, with which we are mainly concerned here, reaches Indian limits as the breeding form in the Frontier hills from Baluchistan to Kashmir.

HABITS AND NESTING. This swallow will first be seen from the car as one winds along the Jhelum Valley, for the rocks and hollows of the excavations on the inner side of the road are exactly suited to its nesting requirements. Unlike the Common Swallow it is not so dependent on human habitations, but likes boulder-strewn valleys, gorges, and hillsides with patches of rocks about which it hawks insects after the usual manner of its family. The flight, however, is rather slow and deliberate and includes a high percentage of volplaning. It loves to take its rest on the telegraph wires by the side of the road, and in the Kishenganga Valley we noted them settling on the rafters and copings of the Forest Rest Houses. Our inquiries there failed to produce any evidence of the birds nesting within buildings or on the rafters of verandas. On the contrary we were assured that the nests would only be found on rocks, and it is true that numbers of such nests may be seen along the Jhelum Valley road as far up as approximately the 68th milestone. Although buildings are eschewed, man-made nesting sites are not altogether avoided, for it is noticeable during the journey from Kohala that pairs are as often as not in the immediate vicinity of culverts and the smaller masonry bridges, under which they undoubtedly breed. In fact, to make quite certain we stopped the car at one such point and were immediately rewarded by finding a half-built nest at the crest of the arch midway through the culvert.

The nest is usually described as retort-shaped. When in a natural site it is placed almost invariably under an inverted ledge of rock, resting in the angle forming the back and the ceiling as it were. It is made of hundreds of mud pellets and is often very large considering the size of the occupants. The only entrance is along a narrow tube about 2 inches in outside diameter. The nest is usually six or eight inches long, exclusive of the entrance-tube which may be anything between four inches and a foot in length, although at times this passage is entirely absent. The interior is lined untidily with straw and a few feathers.

These birds reach their breeding ground in the second or third week of April, but nest-building begins about the first week in May and takes some time so that eggs should be obtainable from approximately the latter half of May to the end of June or early July. These are 3 in number, occasionally 4, and as one would expect in a closed nest, are pure white. They are rather elongated eggs of a very fine texture. They average approximately 20·8 by 14·4 mm.

HODGSON'S PIED WAGTAIL

Motacilla alba alboides Hodgson

(See Plate 42 facing p. 165.)

KASHMIRI NAMES. *Pinskanni, Dobbai.*
BOOK REFERENCES. *F.B.I.*, III, 262; *Nidification*, III, 128; *Handbook*, 243.
FIELD IDENTIFICATION. (2+) An energetic, tame bird of black and white to be seen almost anywhere where water is not far off. Although feeding almost exclusively on the ground, it often perches on buildings or, along the waterways, on the roofs of houseboats and barges. Flies with a dipping motion leaving the ground with sharp calls, *pittit-pittit*, and slightly spreading the long tail so that the outer white feathers show up conspicuously. When it lands the tail is waved energetically up and down in periodic bursts as it trips lightly around the water's edge.

DISTRIBUTION. A widely distributed bird, being very numerous indeed in the main Vale. Ascends the side valleys to at least 10,000 ft in progressively decreasing numbers, breeding most commonly from the level of the Vale up to about 7,000 or 7,500 ft. Hodgson's Pied Wagtail wanders the least of the four Pied or White Wagtails which are to be found in winter within Indian limits. Some, in fact, remain in the Kashmir Valley the whole year round, but the majority move to the foothills and plains adjacent to them, returning to Kashmir in April.

The White Wagtails, in which group Whistler retains Hodgson's Wagtail, are widespread in Europe, North-West Africa, Asia, and extreme North-West America. They are consequently divisible into a considerable number of sub-species. Hodgson's Pied Wagtail breeds in the Himalayas from Kashmir and Ladakh southwards

PLATE 47

European Bee-Eater outside its tunnel

Kashmir Roller with frog for young

PLATE 48

Indian Pied Kingfisher. Male on right

to South-West Tibet. In the winter it drops to the foothills and straggles into the plains as far east as Assam. It is also recorded from North-East Burma, where it may possibly breed, and according to the *Birds of Burma* it is a migrant to Tenasserim.

HABITS AND NESTING. The Pied Wagtail is a familiar bird in Srinagar and indeed throughout the waterways and lakes of Kashmir. Like all wagtails it feeds on the ground, being very quick in its movements, running this way and that in its efforts to seize some winged insect which it has disturbed, occasionally fluttering into the air, a confused conspicuous mass of black and white. It possesses cheerful call-notes and a pleasant song which it will often proclaim from the roof of a building or the ridge of a houseboat, for it by no means spends all of its time on the ground or running about the water-weeds.

Any type of water is to this bird's liking, be it the muddy patches in and around the villages and out in the rice-fields, the swiftly running water-channels, or the sluggish currents of the main river where it is so frequently to be seen perched on garbage and flotsam floating downstream. Away from the Vale it is just as happy by the roaring torrents and tumbling mountain streams. It is particularly addicted to the extensive patches of stones along the lower reaches of the larger side rivers, where old logs have become stranded, twigs and flotsam lie about in heaps, and the larger stones have been piled up into ridges and banks with untidy bushes and prickly stalks growing feebly amongst them. Here, in cavities under the stones, amongst the driftwood, or in the bushes' roots, most nests are to be discovered, but they are by no means confined to such localities. Osmaston records one under the eaves of the church in Gulmarg; they are quite often to be seen on the roofs of houseboats under the flat upper decks; and we once had a pair building —both sexes take part—inside a rolled-up ' chick ' on our houseboat, although we were almost continuously on the move for two of the days during which the birds were employed on the task.

The nest, though often untidy on the outside, is well built of bents, grass, roots, and other odds and ends with a deep neat cup lined with hair. The eggs are 4 or 5, the latter being a very common number. Six are said to be laid occasionally, but we have never seen such a clutch. They are fairly broad eggs as a rule, only slightly glossed, and predominantly grey in tone as the ground colour is a faint greyish-white. The markings, small streaks and specks of brown and brownish-grey, are dense, particularly so towards the larger end. They average 21·9 by 15·6 mm. They breed chiefly in May and June, but also in July.

WAGTAILS

EASTERN GREY WAGTAIL

Motacilla cinerea melanope Pallas
(See Plate 41 facing p. 164.)

KASHMIRI NAME. *Khak Dobbai.*
BOOK REFERENCES. *F.B.I.*, III, 265; *Nidification*, III, 132; *Handbook*, 246.

FIELD IDENTIFICATION. (2) A slender blue-grey bird with pale yellow underparts and a rather long narrow tail. This it continually waves up and down through a considerable arc as it runs about the stones in search of insect food. As it zigzags about the water's edge or betakes itself rising and falling in graceful undulations to the further bank, it utters sharp little notes, *piti, pi-iti, pititi*, or variations thereof. The male has a black patch on the throat and foreneck and his colours are brighter than in the female.

DISTRIBUTION. Immediately on quitting the main Vale, this bird will be met with along all the rivers and more open mountain streams up to considerable altitudes. It breeds up to at least 13,000 ft. In the Kishenganga and Wardwan Valleys it is equally common; in fact it is a widely distributed bird. It occurs rather sparingly in the Vale as it is not as a rule seen by sluggish and stagnant waters. The Grey Wagtail is a migrant to the whole of India and Ceylon, those birds which have bred early at the lower levels of their range commencing the downward journey as soon as the monsoon has well established itself. By the end of August, therefore, Grey Wagtails are to be seen in Northern India. They spread rapidly and soon reach Madras. They return to Kashmir a little earlier than the last species, many of those nesting in the lower reaches of the side valleys having newly-hatched young by the end of May.

The Grey Wagtails—the species has a number of races—breed freely from Europe through Central and Northern Asia to the Pacific. They migrate in winter south to Africa, India and Burma, and Southern Asia generally. The Eastern form nests from the Yenisei eastwards, while southwards it is to be found breeding in Indian limits in North Baluchistan, the Frontier hills and the Himalayas.

HABITS AND NESTING. For breeding purposes the Grey Wagtail favours patches of stones and small boulders near running water. It is therefore found in ground which the next bird generally eschews, so that there is little likelihood of the two Kashmir wagtails which contain yellow in the plumage getting mixed up—not that they are particularly alike in any case, but the Yellow-headed species has a somewhat confusing habit of breeding in different stages of immature plumage. The Grey Wagtail is specially fond of those rivers which have stony banks and scrub-covered islets where logs lie thrown up on the shingly beaches with driftwood and rubbish caught up against them. They will also be found along the banks of those running through forest. The situations of the nests are many and varied, but they are always well concealed. Sometimes a nest will be tucked away in a cavity amongst the stones, in a hole under a small boulder or in an earthy bank, but most often it will be found well hidden amongst the roots of some small bush on an island or near the water's edge on the main bank. At Astanmarg, however, at 11,500 ft, we found one in a tuft of *trollius* leaves in the flat low-lying wet area where the stream runs in many shallow beds and where in a previous year we had seen Yellow-headed Wagtails.

The nest is a well-built cosy cup of bents, weeds and dry grass lined thickly with hair and often wool in addition. In the Kishenganga Valley we did not notice birds carrying building material before 17 May, but breeding must often commence earlier than that as between 5,000 and 6,000 ft in the Bandipur Nullah we have found nest after nest at the end of May either with incubated eggs or young ones. One nest found on 20 May near the mouth of the Sind River contained five fresh eggs. Most eggs are laid before the end of June, even at the higher altitudes, but the Astanmarg nest referred to above contained only two fresh eggs on the last day of that month, while on 5 July we found a nest on an island opposite Inshan in the Wardwan Valley with five fresh eggs in it. Osmaston records seeing a pair with fully fledged young on 10 August at 13,000 ft near Tosha Maidan. Five eggs are frequently laid ; we have seen numbers of 4's, and clutches of 6 are occasionally recorded. They appear an almost uniform brownish-grey or yellowish-grey in colour as the markings of these tints are minute and so profuse as often to hide the whitish ground colour. Eggs are often seen with one or two short and thin wavy black streaks on them. They average 18·2 by 14·1 mm.

WAGTAILS

HODGSON'S YELLOW-HEADED WAGTAIL

Motacilla citreola calcarata Hodgson
(See Coloured Plate II facing p. 40, Plate 39 facing p. 156 and Plate 42 facing p. 165.)

KASHMIRI NAME. *Ledor Dobbai.*
BOOK REFERENCES. *F.B.I.*, III, 274; *Nidification*, III, 134; *Handbook*, 249.
FIELD IDENTIFICATION. (2+) A bright yellow and black slender bird with the characteristic habit of nearly all wagtails of rapidly raising and lowering the rather long tail as it runs about the wet grass or on the weeds of the marshes. In the male the head and underparts are bright lemon yellow, contrasting strongly with the deep black of the nape and back. In the females the whole underside is only tinged with yellow and they breed in at least two colour phases, first-year birds being almost white below while those of two years or over have the underparts much yellower but still very pale. The males too quite often breed in immature dress. They are highly gregarious, many pairs nesting around the same jheels and packing into flocks as soon as breeding is finished. Isolated pairs may, however, be found, particularly at the higher elevations.

DISTRIBUTION. Very common on the marshes of the main Vale. They may also be seen in lesser numbers around the more open lakes such as the Dal lakes and round the eastern and southern shores of the Wular. They are not confined to the Vale. Osmaston, and Magrath, found them breeding at about 11,500 ft near the Gangabal Lake below the Haramukh glaciers, and Wilson took a nest at Sonamarg. We have seen them in small numbers during the breeding season between Sonamarg and Baltal, in the marshy amphitheatre at 13,000 ft below the south side of the Yamhar Pass, near Kolahoi, at Astanmarg (11,500 ft), and in the Sain Nullah which lies between the heads of the East Lidar and Wardwan Valleys. In the winter they migrate to the foothills and the plains of the northern half of India and also to North-East Burma. They return early, Meinertzhagen reporting that they were already abundant in the Vale by 12 March and perhaps earlier.

In India the full breeding range includes North Baluchistan (where it breeds sparingly), the upper Kurram Valley, the Khagan Valley and the north-west Himalayas to Kashmir, Ladakh, and Rupshu. Outside Indian limits it nests in North Persia, Afghanistan, Tibet, and parts of Central Asia. The typical race, with a wider and more northerly breeding range in East Russia, Siberia and Mongolia, also visits India and Burma in winter.

HABITS AND NESTING. Generally speaking, to find this, the most beautiful of the three wagtails which breed commonly in Kashmir, one should repair to the marshes. Whereas the Grey Wagtail is found chiefly along the banks of the turbulent rivers and the open mountain streams, and the Pied Wagtail almost anywhere, this bird is a lover of wet low-lying grassland and weedy marsh. In the Vale it is to be found in great numbers concentrated in such areas as Hokra Jheel, Shalabug, and the extensive marshes extending thence down both sides of the Jhelum to the Wular Lake. Up the side valleys too, it is no good expecting to see them in the drier portions where the ground slopes up immediately from the banks of the rivers, but as one gains altitude and comes up to scored-out vales where retreating glaciers have left broadened flat areas with the streams running through marshy patches, and where tarns with wet grassy margins have formed below the glaciers, then these birds once more make their appearance.

According to the *Nidification* their ' nests and nesting sites are very much like those of the Grey Wagtails '. We do not agree with that statement in so far as it refers to nesting sites in Kashmir, but remain faithful to what we wrote in the *Journal* (**XXX**, 606) in 1925 in describing a visit to Hokra Jheel. We wrote then : ' . . . and it is surrounded by rice-fields and nearer at hand by a considerable strip of low-lying more or less wet spongy ground, covered with soft grass of varying length interspersed with a few scraggy little bushes, mere prickly stalks a few inches to a couple of feet in height. It is on this wet ground that . . . Hodgson's Yellow-headed Wagtails may be found breeding in their hundreds. The nest is rarely, if ever, on really dry ground and its base in fact is often wet. It is generally a well built cup of grass placed in the centre of a tuft or at the foot of a bush, or even in soft grass of a length barely sufficient to hide the nest.' They are not averse to extending their activities to the rice-fields near at hand, where nests may commonly be found in tussocks of coarse grass on the narrow wet bunds separating the fields. Osmaston also describes his nests in the Vale as placed on the ground in tufts of grass a few inches above swamp-level and well concealed, while his Haramukh nests were at about 11,500 ft in an extensive patch of swampy ground near the Gangabal Lake. We also found them breeding in cut-down reed-beds from which the water had receded. Here the nests were amongst a mass of muddy stubble so that their bases were well soaked.

They are bulky nests with a lining of hair or wool, but they would be hard to find were it not for the fact that the birds sit very closely, often fluttering up from under one's feet. The eggs are from 3 to 5, the latter number being by no means uncommon. In colour they vary considerably, but all are of a decidedly greenish hue, mottled or finely streaked with greyish or greyish-brown, but so closely as to appear almost unicoloured. Many of them rather resemble those of the Grey Wagtail, but they are larger and tend to be true ovals. They measure 20·1 by 14·5 mm. In the Vale the birds start laying in May, from the second or third week, and it is difficult to find fresh eggs in the second half of June. Their higher breeding grounds, however, are still snow-bound in May, so June and July are probably their breeding months beyond the main Vale. Both birds build the nest and both are said to assist in incubating the eggs. One of our shikaris found a young Asiatic Cuckoo in a nest of this species.

WITHERBY'S TREE-PIPIT

Anthus trivialis haringtoni Witherby

BOOK REFERENCES. *F.B.I.*, III, 280; *Nidification*, III, 138; *Handbook*, 251.

FIELD IDENTIFICATION. (2) This pipit, though heavily streaked with blackish, is not quite so dark as the next bird, Hodgson's Pipit. It can, however, be more easily separated from it by the black streaks which are distributed over the whole breast and not merely bordering an unmarked area in its centre. Although found in the same areas as Hodgson's Pipit, it is not nearly so common, but it may also be found in open blanks near the upper forest limits. When disturbed in the vicinity of trees, it has a habit of flying up to the lower and middle branches, perching with the tail waving slowly up and down rather after the manner of a wagtail.

DISTRIBUTION. This is a high-level pipit to be found breeding in rather scanty numbers on the margs and slopes above tree-level, being well dispersed in our area over the main Himalayan range up to some 12,000 ft. We have no information about the Pir Panjal, but would expect to find it nesting on those mountains, though perhaps in lesser numbers. Its presence may easily be overlooked owing to the overwhelming numbers of Hodgson's Pipits, but it occurs also at somewhat lower levels, entering the openings in the upper fringes of the forests. In winter it descends to the outer foothills, but apparently seldom to the Plains, which it leaves to the typical race, a more lightly streaked and highly migratory form which goes farther afield to breed. It commences the downward move in September, for Meinertzhagen reports that on the Gilgit road from Burzil Chauki to Tragbal this bird was abundant and in large flocks during the first week of September between 9,000 and 11,400 ft, and was doubtless preparing to migrate as all had just completed their moult. A few were around Gulmarg (about 9,500 ft) in the middle of September.

Witherby's Pipit is a race of the Tree-Pipit of Europe and Central Asia, being the easterly form which breeds from the North-West Frontier (Sufed Koh) to Gilgit, the Khagan Valley of Hazara, and high up in the Western Himalayas to Garhwal. The *Nidification* also informs us that Ludlow got nests of this race in the Tien Shan.

HABITS AND NESTING. This pipit is no doubt often overlooked in Kashmir since most are to be found above the tree line on the upper pastures and grassy slopes which are frequented by such large numbers of Hodgson's Pipits. It has many of the same habits, though it is not so partial to marshy patches of ground, but like that bird it possesses a sweet song which it pours forth in much the same way on the wing and when fluttering to earth. It does, however, enter the forests' limits where blanks and re-entrants occur between the fast dwindling trees. Here when disturbed it retains that characteristic habit of flying into the lower branches of an adjacent tree, where it moves about in a curiously wagtail-like manner, waving the tail up and down. Its nesting season seems to extend throughout June and July. On the Bungus Marg at about 10,000 ft in the Kazinag we took on 12 June a nest containing three well-incubated eggs, but at Astanmarg in early July a nest was found containing well-developed young while at the same time a pair still building were collecting pieces of dry grass from the sides of the sheltered depression in which we had pitched camp.

The nest is a neat cup of grass lined with finer material of the same type, though it is said at times to contain a little moss. The Bungus Marg nest, however, was made almost entirely of pine-needles and was lined with short lengths of the same, a little fine dry grass, and a lone horse-hair. The egg cavity was very deep. Nests are placed in much the same positions as those of Hodgson's Pipits, in a tuft of coarse grass, under a sheltering stone on a hillside, or protected by a straggling bush or weed, but they are often not far from the trees flanking the pastures or creeping up the hillsides. The Astanmarg nest containing young was tucked under the end of a large fallen log. The bird sits very closely, though otherwise it is rather shy. The eggs, 3 or 4 in number, are dark, almost as dark as those of Hodgson's Pipit, and somewhat like those of the skylark, being heavily streaked and mottled with dark chocolate-brown so as almost to obliterate the ground colour. They average 19·3 by 15·2 mm.

BROWN ROCK-PIPIT

Anthus similis jerdoni Finsch

(See Plate 43 facing p. 172.)

BOOK REFERENCES. *F.B.I.*, III, 286; *Nidification*, III, 143; *Handbook*, 252.

FIELD IDENTIFICATION. (2+) A large pipit with an almost uniform sandy-brown plumage found in dry broken ground at low and medium elevations.

DISTRIBUTION. Common, though never much in evidence, on the barer lower slopes around the Vale. We have also seen it along the Jhelum Valley road from Domel upwards and in the lower parts of the Kishenganga Valley. The Brown Rock-Pipit leaves Kashmir in the autumn for the northern and central plains of India.

This Himalayan pipit is a race of the Rock-Pipits found breeding in Persia, Afghanistan, and India. *Jerdoni* nests throughout the Himalayas from Gilgit to Sikkim and South-West Tibet. Meinertzhagen obtained a single specimen of the Brown Rock-Pipit in Afghanistan on 31 May near Jalalabad which he identified clearly as *jerdoni*. In winter, besides northern and central India, this pipit reaches central Burma and the Northern and Southern Shan States. According to the *Birds of Burma*, it also probably breeds in the dry zone of Burma.

HABITS AND NESTING. The Brown Rock-Pipit is the pipit of lower levels which is found commonly on the rough stony hillsides and broken ground at their feet immediately bordering the main Vale, along the Jhelum Valley road, and sparingly in the lower reaches of the Kishenganga. Lack of bushes to perch upon does not trouble it so long as there are coarse grassy tufts, stones and rocks to break up the evenness of the ground ; cultivation and plain grassy slopes are not to its liking. It is larger than the other two pipits of Kashmir, Witherby's and Hodgson's Pipits, both of which

PLATE 49

Indian Pied Kingfisher

Central Asian Kingfisher

PLATE 50

European Hoopoe

White-breasted Kingfisher with insect food for young

breed only at high elevations. The Brown Rock-Pipit does not penetrate above 6,000 ft.

Its sandy nondescript plumage and quiet habits render it an uninteresting bird as well as one whose presence may be overlooked, since the comparatively rough terrain in which it is generally found presents it with plenty of ground-cover to sneak about in. It possesses no song to attract attention, but only a single *cheep* sometimes higher, sometimes lower in pitch. This it utters from a stone or bush-top or as it gets up lazily at one's feet only to land again a short distance farther on, flitting rather aimlessly here and there as if encountering a strong headwind. Magrath talks of a pair hovering ' Kestrel-like ' over the nest before dropping to the ground.

The nest is a collection of coarse grass, bents and grass-roots lined with finer grass. The illustration depicts a typical site. Sometimes it is let into the stones, sometimes into a tuft of grass, but always there is something to protect it, a prickly bush, some overhanging coarse grass, or a projecting rock. The bird begins to lay early, eggs having been taken towards the end of April, but May is the peak period, while numbers of eggs must be laid in June for birds are not infrequently seen in July feeding young still in the nest.

The eggs are typical of the pipits, but the markings are a warm brown in hue as against the dark purplish-brown of Hodgson's Pipit's eggs. The underlying ground colour is white. They average 23·0 by 16·6 mm. Three is said to be the normal clutch with an occasional 4, though at times only 2 are incubated. On 19 May, however, we found a nest containing 5.

HODGSON'S PIPIT

Anthus roseatus Blyth

(See Plate 43 facing p. 172.)

BOOK REFERENCES. *F.B.I.*, III, 295 ; *Nidification*, III, 150 ; *Handbook*, 251.

FIELD IDENTIFICATION. (2) This small dark pipit is only found in the breeding season at high elevations above the tree-level where it is numerous, especially on the wetter ground and around the patches of melting snow. In general the plumage is darker and more heavily streaked than that of the other pipits, but it

PIPITS

may at once be separated from the last bird by the breast which is streaked only at the sides, leaving an unmarked vinaceous area in the middle. It has also some yellow on the underwing, but this is hardly noticeable in the field. It sings on the wing, fluttering lightly to earth like a wind-blown leaf.

DISTRIBUTION. The common pipit at all elevations above the tree line. It haunts the margs and open hillsides on both the Pir Panjal mountains and the ranges of the main Himalayas, being more numerous on the latter. In the autumn it moves down to much lower levels, wintering in the plains of Northern India, though a good many go no farther than the Vale of Kashmir, taking up their quarters around the marshes. It commences its return to its breeding areas in March, but moves up gradually, often in considerable bodies, as the snow recedes and the weather improves. Meinertzhagen mentions that birds in full moult into their breeding plumage were passing through the Vale from 23 March to 9 April. Large flocks may be seen in the beginning of May travelling up the Sind and Lidar Valleys. In May we also found it at Khel in the Kishenganga Valley, and later, at the end of the month, there were still many about the margs and fields of upper and lower Gurais.

In addition to our area Hodgson's Pipit is said in the *Nidification* to breed in Turkestan and Afghanistan as well as in the Sufed Koh and thence to and along the Himalayas to Northern Burma, Kansu, and Yunnan. As regards Afghanistan, there is but a single record of a specimen collected on 10 March. It does, however, nest freely up to 13,000 ft in the Sufed Koh. In winter, as already stated, it moves southwards into Northern India, North-East and Central Burma, and the Northern Shan States.

HABITS AND NESTING. It is impossible to traverse the open valleys above the trees without coming upon this widespread pipit. For widespread it is, especially in the uplands of the main Himalayan range. On the Pir Panjal it is far less numerous, but whether scarce or common it soon brings itself to notice as it is a lively, energetic little bird feeding on the ground on the wet stones around the tarns, or running about like a wagtail over the moist patches and the waterlogged edges by the melting snow-beds. This preference for damp ground is most marked, with the result that at times quite large numbers will be noticed feeding together around the same moist area. The nests, however, are usually found on the slopes, where the crisscross sheep-tracks and jutting-out stones provide little declivities and hollows in which the nests may lie concealed.

The nest is a neat deep cup of dry grass lined with finer grass and sometimes a few horse-hairs, and is always well hidden under a tuft of grass or a straggling plant, in a hole in the hillside, or under the shelter of a protecting stone; so well hidden, in fact, that it is often disclosed only when the close-sitting bird rolls out of it at one's feet. For this bird is very prone to displays of feigned injury, tumbling about and fluttering downhill to draw one away from its treasures. Like the bird, the eggs too are particularly dark, having the ground covered almost completely with deep chocolate and purplish blotches and streaks. Three or 4, often the former, are laid. They are rather broad little eggs with a fair gloss and average 21·8 by 16·0 mm.

We have found the greater number of nests between 11,000 and 12,000 ft, but some are to be found both higher and lower than this. They commence nesting in June and eggs may also be taken in July. The song of this pipit is quite pleasing, somewhat larklike in character, and uttered mainly in the air. It is not, of course, such a sustained song and usually peters out in long-drawn-out *sweet-sweets* as the bird, with rather stiffly upbent wings and cocked tail, sinks unsteadily to the ground like a falling leaf wafted by the breeze.

LONG-BILLED HORNED LARK

Eremophila alpestris longirostris (Moore)
(See Coloured Plate II facing p. 40.)

BOOK REFERENCES. *F.B.I.*, III, 309; *Nidification*, III, 154.

FIELD IDENTIFICATION. (2+) Found only at high elevations running about the barer, more sandy hillsides. The general tone of the plumage is pale, in keeping with the habitat, while the black ' horns ' are quite sufficient to distinguish it from any other bird in such areas. This lark never soars but sings its short and pleasant song from the top of a stone or boulder.

DISTRIBUTION. This in reality is one of the desert larks whose main stronghold is amongst the upland plateaux and bare hillsides across the main Himalayan divide. A certain number, however, are to be seen on suitable ground on the outer side of the Himalayan range and a few even on the Pir Panjal mountains. In summer they are found well above the tree line, seldom below 11,500 ft and more generally a couple of thousand feet higher. A resident species dropping but little even in severe winters. Ludlow, however, saw one in winter in the Vale.

Outside our area it is found from the Sufed Koh to Ladakh and at least as far east as Kulu, Whistler assuming that it breeds along the whole of the Rotang chain. The range of this form as given in the *Fauna* appears to be too wide, Whistler stating that he can trace no authority for the inclusion of Afghanistan while Ticehurst does not mention it in his *Birds of British Baluchistan*.

HABITS AND NESTING. Osmaston remarks that these handsome larks occur chiefly between 11,500 and 13,500 ft on open dry sandy hillsides with scattered tufts of grass and herbaceous plants, especially where there is much artemisia. We have evidence of them around Mount Haramukh ; Loke reports them as comparatively numerous on the open slopes near the Gad Sar and Vishan Sar lakes ; while Osmaston says they are less common on the Pir Panjal range. Others who record them from Kashmir have failed to define the actual localities except for Colonel Phillips, who writes of finding one building on Aphawat in early July, but since this genus is a lover of desert and semi-desert conditions, two species at least being common in Ladakh, it can be taken as certain that nothing is to be gained by looking for them on the green grass and bush-dotted hillsides and margs, but only on the bare screes and drier slopes well above the tree line.

These larks are not infrequently seen as cage-birds in the Srinagar bazaars, but such examples have almost certainly been brought in from Ladakh where they are common birds. The song is not of much account in so far as larks' songs go, but, though short and consisting of few notes, it is loud and pleasing. It is never uttered on the wing, but always from a fixed perch which in the ground it frequents means from a stance on a boulder or even a large stone.

The nest consists of little more than a shell of grass in a small scrape hollowed by the bird in the bare ground. A thin lining of seed-down, other vegetable matter, or hair, is usually added. This is sheltered by a stone or small plant, the whole thing being reminiscent of the nests of the Ashy-crowned Finch-Larks which one finds in the Plains. Osmaston records a nest in Ladakh which the bird had surrounded with a rampart of stones, some of which weighed from a half to one ounce. This we may add is a habit typical of the desert larks. The eggs have a pale stone or yellowish-stone ground profusely blotched and streaked with chocolate-brown, rather lighter in tone than skylarks' eggs, than which they are considerably larger. Many are, however, so closely stippled with light reddish-brown as to be almost unicoloured. Two or 3, often the former number, are laid, which measure 24·5 by 17·4 mm. June and early July are the breeding months.

KASHMIR SKYLARK

Alauda gulgula lhamarum Meinertzhagen

(See Plate 44 facing p. 173.)

KASHMIRI NAMES. *Dider, Didru.*
BOOK REFERENCES. *F.B.I.*, III, 318; *Nidification*, III, 161; *Handbook*, 254.
FIELD IDENTIFICATION. (2+) A streaked brown bird of the open grasslands with a fairly prominent crest. In build it is stockier than the pipits and squats rather than runs away before taking to flight. The sweet pulsating song uttered as it rises straight into the sky on vibrating wings renders it quite unmistakable.
DISTRIBUTION. Common in the meadows and on the karewas of the main Vale, particularly in the waste lands dotted with beds of irises which border the river. It occurs in all the side valleys up to some 10,000 ft and even higher, and was found to be very common in the rough meadows around Sonamarg. It is plentiful enough in Gurais and we took its nest both at Inshan and up at Suknes in the Wardwan Valley. In June we were particularly struck by the numbers heard singing over the blue-hazed linseed fields beneath the foothills of the Kazinag mountains. It is, in fact, a widespread bird extending from the lowest levels in the Jhelum up to at least 10,000 ft in the Wardwan and other large valleys.

This is one of the many races of the Little Skylark, which, besides covering the whole of India, holds sway from Turkestan to Cochin-China and southwards to Ceylon and Tenasserim. The Kashmir race breeds in the Himalayas from Kashmir and Ladakh to Sikkim at elevations from 5,000 to 14,000 ft. In winter it moves down to the foothills where it may be seen in flocks. In the Vale of Kashmir a few birds remain throughout the year.
HABITS AND NESTING. The Skylark's song has drawn forth paeans of praise since the dawn of history, endearing it to all and sundry and rendering it the most publicized bird in the world. Many will tell us that the skylarks of India are but enfeebled

shadows of their counterparts in Britain, whose sustained and powerful song, pouring forth from the heavens, is rightly the envy of all. Nevertheless, our thoughts go back to some pleasant hours lazed away on the roof of a houseboat in listening to the skylarks around Sumbal. For the iris beds, everywhere dotting the waste lands along the river's banks and gay with sheets of little blue irises amongst the tufts of narrow leaves, are most beloved of the Kashmir Skylark, and it is here that in early June a careful search will disclose many nests. It is perhaps true that nowhere throughout the Indian region will the skylarks be heard singing so vociferously, sweetly and for such sustained periods as do our British larks, but in Kashmir many do their very best to emulate them. Not only do they sing while mounting steadily skywards and more intermittently as they drop in successive descents, ceasing only with the final plunge to earth, but they quite often sing perched on a stone or isolated post. Magrath in a note in the *Journal* (XXVIII, 278) wrote as follows : 'The Common Lark is somewhat crepuscular in habits in Kashmir. One delightful bird perched, with its mate, on a stone about 30 yards from my dinner table one evening, and serenaded me till it was almost quite dark.' They never enter the forests, nor do the rice-fields attract ; it is the waste grasslands with tufts of coarse grass, irises or *trollius* plants which they love, or, in such valleys as the Wardwan, the fallow fields and rough ploughed land with the grassy patches and mounds of stones between them.

The nests as a rule are very well hidden within the heart of a grass tuft or in other coarse herbage, at times tucked under a clod of earth in a ploughed field. They are not easy nests to find and considerable parting of the enclosing grass is often required before the contents are laid open to view. The nest is a compact cup of grass lined only with finer grass, and in it from 3 to 5 dark-coloured eggs are laid which match their background to a marked extent. They are white, greyish-white or yellowish-white in ground colour, but this is almost covered up, entirely so at the larger end, with numerous spots, streaks and blotches of varying shades of dark brown. Some clutches have a sepia tinge, others a greyish or purplish hue. They average 21·5 by 16·5 mm., these being the measurements of Osmaston's eggs from the Vale which he points out average smaller than those of the Ladakh birds. They are laid mostly in June, but numbers may also be taken in the latter half of May. In July we have only been able to find nests with young birds in them, even at the higher altitudes.

When endeavouring to photograph the lark, we have always found it very quick off the mark, the wings opening with speed and in such immediate response to the bird's senses that even with comparatively fast exposures blurred wings have often resulted. The bird does not normally alight at, or leave from, the immediate vicinity of the nest, but exercises considerable caution, both coming and going, running for some yards through the surrounding herbage.

WESTERN WHITE-EYE

Zosterops palpebrosa occidentis Ticehurst

BOOK REFERENCES. *F.B.I.*, III, 360; *Nidification*, III, 190; *Handbook*, 264.

FIELD IDENTIFICATION. (1) A bright leaf-green and yellow bird rather more heavily built than a willow-warbler. The short bill is black, slightly curved and sharp, but the most tell-tale character is the conspicuous ring of white feathers around the eye. Usually noted in small parties feeding through leafy trees, the birds calling incessantly to one another.

DISTRIBUTION. This bird is always in evidence in the Jhelum Valley, commonly up to Garhi and probably much higher. In the Vale of Kashmir it may be heard in most years in the groves around Srinagar, but it is by no means common. It has been recorded as breeding on the lower slopes up to about 6,000 ft, though we can trace no specific mention of nests. Osmaston, in a letter, writes that he observed it in and around the Lolab in April. When we moved up the Kishenganga Valley in April and May, it was met with continually from Domel up to the Nilam plateau (5,400 ft) above Keran, where parties were feeding busily in the viburnum and taller scrub. We have no information about its winter movements.

The Western White-Eye is in the main a bird of the Plains, though it is found along the Himalayas to Nepal where it ranges up to 8,000 ft. To the west it is found around Kohat and thence through the plains and continental India to Mysore. It is absent from Sind except for those in the mangrove swamps of Karachi. Other races cover the rest of India. The white-eyes as a family are well spread over Africa, Southern Asia and Australia.

HABITS AND NESTING. The merest glimpse is sufficient to

WHITE-EYE

settle the identity of this bespectacled mite, for the white-rimmed eyes are quite unique, and are responsible for its well-known nickname, in certain quarters, of 'Spectacle Bird'. If any further aid were required, one has only to hear a party making its way through the tree-tops, since white-eyes feed in company in the middle and upper foliage in well-wooded areas. Each bird keeps up a plaintive yet fascinating monotonous whistle, which, in the sum total of many birds continually answering each other in tones of slightly differing pitch, brings with it a welcome feeling of life and bustle. Unfortunately we find it impossible to put white-eyes' notes on paper—perhaps we might say they have a *tswizzing* note—but we know of no other birds in Kashmir which make the welkin ring in such a pleasing manner : most itinerant bands of small birds are composed of different species of tits, warblers, tree-creepers and others, with consequently a multiplicity of widely differing voices. When feeding on the buds and tiny insects they find amongst the leaves, the parties seem to favour the middle and higher foliage, going right to the tops of the tallest trees where they carry on their lively conversations.

The nest on the other hand is often low down—we have seen it within two feet of the ground in a small bush, in potato creeper at shoulder height on a veranda, and in small trees at 12 to 15 ft up, but it is also known to nest at the summits of the tallest trees. White-Eyes, however, are not common in the Vale of Kashmir and we must confess never to having found a nest there, though there is not the slightest doubt that it must breed in some numbers along the Jhelum Valley road and in the lower Kishenganga. Nevertheless, although their appearance in the Vale is intermittent, Osmaston noted it in the Sind Valley up to some 6,000 ft, and we saw a party of four in August at the mouth of the Wangat Nullah, and Ward, in describing it as ' numerous in Poonch and Jammu, less so in Kashmir proper ', adds that it breeds in Kashmir in April.

The nest varies considerably in the materials from which it is built, but in shape it is almost always a dainty thin-walled deep hammock slung by its rim from a slender fork like a miniature oriole's nest. In size it is just about large enough to enclose a golf ball. We have seen a nest composed entirely of thin strands of moss more or less knitted together, but they are commonly made of fine grass and weed-stems, tendrils or fibres, cemented together on the outside with cocoons and spiders' webbing whereby they resemble whitish purses. The eggs, often only 2 in number, but not infrequently 3 and we have twice noted 4, are in colour a pale

PLATE 51

Turkestan Great Horned Owl on nest

Scully's Wood Owl

PLATE 52

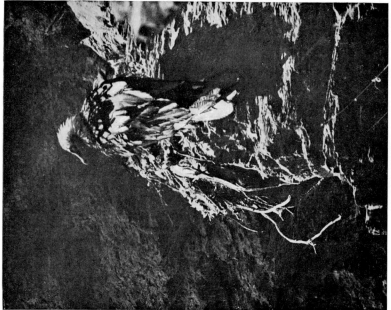

Himalayan Bearded Vulture at nest

Himalayan Griffon Vulture

skimmed-milk blue with little gloss. They are inclined to be oval in shape and average 15·2 by 11·5 mm. Although Ward says these birds nest in April, we feel sure that in Kashmir they will be found nesting in May as well, if not later still.

SCALY-BELLIED GREEN WOODPECKER

Picus squamatus squamatus Gould
(See Plate 45 facing p. 180.)

KASHMIRI NAME. *Koel Makōts.*
BOOK REFERENCES. *F.B.I.,* IV, 7; *Nidification,* III, 272; *Handbook,* 277.
FIELD IDENTIFICATION. (4−) A fairly large green bird with, in the male, the crest and top of the head bright crimson, the female having the crest black. The underparts, except for the grey throat and breast, are whitish, each feather edged with black, which gives the plumage its characteristic scaly appearance. The largest of the three woodpeckers found commonly in Kashmir.
DISTRIBUTION. This fine woodpecker has a wider altitudinal range than the next bird, although it is not nearly so common. It occurs from about 4,500 ft in the Kishenganga Valley, and probably in the Jhelum Valley too, up to at least 10,000 ft, at which altitude we have seen it near Tanin in the East Lidar Valley. We have seen it near Tangmarg and in the winter it is occasionally to be seen and heard near Srinagar. It is a resident bird, being forced out of its higher areas only by heavy snows. It is also resident on Sakesar in the Punjab Salt range.

The two races of this species extend through Transcaspia, North Baluchistan and Afghanistan, and the Western Himalayas to Nepal, our typical form being found eastwards from Kohat, where Whitehead recorded it in winter.
HABITS AND NESTING. The habits of this bird are little different from those of the Pied Woodpecker, which is fully dealt with in the following pages. It prefers mixed forest with clearings, but is found in all types and is also prone to come out into fairly open country well dotted with large trees. It has the usual habit of *chucking* away to itself as it jerks its way up and down the larger branches and trunks of the trees, and of uttering a volley of alarm-notes as it flies off strongly with the usual undulating flight of the family. It also possesses a loud ringing call which carries over a

WOODPECKERS

great distance. The beating wings make considerable noise, thanks to the stiffness of the feathers, but in spite of its comparatively large bulk it is surprising how softly the bird is able to land against the trunk of a tree, often almost noiselessly.

The circular nest-hole is larger than that of the Pied Woodpecker, about $2\frac{1}{2}$ inches across, although it is smaller than one would expect from the apparent size of the bird. The nest may be at any height from the ground, even up to 60 ft or more, but the majority will be found at heights below 30 ft. While Rattray took one at Murree as low as 4 ft, we also found one above a forest path in a rotting stump, the hole being hardly 3 ft above the heads of passers-by. It is often bored into rotten wood so that it opens out into a natural hollow which may be of considerable size and depth.

The eggs are laid on the decaying fragments at the bottom of the cavity and no attempt at a nest is made. They are rather elongated ovals of china white with a high gloss, and up to 6 are laid. They average $31 \cdot 0$ by $22 \cdot 6$ mm. The chief breeding months are April and May, and few if any nests will be found in June which do not contain young.

Magrath in his *Notes on the Birds of Thandiani* made an interesting and amusing note on the habits of this bird. ' When the young are hatched ', he wrote, ' the parents are often to be seen perched on a bare bough keeping up an incessant squarking chuckle, the meaning of which it is not easy to discover. The clamour of the nestlings may be likened to the distant sound of a puffing engine. It seems absurd to see young birds, after they are fully fledged and quite as big as their parents, being fed.' This woodpecker is, in fact, a talkative bird with a considerable repertoire and we have also noticed that when one bird visits the other at the nest, they converse together most freely in subdued soft squeaks. Although the parents are otherwise such vociferous birds, if one taps the tree, the sitting bird does not give away the fact either by sound or flight that the hole is tenanted. On the other hand the young react immediately. In fact they call when being fed and complain bitterly and continuously when hungry. Although many woodpeckers feed their young at least in part by regurgitation, this method alone appears to be resorted to by the Scaly-bellied Green Woodpecker.

KASHMIR PIED WOODPECKER

Dryobates himalayensis albescens Stuart Baker

(See Plate 45 facing p. 180.)

KASHMIRI NAMES. *Hōr Koel, Makōts, Koel Ku-kīr.*
BOOK REFERENCES. *F.B.I.*, IV, 34; *Nidification*, III, 285; *Handbook*, 279.
FIELD IDENTIFICATION. (3) A medium-sized black and white forest bird with an undulating flight and loud call-notes which it often utters on the wing. The black upper plumage with bold white patch on each shoulder and the crimson crest are diagnostic. Clambers up and down the tree-trunks, always keeping the body upright by using the stiff tail-feathers as a support, and raps the bark sharply with its bill in its search for grubs. Nests in holes chipped out by its own efforts with its large chisel-shaped bill.
DISTRIBUTION. Widely distributed throughout the thicker fir and mixed forests normally from about 6,500 ft to 9,000 or even 10,000 ft. To a lesser extent it will be found in more open well-wooded country. In winter it comes down lower and may be seen around Srinagar in some numbers, but on the whole it is a resident species. It is particularly partial to the mixed forests—those containing chestnut, maple and other deciduous trees as well as the pines—which lie between 7,000 and 8,000 ft. In the Kishenganga Valley we met with it at lower altitudes, that is from Keran at about 5,000 ft.

This is a purely Himalayan species with two races extending from Eastern Afghanistan and the Sufed Koh throughout the Himalayas to Kumaon, birds from the Simla States eastwards belonging to the typical form.

HABITS AND NESTING. The Pied Woodpecker (like practically all the woodpeckers) feeds largely on insects and larvæ obtained by chipping away the bark or rotten wood behind which they lie ensconsed. A. E. Osmaston says it also spends considerable time and energy in breaking open the cones of the chir pine to get at the seeds some months before they open naturally to let the seeds escape. The tappings of its bill will therefore probably be noted before the bird itself is actually seen. Every now and then it utters a peevish *chuck* as it moves jerkily up the tree-trunk or slithers clumsily downwards, always keeping the body upright no matter in what direction it may be moving. The stiff pointed tail-feathers are used as an additional support. The rapid drumming of the woodpecker on some hard piece of wood carries a great distance. The method of production and its use are still hotly contested, but we think there is little doubt that its intensity is due to the resonance of the hard wood produced by rapid blows of the slightly open beak. The bird flies strongly with considerable undulations as it more or less closes the wings after every few beats. It often gives forth a volley of ringing calls as it flies from one tree to another.

The nest-hole is circular and less than a couple of inches in diameter, with the chamber about 4 inches across. It is bored by the bird itself sometimes in dead, but more often in living wood, and the ease with which it is accomplished, even in hard wood, is a revelation. After a very few minutes of intense hammering within the half-completed chamber, the carpenter's head will appear at the hole with a beakful of chippings, splinters of wood of anything up to an inch or an inch and a half in length. With a rapid shake of the head the splinters are tossed in all directions. The bird is never content just to drop them sedately to the ground. This manner of disposal of the refuse certainly effects a certain dispersal of the fresh wood chippings, which, if they all fell in a conspicuous pile, might the more easily give away the fact that a nest cavity was in process of excavation. A glance at the ground below a hole, however, can never leave one in the slightest doubt as to whether it is a newly-bored one or not, unless of course it is very high indeed from the ground. The male appears to take a predominant share in the chiselling and works intensively for as much as half an hour at a time, three or four minutes of chiselling being followed by chip-disposal taking about half as long again.

The hole may be drilled at any height up to 40 ft or even more. The lowest nest we have found was not more than 5 ft from the ground. It is usually in the main trunk of the tree though at times the hole may be bored on the underside of a large sloping bough. Pines are by no means the only trees selected; in fact, we are convinced that the Pied Woodpecker prefers deciduous trees, particularly chestnuts where these occur.

They breed from mid-April to June. By early July we have only been able to find nests with young, and but few of these, three to be exact, all in the vicinity of Pahlgam. The number of eggs varies from 3 to 5, the latter number being very rare. They are china white ovals with a considerable gloss, and measure 26·5 by 19·1 mm. It is easy to trace a nest once the young are hatched, as they soon become noisy. The parents also become extremely fussy, *chucking* away angrily when one is in the vicinity, and meeting one often as much as a hundred yards from the nest.

BROWN-FRONTED PIED WOODPECKER

Dryobates brunifrons (Gould)

(See Coloured Plate IV facing p. 232.)

BOOK REFERENCES. *F.B.I.*, IV, 42; *Nidification*, III, 290; *Handbook*, 279.

FIELD IDENTIFICATION. (3—) Easily distinguished from the Kashmir Pied Woodpecker by the black-and-white barring on the back and wings and the whitish face with its brown forehead. The underparts are streaked with black. In the male the crest runs from golden-yellow in front to crimson behind; the female's crest is yellowish throughout. It is a bird of lower levels and more given to frequenting light mixed forest, fairly open country, and gardens.

DISTRIBUTION. Of Kashmir it is safe to say that this woodpecker represents the family only at the lower elevations. Colonel Phillips, *in epistola*, writes : ' A fair number is resident in the Vale ; one comes across the bird in Srinagar gardens throughout the winter. Their numbers visibly increase in March and April, and then all the gardens of any size have their pair. Many nests have I found in Srinagar itself ranging from 5 to 20 ft. Pampur, Telbal, Ganderbal and Shalabug have also produced nests, so one can say they are general throughout the Vale.' Although Osmaston says they occur uncommonly on the lower wooded slopes in and around the Vale, they do not appear to ascend as high as Gulmarg. Outside the main Valley we noted a single bird in June near Kangan, while in the Kishenganga Valley we found it as low as Pateka, elevation 2,670 ft, and again at Keran.

Outside Kashmir this woodpecker becomes a common representative of the family in all the hill-stations of the outer ranges of the Western Himalayas as far east as the Nepal border. To the north-west, it extends into Afghanistan and along the hills of the North-West Frontier to the Samana and the Kurram Valley.

HABITS AND NESTING. Although it would appear that this bird is not particularly numerous within our area, it must be remembered that it is a species of the foothills and lower levels. We field-workers have always been in too much of a hurry to reach the Vale of Kashmir, with the result that much of interest has been overlooked in the tract of country bordering the motor road from Kohala to Baramullah. This woodpecker may therefore well turn out to be the main representative of the family near such places as Rampur, Uri and Garhi ; especially as the pines of the lower levels are the long-leafed pines and, according to an observation of A. E. Osmaston's made in Garhwal, an article of their diet—a surprising article for a woodpecker—is the seed of the chir pine (*Pinus longifolia*). Elsewhere its nest has been taken as low as 2,000 ft and it is said not to occur much above the 6,000-ft level, though A. E. Jones says it is the common woodpecker around Simla. It is on the whole a quieter bird than the Kashmir Pied Woodpecker though we heard one at Keran using much the same *chucking* call-note. They are not shy birds, but it certainly surprised us to see the bird in the Chenar Bagh, within a few feet of the houseboat's veranda, unconcernedly hammering its way up one of the wooden cable standards.

They are said to lay in April and the first half of May, boring their nest-holes in firs and other trees at heights from the ground varying from 5 to 40 ft, according to Colonel Phillips, choosing in the Kashmir Valley trees on bunds, round villages, and in open gardens. The bird at Keran was close to a hole approximating to the latter figure. No extraneous material is introduced into the hole, the 4 eggs being laid merely on the fallen chips. These have the usual pure white shell with a considerable gloss and are a little smaller than those of the last bird, averaging approximately 23·4 by 17·4 mm.

JAPANESE WRYNECK

Jynx torquilla japonica Bonaparte

(See Coloured Plate III facing p. 136.)

KASHMIRI NAME. *Viri Mōt.*
BOOK REFERENCES. *F.B.I.*, IV, 100; *Nidification*, III, 322; *Handbook*, 286.
FIELD IDENTIFICATION. (2+) In general tone matches the bark of a lichen-covered tree, being mottled above in grey and dark brown and closely barred below in the same shades. Though not unlike a woodpecker in habits, progresses well on the ground, to which it freely descends to feed upon ants; and where the broad black markings down the back show up prominently. With its rather long tail a perfunctory glance may then lead one to take it for a pipit. When clinging to the trunk or branch of a tree, has a peculiarly grotesque manner of twisting round its head, whence is derived its trivial name.

DISTRIBUTION. The Wryneck is a not uncommon bird in the willow groves of the Vale. It also ascends the side valleys to moderate elevations, frequenting the better wooded portions of the valley bottoms and the edges of the thinner mixed forests. It was comparatively common and very noisy in mid-May in Gurais and we have met with it at the same elevations, approximately 7,000 to 8,500 ft, in the Sind and Lidar Valleys. We also noted it at Inshan and Basman in the Wardwan. Osmaston's find of a nestful of young ones in a birch tree above Wanhoi near the head of the Wardwan Valley at an elevation of at least 11,000 ft extends its altitudinal range considerably above what it is normally considered to be. It of course penetrates the many smaller valleys at the north-eastern and southern end of the Vale in the same way, such as the Bringh, Ahlan and Naubug Glens.

The Japanese Wryneck breeds from Japan, Manchuria, and Central Asia to Chitral and Kashmir. It occurs, but is rare, in Ladakh, while Ludlow says it arrives in Lhasa in April where it is fairly common in the parks. In winter migrants reach South China, Burma, and much of Eastern India. The European, typical, race also enters India by way of Baluchistan and Sind, confining itself mainly to the western side and southern part of the peninsula.

HABITS AND NESTING. The Wryneck arrives in Kashmir from its winter quarters in India in March, but does not get down to nidification for some weeks after its arrival, contenting itself by advertising its presence vocally throughout the remainder of that month and April. It does not possess a song, but only loud call-notes consisting of a harsh squeak, a quick *teeonk-teeonk-teeonk* repeated about five times, which it proclaims very freely until the serious business of nesting has begun. We see that Osmaston credits it with excavating its own nest-cavity, a fact which would, of course, account in part for the hiatus between its arrival in its breeding area and the deposition of the eggs, but we have been unable to trace any other record of this fact. Our own view is that the bird chooses a ready-made hollow for its nest and then proceeds to tidy it up to meet its requirements by picking away pieces of the decayed wood ; we have watched a bird doing this.

It is on the whole a quiet unobtrusive bird when once it is nesting and easily escapes notice, particularly as its coloration blends so well with the bark of the willows, mulberry and other trees which it frequents. In its hunt for insect food it moves about the branches and tree-trunks in the same manner as a woodpecker, but it is altogether more agile on its feet and moves well on the ground, to which it descends whenever a chance occurs of gorging itself on a swarm of ants.

A natural hollow, usually at the end of some rotten branch and often a considerable height above ground, is appropriated and in this the oval glossless white eggs are deposited on the debris at the bottom. The cavity may be one of those many holes so often found in willows or one caused by a deep split in a broken and rotting branch, or yet again the disused boring of a woodpecker. Davidson took a nest of the latter character in the stem of a walnut tree only about 7 ft from the ground. It contained 7 fresh eggs on 24 May. Eight seems to be the maximum number of eggs to be taken, for Wilson found that number at Sonamarg, presumably on 13 June, in a hole only 10 ft from the ground in an old tree, while Osmaston records two clutches of 6 and 7 nearly fresh eggs. Osmaston's clutches were taken on 23 and 24 May. We found a nestful of young ones in the rotting end of a snapped branch of a tall tree on an island in the Wardwan River near Inshan on 8 July. It would, therefore, appear that the eggs are laid in the latter half of May to mid-June, though reference must again be made to Osmaston's high-elevation nest at Wanhoi which still contained young on 29 July. In size the eggs average 21·1 by 15·5 mm.

ASIATIC CUCKOO

Cuculus canorus telephonus Heine

(See Plate 44 facing p. 173.)

KASHMIRI NAMES. *Shāh Kuk* or *Kukū*, *Zoeb Kuk*.
BOOK REFERENCES. *F.B.I.*, IV, 136; *Nidification*, III, 341; *Handbook*, 318.
FIELD IDENTIFICATION. (4) The familiar notes of the Cuckoo are quite unmistakable, but *Cuck-coo*, or *Cuck-cuck-coo*, is the prerogative of the males, the females usually uttering a less well-known but distinctive bubbling call. Both sexes are capable of some harsh cackling notes. In flight the Cuckoo is very hawklike, but the head is held high with the back hollowed, giving the bird a slightly arched appearance. Even at rest the resemblance to a Sparrow-Hawk is striking, the upper parts being mainly dark ashy. The sides of the neck and breast are paler ashy, while the lower parts become white narrowly barred with blackish. The tail is long, slightly graduated, notched and narrowly tipped with white.

Himalayan Bearded Vulture in flight

Large White Scavenger Vulture

PLATE 54

Pallas' Fishing-Eagle, the lower illustration showing the broad white band in the tail

DISTRIBUTION. The Cuckoo is a summer visitor to Kashmir, arriving in April. At the end of July the call ceases, although the birds probably do not start their journey to winter quarters in the plains and hills of Continental India until September. In the breeding season they have a wide altitudinal range from about 4,000 ft in the Jhelum and Kishenganga Valleys, throughout the main Vale, the side valleys, the lighter forests, and even the higher margs of both the Pir Panjal mountains and the main Himalayan range up to well over 12,000 ft. In the Vale they are particularly numerous about the lakes and willow groves.

This race of the familiar Cuckoo extends from Northern Asia to Japan and down to the Himalayas. Within Indian limits breeding birds in the extreme north-west and Baluchistan may be the typical race, while those occurring in the hills south of the Brahmaputra and eastwards through Burma to Western China belong to the Khasi Hills form *Cuculus canorus bakeri*. In winter these cuckoos are great wanderers and are said even to reach Australia, but there is still a great deal to be learnt in regard to the movements and status of the different races.

HABITS AND NESTING. So much has been written on the subject of the Cuckoo that it seems almost presumptuous on our part to add more, beyond extolling the occurrence of this bird in Kashmir. A perusal, however, of Stuart Baker's *Cuckoo Problems* will bring home to the reader how little we really know of this ubiquitous creature's inner life, and in particular its status in our area.

In Kashmir the Asiatic Cuckoo has an exceedingly wide range. It is common in the Vale, and we have heard and seen it well below 5,000 ft in the Jhelum and Kishenganga Valleys, yet it is to be found at 13,000 ft where only a few stunted birches remain to serve it as vantage-points from which to observe the movements of its would-be dupes, for the Cuckoo imposes upon less cunning birds the duty of hatching out its eggs and bringing up its voracious young. Indeed, as Miss Frances Pitt so aptly puts it : ' There is no more scandalous bird than the Cuckoo. By no standard of conduct, whether human or animal, can its habits of life be condoned. Whereas the majority of birds pair respectably, often marrying for life, the Cuckoo knows naught of marital ties ; and as for its offspring, we are all aware how it foists its eggs upon other birds and leaves them to bring up the young ones.'

The photograph of a chick in a Blue-fronted Redstart's nest in the shattered end of a birch stump was taken at over 12,000 ft at Astanmarg. This cuckoo is perhaps commonest between 6,000 and 7,000 ft ; in fact nowhere have we beheld so large a population

CUCKOOS

of this and the Himalayan species as in that tiny forest-encircled vale of Rampur-Rajpur, 1,500 ft above the Wular Lake. This glen suits the Asiatic Cuckoo down to the ground, for, in addition to the undulating valley bed, dotted with trees and presenting a patchwork of scanty fields divided here and there by rose-bedecked hedges or rain-worn straggling ditches, in the surrounding forests are grassy clearings of varying size. The whole supports a thicker population of Dark-grey Bushchats, Meadow-Buntings, and Stonechats than we have seen elsewhere, and some pairs of Rufous-backed Shrikes, all amongst the chief fosterers of the Cuckoo. These are not, of course, the only birds duped, but the last-named is a widely distributed and easily traced bird from Valley level to some 8,000 ft of elevation. It also lays an egg to which the Cuckoo's has become so attuned that rejections by the fosterers are probably less than those by any other species cuckolded.

We have always noticed the presence of Cuckoos around Hokra Jheel. Here vast numbers of Yellow-headed Wagtails place their nests in the soft grasses, and although we have only a couple of such records and Stuart Baker's comprehensive collection contains few Kashmir eggs from wagtails' nests, two only to be exact, we feel that this bird and Hodgson's Pied Wagtail should be common fosterers. On leaving the Vale, Western Spotted Forktails appear to be frequent victims, but Cuckoos which victimize this bird surely have other strings to their bow, Plumbeous Redstarts or shrikes, for the Western Spotted Forktail does not appear to be sufficiently numerous to satisfy the egg-laying capacity of a Cuckoo which can lay from a dozen to twenty eggs in a season, each of which is placed in a separate nest. At the higher elevations rubythroats, pipits and redstarts are the main fosterers. There are yet others, for Stuart Baker's list also contains accentors, bluechats and Streaked Laughing-Thrushes which, with those already mentioned, provide suitable fosterers for the three most highly developed types of Cuckoo's eggs, viz. an immaculate blue egg, one reproducing the distinctive pattern and coloration of the Meadow-Bunting's, and thirdly an egg often closely resembling that of the Rufous-backed Shrike. It is noticeable that out of Stuart Baker's 93 Kashmir-taken Cuckoo's eggs, no less than 20 are of the latter type from shrikes' nests.

One would expect the Indian Great Reed-Warbler to figure prominently, for this bird swarms in the Vale, yet the above collection contains only one egg from the Reed-Warbler's nest. Is it possible that so many young cuckoos have found a watery grave that the evolution of a Reed-Warbler Cuckoo has proved impossible? This seems unlikely, for in the Balkans and Eastern Europe the Great Reed-Warbler is the cuckoo's favourite and there an egg has been evolved matching the Reed-Warbler's to perfection. We have personally twice found a Cuckoo's egg in Indian Great Reed-Warblers' nests, but in each case the egg had a reddish ground, very different from the grey or greenish ground of the Reed-Warbler's egg. It might be inferred from this that no assimilation has taken place between the eggs of the Asiatic Cuckoo and the Indian Great Reed-Warbler, still further bearing out the contention that this warbler figures only amongst the casual fosterers. On the other hand, a cuckoo's egg taken on 2 June from this Reed-Warbler's nest on the Anchar Lake provides much food for thought. This egg is described by J. E. Scott, who took the nest, as 'in ground colour dingy cream or grey-stone, markings olive-brown and yellow-brown confined to large end and in small speckles widely scattered except at top where they form a cap somewhat to one side and composed of one large blotch three-eighths of an inch in diameter and other confluent spots continuing the cap or ring; a few lilac secondary spots. Compared with the fosterer's eggs, the colour scheme is a passable imitation. The size is almost identical, but the markings differ in being confined to one end and lack the very dark flecks of the Reed-Warbler.' The size of this egg is 20·5 by 14·75 mm. Stuart Baker, to whom it was sent, was of the opinion that it was not an egg of the Asiatic Cuckoo, and suggested, presumably on account of its size, that it might be that of *poliocephalus*, the Small Cuckoo. We naturally hesitate to contradict such an authority as Stuart Baker where cuckoos' eggs are concerned, but we have never seen or heard *poliocephalus* in the marshes. In Kashmir the Small Cuckoo is a bird of the better wooded areas fringing the Vale, and of the forests and parklands of the side valleys. Besides, what of the coloration? Might not this egg be an abnormally small egg of the Asiatic Cuckoo? After all, Stuart Baker gives the measurements of the smallest of a series of some 600 eggs of typical *canorus* as 20·0 by 15·5 and 20·7 by 14·7 mm.; *telephonus* admittedly averages larger. As typical *canorus* has in Hungary produced an egg showing advanced assimilation with its Great Reed-Warbler fosterer both in size and colour, it might reasonably be argued that *telephonus* has done so in Kashmir. Both cuckoo and fosterer are sufficiently numerous and concentrated for such assimilation to have taken place, though surely were this the case in such a well-worked area, the fact would long since have become known. This particular egg might, of course, belong to some cuckoo not as yet recorded from

CUCKOOS

the Vale. For example, the Plaintive Cuckoo, *Cacomantis merulinus passerinus*, is known to occur in the Jhelum Valley up to at least Garhi. We do, therefore, suggest that visitors to the marshes make a point of assuring themselves that cuckoos' eggs of this and the known *canorus* types are not being overlooked in the many Reed-Warblers' nests they are certain to find.

In Kashmir the Cuckoo begins to arrive towards the middle of April and from then until the end of July is very vociferous, the *cuck-coo*, sometimes *cuck cuck-coo*, of the males and the bubbling notes of the females being heard from all sides in most types of country except thick forest, though cultivation and bush-dotted country are the most to its liking with here and there tall trees to provide it with good lookout points from which it can spy out the positions of the necessary nests. It calls on the wing as well as at rest.

To deposit its eggs the Cuckoo either sits in the nest in a normal manner, or if the site or shape of the nest precludes this, it clings to the surrounding material and ejects the egg into the nest. In this it is not always successful, and eggs not infrequently fail to reach the egg cavity. The incubation period of a Cuckoo's egg is short, 12 to 13 days, to ensure that the chick will be able to get rid of the remaining eggs or newly-hatched young of its fosterers. This object it attains by working them into a peculiar hollow depression in its back and then heaving them out of the nest; an amazing provision of nature, for the fosterers are often hard put to it to bring sufficient food even for their sole remaining charge, the young Cuckoo.

The coloration of the egg is variable in accordance with the development towards assimilation with those of the fosterer, but most of the marked Kashmir eggs have a whitish or pink background mottled or spotted to a varying degree, and with a strong range in depth, with brownish or yellowish red. In addition the Meadow-Bunting type must not be forgotten. In this the markings, even to the hair-lines, are well followed, though the ground colour usually shows a pinkish tinge. They are blunt ovals with a heavy shell and a faint gloss. Stuart Baker gives the average measurements of 100 eggs as 23·6 by 18·1 mm.

HIMALAYAN CUCKOO

Cuculus optatus optatus Gould

(See Coloured Plate IV facing p. 232.)

BOOK REFERENCES. *F.B.I.*, IV, 140; *Nidification*, III, 344; *Handbook*, 319.

FIELD IDENTIFICATION. (4) Closely resembles the Asiatic Cuckoo in appearance and size, but has distinctive notes easily mistaken for the *hud-hud-hud* of the Hoopoe. The black barring on the underparts is perhaps slightly coarser and the upper plumage is somewhat darker. It is purely a forest bird.

DISTRIBUTION. We have heard or seen it throughout practically our whole area, except in the lower Kishenganga and Jhelum Valleys and above the tree line. It occurs not uncommonly from about 6,500 to at least 10,000 ft in all the side valleys, in the Kazinag hills, around the Lolab and west of the Wular Lake, in Gurais certainly down to Taobat, and in the Wardwan Valley. We noted its call occasionally as low as 5,500 ft near Handowar.

The full distribution of this race as given by Stuart Baker embraces the whole length of the Himalayas to Assam (including the hills south of the Brahmaputra) and North Burma, while outside Indian limits to the north and east this or some other form breeds over the greater part of Central Asia, Siberia and the mountain ranges to North China. In winter numbers wander far to the south; in India to about the latitude of Lucknow, but farther east it has been recorded from Maymyo in Burma and it wanders through Malaya, even reaching Australia.

HABITS AND NESTING. The habits of the Himalayan Cuckoo are in the main those of its better-known relative, but, being most retiring by nature, it is far more often heard than seen. It is not that it is particularly shy, but it is not prone to over-much movement and sticks closely to a forest habitat, being found even in the thickest pine woods, where consequently only its distinctive calls lead to knowledge of its presence. At close quarters a feeble high-pitched *cuck* can be heard preceding the booming hoopoelike notes. These are generally four in number though sometimes only three. This bird at times indulges in volleys of gurgles and groans of quite a startling character and the female has the usual bubbling call.

Considering how widespread this cuckoo is in Kashmir, and the Himalayas generally, surprisingly little is on record about it from our area and comparatively few of its eggs appear to have been listed. For fosterers the Himalayan cuckoo favours the willow-wrens, the majority of Kashmir eggs having been found in nests of the Large Crowned Willow-Warbler. Other recorded fosterers found in our area are Hume's and Ticehurst's Willow-Warblers, Meadow-Buntings and Red-flanked Bush-Robins. It is a moot point as to how the egg is introduced into some nests. One egg, taken by us on 6 June in a nest of the Large Crowned Willow-Warbler in a cavity in a perrottia, had cracked one of the warbler's eggs in its fall. There is nothing, of course, to rule out the possibility that the egg had been ejected directly from the oviduct while the bird clung precariously to the trunk of the sapling, except that in this case surely the cuckoo's egg or those of the fosterer would have been smashed to bits. It is dangerous to assume with our present knowledge that the bird at times lays its egg in safety at a distance and deposits it with its bill.

The Himalayan Cuckoo's egg is a very elongated oval, white or skimmed-milk blue in coloration. Two exceptions in Stuart Baker's collection have the ground colour respectively faintly tinted green and equally faintly tinged with pink. The markings are never profuse and often all but absent. They consist of minute purple-grey or blackish specks. Stuart Baker states that, though sometimes more numerous, they are never in any way dense. The average of 58 eggs in his collection is 20·11 by 14·28 mm., which means that this bird has evolved an extremely small egg for its size. May and June are the principal breeding months, corresponding with the nesting of the above fosterers. Brooks took an oviduct egg on the Ratan Pir on 17 June.

SMALL CUCKOO

Cuculus poliocephalus poliocephalus Latham

BOOK REFERENCES. *F.B.I.*, IV, 142; *Nidification*, III, 346; *Handbook*, 319.

FIELD IDENTIFICATION. (3) In appearance this bird is a miniature of the Himalayan Cuckoo, being approximately two-thirds of its size. It is, however, shy and little seen, but reveals its presence by noisy calling, by night as well as by day, in unmistakable harsh tones which may be syllabized as *So take your choky pepper—your choky pepper*.

DISTRIBUTION. Its distribution in Kashmir coincides largely

with that of the Asiatic Cuckoo in its middle and higher ranges. It does not occur in the central parts of the Vale, but we heard it at Achabal and to the west of the Vale near Handowar and at Bandipur. It is met with more frequently from about 6,000 ft upwards in the side valleys, spreading on to the margs above tree-level. It is also to be heard on the wooded slopes of the Pir Panjal, but is scarcer on those mountains than on the main Himalayan range. It is more of a forest bird than *canorus*, but less so than the Himalayan Cuckoo, and will be found in any well-wooded country as well as in the mixed forests and where there are blanks in the thicker woods.

Its full breeding range takes it through the Himalayas and the hills of North Burma and China to Japan, while in winter this form further extends through the Indian peninsula to Ceylon, and throughout Burma, Siam, and the Indo-Chinese countries to Malaya.

HABITS AND NESTING. This little cuckoo is as much a bird of the night as of the day, its harsh cacophony of grating calls ringing out at any hour up to midnight and increasing again about 3 o'clock in the morning. For this reason its presence can hardly be overlooked. Were it not for this insistent calling at times when most other creatures are silent, its wide distribution throughout the better-wooded parts of Kashmir might easily be overlooked, for it is a wary self-effacing bird in other respects.

Its character is admirably summed up by Colonel Magrath in a *Miscellaneous Note* in the *Journal*, from which the following is an extract. 'This Cuckoo is fairly ubiquitous in Kashmir, occurring at all elevations, and in the same month, from the Valley itself, 5,100 ft to 11,000 ft and over, though it is, perhaps, not so common as *C. canorus*. In some favoured spots these little Cuckoos collect, and here they make both day, and a good part of the night, "hideous" with their very extraordinary notes, although perhaps not more than half a dozen individuals may be present. The number of birds in a locality must always be hard to compute from the notes. At Lidarwat, 9,500 ft, I was encamped in a clearing in silver fir forest, and for one whole day one solitary individual of this species kept up an almost incessant cackle, not being silent for more than half an hour from 8 a.m. till dusk. It constantly changed its perch from tree to tree in the forest around my camp, and sometimes called on the wing, and to anyone not versed in the ways of this freak amongst birds, it would certainly have appeared that there were at least a dozen or more birds calling. This particular bird so exhausted its syringeal muscles by its performance this day that it remained silent throughout the night, but recommenced at 7.30 a.m. next morning, and called till about 9 a.m.; then finding, I suppose, that there were no responsive females in the neighbourhood it shifted to another part of the valley. But it returned in the evening, and called at intervals during the night. It remained in the vicinity calling daily, and intermittently at night, till I left on the 23rd July.

'The following words may serve to convey to the ear the cadence, and to some extent, the sound of the notes, etc., viz. " That's your choky pepper ". . . On an open hillside it can be heard a quarter of a mile away or more. The note of the female is not unlike that of the female *canorus*, but is more slowly repeated, and perhaps more nearly resembles that of the female *C. saturatus*, but is less loud. . .

'Unlike birds of the passerine families and indeed others of its genus, the Small Cuckoo does not call at dawn, but generally commences some two hours after sunrise. It becomes active in this respect from about 8 or 9 a.m. till noon; then again in the evening; again at dusk; again, moderately so, after dark, and even at midnight, and again commonly at about 3 or 4 a.m.

'This is not an easy bird to spot or collect. One hears its notes all round, but they are ventriloquial in character, and when in a tree and calling, like others of its genus, it keeps very still. Under favourable circumstances, however, it may be seen perched on the topmost twig of a tree, uttering its call, or may be found towards evening coming to the ground to hunt for caterpillars in the undergrowth.'

It penetrates to high elevations, and we have found it at Astanmarg at 12,000 ft where but for juniper and a few patches of other bushes only isolated birches remain. Here it perhaps cuckolds Tickell's Willow-Warbler. It seems common enough in the lower mixed forests and on the fringes of the heavier pine woods. It is, however, to be found in comparatively open country where the trees are in isolated groups. Of its nesting, or rather laying, in our area not much is on record. The full distribution of this cuckoo extends throughout the Himalayas and the hills of Burma and China to Japan. In this huge area it lays two widely divergent types of eggs. In Japan a deep terra-cotta egg is invariably produced; over most of the central area both white and terra-cotta eggs may be found, the latter the more common; but in the Western Himalayas none but the white egg, and only few of these, have so far been taken. These have had for their fosterers mainly willow-wrens and flycatcher-warblers which lay white or spotted white eggs. Somewhat farther east a large proportion of red eggs

PLATE 55

Nest and eggs of Impeyan Pheasant

Himalayan Rufous Turtle-Dove on nest

PLATE 56

Chukor on nest

Nest and eggs of Chukor

have been taken in nests of *Homochlamys fortipes*. As the Pale Bush-Warbler (*Homochlamys pallidus pallidus*) is common in our area, and its distribution comes well within that of the Small Cuckoo, it seems strange that the red egg has not been taken from its nest. A lookout for such eggs should certainly be kept, although Osmaston, who has taken many nests of the Pale Bush-Warbler, says he never found an egg of this cuckoo of either the white or red variety. It is also worth mentioning that another fosterer might well be the exceedingly common Rufous-tailed Flycatcher, in whose nests the white egg should prove acceptable.

In spite of its being much smaller, it lays an egg almost identical in size with that of the Himalayan Cuckoo, Stuart Baker giving the average measurements as 21·0 by 14·2 mm. It is therefore at times difficult to tell the unspotted white egg from that of the Himalayan Cuckoo. Most of the eggs have been taken in May and June, a few in July.

NORTHERN PIED CRESTED CUCKOO

Clamator jacobinus pica (Hemprich & Ehrenberg)

(See Coloured Plate IV facing p. 232.)

KASHMIRI NAME. *Hōr Kuk*.
BOOK REFERENCES. *F.B.I.*, IV, 167; *Nidification*, III, 356; *Handbook*, 324.
FIELD IDENTIFICATION. (3+) Though a typical cuckoo in its parasitic habits, in appearance it is quite different from the three preceding birds, being black above and white below. There is a white patch in the wing, conspicuous in flight, and the long graduated tail-feathers are tipped with white. The uptilted crest is well-defined. It possesses a double screaming call. This sleek cuckoo is not shy and may be observed at fairly close quarters as it perches on the top of a bush or creeps about the lower foliage in search of insects.
DISTRIBUTION. Confined to the main Vale and the bush-clothed slopes on its rim, especially in the Ganderbal area and around the southern and eastern extremities. Occurs in small numbers only, and is a summer visitor whose movements are still somewhat shrouded in mystery, but it probably winters in Eastern Africa. It arrives in June and leaves again in August or September.

At the commencement of the rains this species spreads rapidly

CUCKOOS

throughout the greater part of India down to the Madras province where it meets a resident, Ceylon, race. It also penetrates into Burma.

HABITS AND NESTING. The Pied Crested Cuckoo is an uncommon bird in Kashmir, but with the advent of the south-west monsoon it suddenly makes its appearance in limited numbers in the Vale in June, about the same time as elsewhere in Northern India. We have noted it halfway through that month on the scrubby hills below the Banihal Pass, while one of its chief strongholds is the bush-covered delta of the Sind River between Ganderbal and Shalabug. Where our other cuckoos prefer the tall trees, the Pied Crested Cuckoo delights to perch on a bush top or to creep about the foliage near ground level. Although it is generally understood that it does not descend to the ground for feeding purposes, it will on occasion do so to seize a caterpillar or succulent grub. Its call is a somewhat unmusical whistle, almost a shriek, *pie-ou*, becoming more metallic as the breeding season draws to a close.

In the Plains this cuckoo places its eggs in the nests of certain species of babbler (*Argya* and *Turdoides*) which lay blue eggs. With these the Pied Crested Cuckoo's eggs match to perfection, the cuckoo's eggs being a pure greenish blue with a hard satinlike shell. They are blunted symmetrical ovals measuring $23 \cdot 9$ by $18 \cdot 6$ mm.; rather beautiful eggs which match up best in our area with the clear blue eggs of the Simla Streaked Laughing-Thrush. It is, in fact, from these bush-loving skulkers' nests that in late June and in July the eggs have alone been taken, and it appears that in the extreme Western Himalayas no other fosterer has so far been systematically exploited. With the egg so perfectly modified to agree with the eggs of the plains' babblers, Stuart Baker suggests (*Cuckoo Problems*, 82 and 83) that the Pied Crested Cuckoo is to all intents but a recent excursionist into the hills. What he says is this: 'All the eggs of this Cuckoo referred to in Hume's *Nests and Eggs of Indian Birds* (III, 388-91) were found in the Plains of India or in the lower ranges of the Nilgiris or sub-Himalayas, and all without exception (omitting oviduct eggs) were deposited in the nests of the two genera referred to.

'When we come to the Hills, however, we find the Northern Pied Crested Cuckoos placing their eggs in a great range of birds' nests, though in most cases these are nests of the larger Laughing-Thrushes, nearly all laying blue eggs with which the eggs of the Cuckoos do not contrast. The normal fosterers here are undoubtedly the Necklaced and Black-gorgeted Laughing-Thrushes in the

Eastern Himalayas and the Striated Laughing-Thrush in the Western...

'The Pied Crested Cuckoo seems now to have no definite limits to its elevations for breeding. In Assam it certainly breeds up to 6,000 ft and probably higher than this in the Central Himalayas. At the same time it is possible that its breeding in the hills is a more or less modern extension of its breeding habitat. In the Plains, as I have already written, its normal fosterer, or group of fosterers, is so completely established that exceptions are very very few. In the lower hills, the cuckoo adheres closely to the Necklaced and Striated Laughing-Thrushes, but above the normal elevation of the breeding areas of these birds, or where these birds are not found or are rare, it launches out into the use of all kinds of nests which bear some resemblance to those they usually cuckold.'

In other words, in its penetration into the inner ranges it has not yet fixed upon certain fosterers to the exclusion of all others. Nevertheless, so far as we can ascertain, in Kashmir the Streaked Laughing-Thrush is the only bird regularly victimized, that being the commonest bird in our area with which the Pied Crested Cuckoo's immaculate blue egg agrees sufficiently closely as to avoid rejection by the fosterer. The Indian Bluechat could conceivably be another fosterer, but so far we have no information whatsoever in its connexion.

HIMALAYAN SLATY-HEADED PARAKEET

Psittacula himalayana himalayana (Lesson)

(See Plate 46 facing p. 181 and Plate 76 facing p. 301.)

KASHMIRI NAMES. *Tota, Shoga* (in the Lolab).
BOOK REFERENCES. *F.B.I.*, IV, 206; *Nidification*, III, 380; *Handbook*, 336.
FIELD IDENTIFICATION. (3+) A bright green bird with an outsize slaty head and yellow tip to the long and narrow blue tail. Often in small flocks of half a dozen birds which fly swiftly through the trees uttering shrill screams. The immature bird is dull green with no adornments, attaining its full colours after the first moult.
DISTRIBUTION. During its breeding season the Slaty-headed Parakeet is very locally distributed in Kashmir. In the Vale, except for the immediate vicinity of Baramullah where it is not common, we have only seen it on the wooded slopes above the

PARAKEET

Wular Lake at elevations from about 6,000 ft. These particular birds come from the Lolab Valley, of which it is a very common inhabitant. There, and in the Kazinag range, in fact throughout the area drained by the Pohru River to its junction with the Jhelum, it occurs in some numbers. We have met with it in the Kishenganga Valley between Salkalla and Doarian. It seemed most numerous there around Keran, which lies at 4,900 ft. On a few occasions we have seen it near Rampur on our way up the Jhelum Valley, but in days of more leisurely travel, 1896 to be precise, Davidson recorded seeing it in flocks as well as in pairs up to within 7 miles of Baramullah. It is a noticeable fact that these are areas in which the slopes are mainly afforested with that valuable and beautiful cedar, the deodar, which occurs in quantity in no other parts of Kashmir. The Slaty-headed Parakeet, in fact, gives one the impression that it is pre-eminently a forest bird, and except for the purpose of making raids into gardens and orchards, during the breeding season it remains within the forest areas, preferably those in which the deodars predominate. It is later in the year that it strikes farther afield. Osmaston records seeing it about Gulmarg in August, and Ward remarks that in the autumn numbers venture into the side valleys—Sind, Lidar, etc.—up to about 7,000 ft. The reason underlying this expansion is, of course, food-supply. It is then that the maize is ripe, while Unwin remarks, ' In autumn it plunders the walnut groves at Atawat '. In the spring these parakeets attack the apple blossom. They can at most be called local migrants wintering in a lower zone down to about 2,000 ft, although a few find their way into the northern plains.

This species in two races extends throughout the Himalayas to Burma and Yunnan, except in the plains, and down to Tenasserim. Whitehead found a small nesting colony in the Kurram Valley and there are a couple of records of its occurrence in Afghanistan, so it may well breed in some numbers in the Sufed Koh.

HABITS AND NESTING. We doubt that this very striking bird lives in the summer at a higher elevation than 7,000 ft. There is a low pass, about 8,000 ft, leading out of the Lolab Valley which provides a well-worn route for these birds to the slopes and little vales overlooking the Wular Lake and the lower Pohru River. At any time of day, but particularly in the early mornings and evenings, parakeets, sometimes in ones and twos, sometimes in small bands, may be seen flying up to the pass only to curve down over the crest and be lost sight of in their rapid descent: we have never seen them pause on the summit. They are far more forest birds than most Indian parakeets and are usually seen in mixed

fir and deodar woods.   In the sombre shade of the pines, a flock
of half a dozen of these brightly hued parrots provides a contrasting
splash of colour not easily forgotten as they clamber about the
dark branches using both bill and feet as is their wont to lever
themselves from perch to perch.   Besides the wild rasping screams
which they utter on the wing, and which are reminiscent of the
shrill squeals of the marmots, they also possess a purer exceedingly
high-pitched shriek used at rest, while for conversation some quite
pleasant subdued cries are employed.

They nest in the woods, where to meet their requirements they
either enlarge an existing hole in a tree or chisel out a new one.
Unfortunately there appears to be little on record about their
nesting, but it undoubtedly starts early in the year, according to the
Nidification from late in March to the end of April ; Hutton (*Stray
Feathers*) says April and May.   At Keran, where we occupied the
Forest Bungalow between 29 April and 6 May, we marked down
two pairs in the woods above us which we suspected to be feeding
young in the nest.   Of Murree, Marshall, who gives late April as
the nesting time, wrote that this parakeet usually selects trees of
great height, and this is in general the impression we have gained
from our experiences in Kashmir.

The nest which figures in Plate 46, however, was not above
18 ft from the ground and we have noticed a few holes even lower.
This nest was near Sanzipur in the Kazinag, and though the date
of its discovery was 10 June, it still appeared to hold eggs.   It
will be noticed that the two birds on the perches are in immature
plumage.   There were five birds clustered round the hole while
one of the owners was within, so all, except perhaps the adult
clinging to the entrance, had no proprietary rights in the abode.
We already had reason to believe that these birds are colonial
in their habits.   Our experiences a few days later near Panzal,
9 miles north-west of Baramullah and a few hundred feet above
the level of the Vale, confirmed this conclusion.   Here the path
to Baramullah skirts the lowest rim of the forest, cutting through
a little tongue—what in the Plains might be termed a *bagh*—
which consisted mostly of long-leafed pines with about half a dozen
tall deciduous trees scattered amongst them.   The latter were
tenanted by colonies of Slaty-headed Parakeets—unfortunately we
omitted to inspect the pines.   In one trunk alone we counted five
holes on the side facing us, three of which were definitely occupied
and hardly two yards apart.   Many of the holes were far from
perfect circles, a few of them being unsymmetrical ovals with the
long axis at no particular angle.   They varied considerably in

size, from little bigger than that of a Pied Woodpecker, through which the birds squeezed with difficulty, to about 4 inches across. The lowest was about 20 ft from the ground while some we estimated to be 60 or more feet up. We saw three or four birds entering and leaving holes while all the time there was an incessant screaming from birds conversing with one another or from small bands weaving swiftly in and out of the trees. These bands contained numbers of immature birds and seemed to be family parties whose nesting for the year was over and done with. The flight is indeed swift and very sure; it is a remarkable sight to see these tearing bands weaving and banking through the sombre tree-trunks in perfect unison. Some parties will pass above the tree-tops, but most prefer to fly low, all the members suddenly swinging upwards, as if by word of command, to alight in the denser foliage.

No actual nest is constructed, the eggs being deposited on the chips and rotten fragments in the bottom of the cavity, which is often of great depth. The eggs, 3 to 5 in number, generally 4, are of course the usual all-white, glossless broad ovals of the family. They average 28·3 by 22·2 mm.

KASHMIR ROLLER

Coracias garrula semenowi (Loudon & Tschusi)

(See Plate 47 facing p. 188.)

KASHMIRI NAME. *Nīlakrāsh.*
BOOK REFERENCES. *F.B.I.*, IV, 222; *Nidification*, III, 387; *Handbook*, 294.

FIELD IDENTIFICATION. (4—) The raucous cries and the tumbling aerobatics during courtship of this grey-blue bird are its most striking points. Lacks the deep blue wing-patches and tail-band so characteristic of the Blue-Jay of the Plains. The rather slow beat of rounded wings, the large head and heavy bill are also points to note. Descends from a view-point to seize large insects and frogs on the ground.

DISTRIBUTION. Osmaston records seeing a straggler at 10,500 ft on Killenmarg. A couple of pairs certainly resided in that lovely little forest-encircled valley of Rampur-Rajpur of which the elevation is approximately 7,000 ft, but above that elevation this bird will not normally be found. Their chief stronghold is undoubtedly the main Vale, more particularly the drier portions of it, while they ascend the lower wide reaches of the side valleys, such as the Lidar, Sind, Erin, and Bandipur Nullahs, to where they begin to narrow down and up to where the more lightly-wooded parkland and cultivated areas give place to enclosed country and forests. This roller is not uncommon along the Jhelum Valley road, where we have seen it in summer as far down as 7 miles from Kohala. It is seen in Kashmir mainly between April and September. Meinertzhagen states that none had arrived in Kashmir by 24 March, the date he left the Vale, and that although a few remained around Srinagar in early September, all had gone by the 25th. Its migration to winter quarters in North Africa probably starts early but is well drawn out. In Baluchistan its numbers are considerably increased at the end of July, while Sálim Ali notes in the *Birds of Kutch* that it also passes over Kutch, a few in August but many between the second week and end of September. This southern route figures only on the outward journey.

Stuart Baker gives the range of this Eastern race of the European Roller as Transcaspia and Turkestan to Kashmir, Baluchistan, Afghanistan and the North-West Frontier. It is also a common breeding bird in Iraq and Persia.

HABITS AND NESTING. In comparison with the brightly-hued Blue-Jay of the Plains, which is clothed, as Whistler so aptly puts it, in the combined colours of Oxford and Cambridge, the Kashmir Roller is more sober in his dress, flaunting his allegiance as is right and proper to but one seat of learning, and that Cambridge. In habits, however, this roller differs little from his Plains' cousin, though his guttural croakings are in our opinion not quite so harsh and his behaviour not so wild, carefree, and demonstrative. In the excitement of mating, or on later occasions when the even tenor of their ways is upset, the same aerial contortions are indulged in. Rollers so obviously glory in wild aerobatics, whence of course they derive their name, proclaiming their pleasure in carefree antics and abandoned flying by a series of most harsh croakings which are as displeasing to the ear as their colour and evolutions are pleasing to the eye. A wild upward sweep ends with the bird gracefully upending itself as it reaches the stalling-point, to lurch earthwards with bent-back wings till it pulls out of the headlong dive as it once more gathers speed. The rollers are said to somersault in the air. Certainly many of their evolutions are startling and complicated, but to be quite truthful we find we have recorded only on one occasion seeing a roller complete a real loop, and that was such a perfect example that there was not the slightest doubt about it.

ROLLER

In sober mood the Roller flies with a rather slow continuous flapping of the broad wings and a direct purposeful flight. In between times it sits quietly on some good view-point at any height from the ground, whence it can survey a goodly area around it, be that point the tip of a gaunt dead branch of some tall tree, the rounded top of an earthy mound, or merely a stone or weed almost on the ground itself. In seizing its prey it usually planes down to alight beside it and then snaps it up, be it a slimy frog, a grasshopper, or an unappetizing armour-plated beetle. And sometimes it has to carry out a series of awkward fluttering hops to complete its capture.

For the nesting site some natural hollow is always chosen; at times a hollow in a tree from a few feet to even 30 or 40 ft up, but perhaps the most favoured site is a cavity in a sand-bank. As the face of a bank weathers, nest-cavities of bee-eaters and kingfishers' disused tunnels become exposed and often enlarged by time and water-action into considerable caverns of irregular shape. Such places are those most favoured by the rollers, being in Kashmir the sites chosen three times out of four in preference to a hole in a tree. The sandy bank may be that in a quarry, on the side of a dry nullah, or the corroded bank of one of the larger rivers with the roaring swirling water dashing past within a few feet of the sitting bird. In such a situation near Sonarwain, on the Madmatti River, we found two nests within 20 yards of one another, while Davidson records having taken eight nests in the bank of the river near Baramullah all on one day, the 26th of June, when they contained either incubated eggs or young. Although the finding of such numbers in a single day may warrant the application of the word common, Stuart Baker's opening remark in the *Fauna* that the Kashmir Roller is one of the commonest birds in Kashmir in summer is very far from correct.

Never yet have we found any nest material beneath the eggs. On the rotten wood or amongst the sand on which the eggs rest, rejected parts of the insect food may be found, such as the hard wing-cases of beetles, but there is never any semblance of a special bed for the eggs.

As the Kashmir Roller is migratory, the majority leaving India altogether, breeding does not commence very early and few eggs will be deposited before the last week of May, the young hatching out towards the end of June or in July. Stuart Baker states that both birds take part in incubation and certainly both share the task of feeding the young, for we photographed a pair which were bringing their chicks frogs. The parents entered the nest-cavity in strict rotation with their provender, which was being obtained

PLATE 57

Northern Ruddy Crake on nest

Turkestan Water-Rail on nest

PLATE 58

Eastern Baillon's Crake repairing nest canopy.
Note the curious white smears on the upper plumage

from a boggy patch nearly half a mile away. Rollers' eggs, like those of the bee-eaters and kingfishers, are almost spherical. They are, of course, pure white and are large eggs measuring 36·2 by 29·0 mm. Four, or occasionally 5, are laid.

EUROPEAN BEE-EATER

Merops apiaster Linnæus
(See Plate 47 facing p. 188.)

KASHMIRI NAMES. *Tuleri Kāv, Rodabubru.*
BOOK REFERENCES. *F.B.I.*, IV, 233; *Nidification*, III, 393; *Handbook*, 297.
FIELD IDENTIFICATION. (3—) The illustration gives a good idea of the streamlined appearance of this bird. It is considerably larger than the Common Green Bee-Eater of the Plains and is at once recognizable by its bright yellow throat. Beneath, it is largely blue; brown and yellow on the upper plumage. When seen in pairs or small parties hawking flies and *hymenoptera*, its flight is particularly graceful, consisting of a high proportion of gliding on motionless wings. In the evenings collects high overhead into flocks preparatory to roosting communally in the leafy foliage of tall trees.
DISTRIBUTION. Confined to the main Vale and lowest reaches of the large side valleys where it arrives in April or early May, leaving again in September. It is a migratory species, Kashmir birds possibly wintering in the Northern Punjab.

The European Bee-Eater, as its name implies, breeds extensively in Southern and Central Europe, but it is equally common through much of the western half of Asia, nesting in Indian limits from Baluchistan to the Western Himalayas. Stuart Baker says that in winter it is found over most of North-West India, but most birds probably pass on through Arabia to winter in Africa.
HABITS AND NESTING. To our minds the bee-eaters are perhaps the world's finest exponents of graceful flight. To lovely colours they add great beauty of form. For when soaring their outline is that of a Cupid's bow, thanks to the delicate curve of the tapered wings. The slender sharp bill and the central tail-feathers prolonged into fine needles enhance their attractiveness. The Green Bee-Eater of the Plains, so often seen in rows on the telegraph

BEE-EATER

lines, glows a brilliant golden green in the strong light and is often referred to as the Sunbird. The European Bee-Eater is larger, and although possessing a greater range of colours, does not catch the eye so readily. It is also more retiring. Its tinkling yet fluty notes reach the ear from afar, for near nightfall numbers of these birds collect high overhead prior to departure to a common roosting-place in some tall leafy tree such as a chenar or walnut. It is then that their mellow notes mingle so pleasantly together and that their easy flight is so striking. For they swing and dip and rise on gliding wings, their bills snapping audibly together on the fleet *hymenoptera* and other insects which are no match for their practised aerobatics.

Though considerable numbers assemble in this manner, they do not nest in compact colonies. Each sandy bank or workable declivity spread over a wide area sports a few of their borings, only one or two of which prove to be occupied. The hole is oval or roughly rectangular with the long axis vertical, and has a slight upward trend for two to four feet culminating in a widened egg-chamber of considerable size. Boring takes time, and numbers are discarded before the final working is chosen, so that eggs are not generally deposited before mid-June. The drier bare sandy declivities such as the moundlike hillocks between Shadipur and the main road, and the spurs on the right bank of the river between Shadipur and Manasbal, are most to their liking, but they are widespread birds at Valley level and provided the necessary banks or quarries are available, they will be found anywhere around the Vale's rim. They nest on the Takht and along the lower slopes round the Dal lakes. We have also found them at Kyunus and on the other promontories around the Wular's shores, while no doubt the undulating downs between Baramullah and Patan are well patronized. Colonel Phillips records that during two successive summers a pair made use of one of the bunkers of the Srinagar golf course. He also appears to have found their tunnels drilled in level ground.

Fresh eggs may be taken up to the end of July. They are almost spherical glossy ovals, in colour pure white. Very occasionally a clutch is found in which one or two eggs bear some bold purple-black spots, Ward having sent Stuart Baker one such clutch and Buchanan having found two others. Five or 6 is the normal number laid, but up to 8 have been recorded. They average $25 \cdot 8$ by $21 \cdot 9$ mm. They are laid direct upon the bare floor of the chamber, no materials of any kind being introduced.

In August the birds begin to congregate preparatory to migration.

Unwin relates that ' the migration commences almost at once and by September scarcely a Bee-Eater is left in the Valley. In early September 1891 I saw a large flock crossing Gulmarg and passing South-West.'

INDIAN PIED KINGFISHER

Ceryle rudis leucomelanura Reichenbach

(See Plate 48 facing p. 189 and Plate 49 facing p. 196.)

KASHMIRI NAMES. *Hōr Kola Tōnch, Sufaid Tōnt, Duddru.*
BOOK REFERENCES. *F.B.I.*, IV, 246; *Nidification*, III, 402; *Handbook*, 299.

FIELD IDENTIFICATION. (3) Boldly patterned in black and white. The upper parts are barred with black while there is a broad collar on the breast with, in the male, another indefinite one just below it. The feathers on the crown of the large head are slightly elongated into the semblance of a crest. The bird fishes by plunging into the water from a height of 10 to 20 ft, first hovering in mid-air with its heavy pointed bill directed vertically downwards.

DISTRIBUTION. Common in the Vale at all seasons over almost any water except shallow running streams. May be seen in the side valleys only up to about 6,000 ft; for instance, in the Bandipur Nullah to a little beyond Sonarwain, in the Sind to about Kangan, and in the Lidar Valley to Aishmakam. This bird remains in the Vale throughout the year and commences to breed very early in the spring.

The species as a whole has a range from Egypt to China, the Indian race being found throughout India and Burma, Yunnan, South China and the Indo-Chinese countries. It is everywhere a resident.

HABITS AND NESTING. This widely distributed bird, which is spread throughout the length and breadth of the Indian region, is in its element chiefly over the still waters of lake and marsh and the undisturbed slow-flowing waters of wide canals and the main river. It does, however, patronize the banks of the larger side rivers for a short distance after they become noisy torrents of heaving water. Here, however, the birds will only be seen fishing in the quieter backwaters and in from the banks over channels and pools amongst the adjacent fields.

KINGFISHERS

The method of fishing is most arresting. Suddenly checking its flight on sighting some small fish near the water's surface, the bird hovers on rapidly vibrating wings, the body inclined upwards at an angle of about 45°. The bill, however, is directed vertically downwards while the bird corrects its aim at its intended victim. Suddenly the head drops, the body tilts over, the wings close, and the bird drops like an arrow into the water below. Sometimes its hovering ceases and it moves on a yard or two to retake its aim ; sometimes it checks its dive almost as soon as it has started, and flies off chittering querulously to try its luck elsewhere. On two separate occasions in different years birds were seen to skim over the water cutting the surface with the slightly opened bill just as the Indian Skimmer does. The birds flew upstream cutting the surface for 50 yards or so at a stretch.

The nest-hole is usually over water in some vertical sandy bank, but it may at times be a considerable distance away in the side of a borrow-pit or quarry, Colonel Phillips recording one such site as being over two miles from the nearest water. In the same article (*Journal*, 46, 94) he speaks of ' many nesting colonies in the vicinity of the Lakes each used by not more than about a dozen couples. Many of these localities appear to have been used over a period of years. The bank faces are riddled by their tunnellings and it is evident that the burrows are used as shelters during the winter months as, nearing sunset, many birds congregate in the area and have been seen entering the holes.' This bird is not, in fact, colonial in its nesting habits in normal circumstances, but where particularly attractive sites occur, such as the sandy banks referred to by Colonel Phillips, a number of pairs are apt to take advantage of the accommodation offered. In depth the tunnel varies a great deal, from as little as 12 inches to half as many feet. It opens out into a wide flattened chamber in which the eggs will be found lying surrounded by the remains of undigested fish-bones disgorged by the owners. Between 4 and 6 eggs are laid, which are very broad ovals, at times almost spherical, of pure glossy china-white. They measure 30·3 by 23·6 mm. They are deposited chiefly in April and May, but fresh eggs may at times be found in early June. On the other hand Osmaston records taking eggs on 1 March when snow was still lying about the ground. We possess no proof that this bird is double-brooded, but in view of this extended nesting period the possibility should be borne in mind for investigation.

PLATE IV

1. Black and Yellow Grosbeak (adult male). 2. Brown-fronted Pied Woodpecker (adult male). 3. Northern Pied Crested Cuckoo. 4. Black-throated Jay. 5. Himalayan Pied Kingfisher. 6. Indian Grey Drongo. 7. Himalayan Black Bulbul. 8. Indian Blue Rock-Thrush (adult male). 9. Himalayan Cuckoo. 10. Larger Spotted Nutcracker

HIMALAYAN PIED KINGFISHER

Ceryle lugubris guttulata Stejneger
(See Coloured Plate IV facing p. 232.)

BOOK REFERENCES. *F.B.I.*, IV, 248; *Nidification*, III, 404; *Handbook*, 300.

FIELD IDENTIFICATION. (4+) Easily distinguished from the Indian bird, not alone by its larger size, but by its different habits and the pattern of its plumage. In this kingfisher the entire plumage is heavily and closely barred with narrow bands of black which give to it a speckled rather than a pied effect as the amounts of black and white are just about equal. The elongated loose feathers of the head are in the nature of an untidy mop rather than a crest. Lastly the Himalayan Pied Kingfisher does not hover, but drops on its prey from a shady perch.

DISTRIBUTION. Although it is said to be found in Kashmir up to about 7,000 ft, this fine kingfisher will only be met with in any numbers at all at less than half that altitude. It is comparatively common in the Kishenganga Valley from Ghori at about 2,400 ft to Mirpur at 3,600 ft. We saw a nesting pair at Salkalla and also noted signs of them at Keran, which lies at just under 5,000 ft. Of the Jhelum Valley we can find no records, but one would expect to find it up to at least Rampur. It does not occur in the Vale except perhaps in that quiet labyrinth of shady channels between Shalabug and Ganderbal. It has been seen at times in secluded reaches of the larger side rivers up to between 6,000 and 7,000 ft, where it would probably be more numerous were it not for discouragement on account of its depredations amongst young trout. Captain Livesey found young ones in a nest in the Lidar Valley.

This race of the Great Pied Kingfisher is found throughout the lower Himalayas to Assam and on to Burma and Yunnan. A second race extends the range of the species still farther east to Japan.

HABITS AND NESTING. The Himalayan Pied Kingfisher is a resident bird of medium elevations along the larger streams and rivers which run through forest and well-wooded country. Bare rocky banks are not to its liking, for it loves to sit on some shaded branch over the water waiting for its prey to appear below it. For this reason its presence may well be overlooked so it may, in fact, be less rare in the Kashmir valleys than it appears to be. Osmaston remarks that it would undoubtedly be commoner if it were not for

KINGFISHERS

the ruthless persecution to which it is subjected in the interests of fish-preservation.

Great volumes of rushing water hold no terrors for it, and it will unhesitatingly plunge into foaming rapids. As far as Kashmir is concerned, it is wrong to say—as the *Nidification* has it—that it 'breeds in the banks of streams . . . always where the water is flowing bright and clear, in rapids and pools, but not in a rushing torrent'. We know of no more turbulent river than the lower and middle Kishenganga, whose waters are always brown with silt. Along its banks the nest-holes of this bird are common, whereas we failed to find their traces along the lesser side streams. The waters of the Lidar, too, can hardly be called bright and clear.

The nest-hole is large and from all accounts varies much in depth. We came upon a pair making their tunnel in the wall of a sandy excavation by the side of the track where it runs close to the water's edge near Salkalla. When first noticed, it was hardly a foot deep, but twenty hours later another three and a half feet of sand had been excavated and the birds were still at work. Four or 5 large white eggs of the usual almost spherical type are laid, measuring 38·5 by 32·5 mm. The usual breeding season is stated to be March and April, but our Salkalla birds were still tunnelling on 27 April.

CENTRAL ASIAN KINGFISHER

Alcedo atthis pallasii Reichenbach

(See Plate 49 facing p. 196.)

KASHMIRI NAMES. *Kola Tōnch, Chotta Tōnt.*
BOOK REFERENCES. *F.B.I.*, IV, 253; *Nidification*, III, 407; *Handbook*, 301.
FIELD IDENTIFICATION. (2+) A brilliantly coloured dumpy little bird with large head and equally large bill. The upper parts are mainly blue with some white on each side of the neck, and the underparts are chestnut. Usually fishes from a low perch, dropping into the water with quite a plop. Skims the water's surface in swift direct flight uttering shrill squeaks as it flashes by.
DISTRIBUTION. An exceedingly common bird in the Vale, numbers remaining throughout the year, where it will be seen along the irrigation channels, canals, rivers, marshes, and even around the shores of the largest lakes; in fact everywhere where

water is not far off. It does not penetrate deep into the side valleys, about 7,000 ft being its normal limit. It may be met with in Pahlgam, and we saw it on one occasion in Gurais at about 8,000 ft, while it is recorded not infrequently from Gulmarg. A pair was also noted regularly between Inshan and Tsuidraman in the Wardwan Valley, elevation about 8,400 ft.

The Central Asian Kingfisher occupies the plains of Baluchistan in winter and breeds from the hills of the North-West Frontier to the Western Himalayas. Outside Indian limits it is the common kingfisher northwards to Western Siberia and westwards to Afghanistan and Persia. It is, of course, one of the many races of the widespread Common Kingfisher of Europe, North Africa, and Asia generally.

HABITS AND NESTING. Since the days of the Mogul Emperors the beauty of form, the brilliant hues, and the trustful familiarity of the kingfisher have drawn the admiration of the Kashmiris from the humblest cultivator and artisan to the highest in the land. No branch of the many beautiful crafts and handiworks practised in the bazaars of Kashmir is without its representations of this common but none the less distinctive bird of the Vale. It appears in the patterns on the carpets, the numdahs and embroideries, on the exquisite papier-mâché bowls, on the intricately worked silverware, and on the remarkably fine wood-carvings. And is not this just as it should be, for this little gem of blue and chestnut is part and parcel of the life of the Vale? It is quite impossible to visit even the busiest waterways without seeing an industrious little fisherman seated on the gunwale of a passing dunga, quite oblivious of the Manjis as they pole their deeply-laden craft crying rhythmically upon God and the prophets to aid them in their heavy task. Others will be seen on slender drooping branches of the willows or even on the prow of a houseboat, beating the life out of a freshly caught minnow.

In their fishing they employ two methods, the usual one being to watch the waters from a fixed perch whence they drop on their prey at an angle, a sharp little plop resulting as they break the surface. Not infrequently, however, when they fish the open waters or hunt for frogs in wayside pools, they will hover in mid-air, even as the Pied Kingfisher does, with bill directed almost vertically downwards aiming at their victim. When passing rapidly by in swift direct flight, uttering sharp *peets* as they go, they will also at times drop without pause upon some hapless creature which catches their attention.

Sometimes they will skim considerable distances overland from one stretch of water to another or on the way to their nests. For the Central Asian Kingfisher does not always drill out his tunnel in the bank of a stream. Often it will be in the sandy shaft of some shallow well, in the bank-side of a dry nullah, or yet on a hillside where rocks have been quarried from the slope, and we have found them even in the earthen walls surrounding a village garden whither the birds had to carry the food for their young 300 yards from the nearest water-channel. True it is that the majority of nests will be found in vertical sandy banks overlooking streams and rivers, and every year numbers of nests in these last sites are drowned out by flooding.

The hole, in the excavating of which both birds take part, is usually slightly oval in shape with the longer axis vertical and it may be anything from 2 or 3 to 5 ft deep, ending in a rounded chamber about 6 inches across. In this 5 to 7 pure white, almost spherical eggs are laid. They have a high gloss and measure 21·2 by 17·6 mm. In nests which have been in use for any length of time dried-up fish-bones will be found in varying quantities, but no nest is made and these are merely the accumulated disgorgings of the sitting birds. It is disappointing to find that such beautiful creatures are so lacking in all ideas of sanitation, for from the tunnel of a kingfisher's nest with young there oozes an evil-smelling mass of ordure and fishy slime. Whistler goes so far as to state that the entrance-tunnel starts to grow dirty with the excreta of the parent birds before the eggs are even laid. The food is not restricted entirely to a fish diet; they will eat most water insects and amphibiæ to satisfy their voracious appetites, while frogs seem to find favour amongst those with large young ones to feed. The birds in the plate were feeding their brood exclusively on frogs, which they brought to the nest throughout the hours of daylight at intervals of about fifteen minutes.

The nesting season is very prolonged. Full complements of eggs may be seen by the end of April, yet we have noted birds still entering their tunnels with food well on into September. It is in May, June and July, however, that nesting is at its height.

PLATE 59

Moorhen on nest

Painted Snipe, male, on nest

WHITE-BREASTED KINGFISHER

Halcyon smyrnensis smyrnensis (Linnæus)
(See Plate 50 facing p. 197.)

BOOK REFERENCES. *F.B.I.*, IV, 268; *Nidification*, III, 418; *Handbook*, 303.

FIELD IDENTIFICATION. (3+) Slightly larger than the Pied Kingfisher, but of dark coloration, having greenish-blue patches on the wings and back, and dark rufous underparts with a conspicuous irregular white area on the chin, throat and breast. The heavy stout bill is dull red and waxlike. In flight a large spot of white shows up prominently on each wing. It possesses a plaintive ringing call.

DISTRIBUTION. Until lately this bird found a place in this book solely by virtue of its presence in the Jhelum Valley, being comparatively common up to Domel. It has also been recorded at Garhi and recently we saw a single bird near Chenari. It occurs for a few miles up the Kishenganga River from its junction with the Jhelum, but its penetration there appears to have been stopped by the first of the gorges. As far as the Vale of Kashmir is concerned it is extremely rare. In 1932 we saw a single bird in some pits near the Nasim Bagh where it was perhaps nesting. Lately, however, we photographed the bird at a nest in a quarry at Ganderbal, and Colonel Phillips, *in epistola*, remarks that they now appear to be gaining ground in the Vale. Further, in a recent article in the *Journal* he writes of their sporadic appearance at all seasons of the year.

This beautiful kingfisher has a range from Asia Minor eastwards throughout all the southern countries of Asia. It is divided into a number of races, the form found throughout India, excepting the extreme south, belonging to the typical race. It is a resident species wherever found.

HABITS AND NESTING. The habits of the White-breasted Kingfisher differ very considerably from those of other members of the family for the immediate presence of water has long since become of secondary importance in its economy. It is now content to haunt quarries and gardens and almost any country sufficiently well wooded to provide it with shade and perches. Unlike our other Kashmir kingfishers, it never hovers but planes down to seize its prey off the ground, flying back to some shady perch to devour it. Well content to subsist on a diet of frogs, lizards, grasshoppers and other insects, it has practically given up fishing and it is for this reason that a shady garden attracts its presence rather than the more open banks of rivers and streams. It will, however, catch small fish upon occasion, and the Ganderbal pair were certainly regaling themselves with a mixed diet of minnows, frogs, and aquatic insects from a pond adjacent to their quarry, as well as on drier game.

For nesting purposes a vertical bank is a necessity, although in Assam Stuart Baker found loose oven-shaped nests of moss pushed into exposed roots of trees overhanging the banks of streams. Normally, a hole of considerable diameter, and often of some depth according to the hardness of the material, is bored into the face of a vertical bank, be it that of a river or stream, the side of an unlined well, or the face of a quarry or other vertical wall well away from water. The quarry face at Ganderbal was of hard sandy grit in which were mixed pebbles and larger stones, with the result that the shape of the tunnel was very irregular. The tunnel contained, in fact, two large stones near the entrance, too heavy for the birds to eject.

A high-pitched ringing call is often uttered on the wing, while the song, if song it can be called, is a rolling crescendo of querulous screams uttered from the perch with the bill thrown upwards. Both carry a considerable distance so that the presence of this bird can hardly remain unnoticed by anyone who knows it.

In the Plains the nesting season is prolonged, those boring their nest-holes in the banks of rivers and streams liable to seasonal flooding when the snows melt, commencing in February and March, while eggs have been taken as late as August. The Ganderbal nest was discovered on 12 May when it probably contained eggs. We photographed the birds on 7 June when their behaviour indicated that the young had recently hatched out. Both birds were then bringing food so minced up that it was often impossible to judge of its character, but frogs and crickets figured conspicuously.

The number of eggs laid is from 5 to 7, the usual number being 6. Stuart Baker remarks upon the peculiar 'watered-ribbon marks' to be seen in the eggs of this and certain other kingfishers if held up to the light when freshly blown. They are the usual spherical immaculate eggs averaging $29 \cdot 4$ by $26 \cdot 2$ mm.

EUROPEAN HOOPOE

Upupa epops epops Linnæus

(See Plate 50 facing p. 197.)

KASHMIRI NAMES. *Satūt, Hud-hud.*
BOOK REFERENCES. *F.B.I.*, IV, 308; *Nidification*, III, 442; *Handbook*, 309.
FIELD IDENTIFICATION. (3) The long curving bill, the fanlike crest, and the barred black and white wings and tail on an otherwise fawn body, render this handsome bird quite unmistakable. Other distinctive features are its loud hooting call, *hud-hud-hud*, whence the bird gets both its scientific and trivial names, and the butterfly flight as it rises unconcernedly before one when approached while busily probing the earth for larvæ and grubs.
DISTRIBUTION. The Hoopoe is an abundant summer visitor to the Vale of Kashmir and in lesser numbers penetrates the larger side valleys up to 9,000 and even 10,000 ft. At this elevation it breeds around Sonamarg, and we have noted odd pairs nesting in the vicinity of Pahlgam. Osmaston records a few in Gulmarg compounds in July and August, and they occur up to about the same elevation on other margs of the Pir Panjal range. In the Kishenganga Valley we met with them at Keran, Sharda, and in Upper Gurais, but they were nowhere common. They seem fairly plentiful along the motor road below Baramullah. The European Hoopoe is migratory, arriving from the Plains of Northern and Western India in March and leaving in September or October. A few brave out the winter and may occasionally be seen then in the Vale. It should be mentioned that Colonel Meinertzhagen, in his 'Birds Collected in Ladak and Sikkim' (*Ibis*, 1927, 603), states that three breeding hoopoes obtained in March near Srinagar proved to be *indicus*, the Indian Hoopoe, and not *epops*. A male from Pampur obtained on 27 September was typical *epops*.

The hoopoes range throughout Southern and Central Europe, reaching the British Isles and South Sweden in small numbers, Africa, and much of Asia. There are three races in India, the typical race breeding in the Western Himalayas, but wintering also in the Plains of North India and the United Provinces.
HABITS AND NESTING. It is obvious from the many legends regarding the acquisition of the Hoopoe's unique crest that this bird has been an object of interest for centuries. That is not surprising, for no one can possibly help taking notice of a Hoopoe.

HOOPOE

It is a tame and confiding bird, little put out by the proximity of man : in fact it takes every advantage of his achievements, grubbing about on his lawns and in his flower-beds, dust-bathing on the sandy roads, and nesting in his outhouses and cavities in walls. It will hardly stir from his path, running a few feet at the last moment to get out of his way before resuming its vigorous digging for the larvæ it senses beneath the earth. The slightly curved bill is used like a pick-axe to break through the surface, the analogy being further strengthened by the manner in which the flattened crest continues the line of the bill behind the head. When the Hoopoe alights, however, or is in any way excited, the crest swings forward to form a wide-open fan, each black-tipped feather being set off by a narrow white patch beneath it.

The flight of the Hoopoe is easy and graceful, the broad wings, heavily barred with black and white, bearing the bird aloft in a manner reminiscent of a butterfly, as the wings are often closed to the sides so that the flight is dipping and intermittent. When leaving the ground the Hoopoe often utters a short snipelike *pench !* The *hud-hud-hud* of the Hoopoe is most distinctive but easily confused with the call of the Himalayan Cuckoo, whose notes may also be heard emanating from the woods in the side valleys. This cuckoo, however, usually gives vent to a series of four hoots to the Hoopoe's two or three. To produce his call the Hoopoe drops the bill towards his chest, at the same time depressing the tail. When using the rasping note the bill is raised upwards. In both cases the mandibles apparently remain closed.

Osmaston says the staple diet of the Hoopoe in Kashmir is the grub of the cockchafer and there is little doubt that the larvæ of beetles are most sought after by them. In fact, when photographing this bird at the nest we have noticed that on every occasion cockchafer grubs alone were being brought to the birds within. They commence to nest as early as April and by the third week of May many eggs are hatched. But the season is prolonged, for we have often noted young still in the nest in July, so it is possible that two broods are raised each year.

The situations are varied, but many choose cavities in the mud and stone foundations of the village houses, while away from habitations—they are equally at home in well-wooded country and even in light forest—hollows in trees are well patronized. At times most curious sites are recorded, such as the bare floor in the corner of a darkened room entered by some small hole in the floor or wall ; we have personally found two such nests. Only a scanty collection of waste materials is scraped together to hold the eggs, consisting of bits of grass, leaves and feathers, and we have at times found small pieces of dirty rag.

Once incubation has fairly started, the male shows considerable anger if the female leaves the nest and he soon drives her back to her task. She is, therefore, often incarcerated for something like three weeks. During this time the nest-hole becomes much fouled so that it smells rather like an ill-kept hen-house. At this period the male keeps his spouse supplied with food. We once came upon a bigamist whose two wives were occupying separate holes in the eaves about 10 ft apart. That immoral bird was perhaps to be pitied for he had a hard time of it keeping both of them supplied and policing the entrance-holes. Yet his anger when one or other attempted to leave the nest to fend for herself knew no bounds, and the offender was soon driven indoors with the most dreadful scolding, for when really angry the Hoopoe marks his displeasure with a wheezing noise, at the same time puffing out the throat-feathers.

The eggs, up to 8 in number, are a faint blue, white, or buff when fresh, but they rapidly become discoloured to a dirty brown. A considerable percentage of eggs fail to hatch out judging by the numbers of addled eggs we have found in disused nests at odd times. They are rather long true ovals without gloss and measure 26·2 by 17·6 mm. There is probably considerable mortality amongst the chicks thanks to the insanitary conditions in which they are raised. It is noticeable that the very young chicks, which are clothed in a bluish-grey down, have a short comparatively thick bill not unlike that of a young myna, the commissure being of an ivory white colour.

ALPINE SWIFT

Apus melba melba Linnæus

BOOK REFERENCES. *F.B.I.*, IV, 324 ; *Handbook*, 312.
FIELD IDENTIFICATION. (3—) In spite of its great speed, this swift is more heavily built than the next species, being much wider in the chest. The body, in fact, is almost pear-shaped—presumably to accommodate the powerful pectoral muscles—rather than narrowly streamlined like that of the commoner swifts. The narrow sickle-shaped wings, and the white underparts divided by a wide brown band across the breast, are quite distinctive features.

SWIFTS

Usually seen in parties hawking flies at great heights, but at odd times they come down quite low. Owing to their great powers of flight these bands are seldom in sight for more than a few minutes.
DISTRIBUTION. It is hard to estimate either the numbers or define the permanent habitats of these fine birds, for their abnormal speed and powers of endurance take them anywhere from quite low to the highest elevations and passing bands may be hundreds of miles from their permanent quarters. They are to be seen at odd times anywhere in our area, but once only have we been lucky enough to discover a roosting or nesting place. We cannot say for certain which it was, nor can we state categorically that they nest where they roost. On discovery, on 23 April, birds were entering at short intervals clefts in a rock-face from which intermittent squeaks came almost continuously; yet the same evening well before dusk the place seemed quite deserted. The cliff was in the Tithwal gorge in the lower Kishenganga Valley at an elevation of about 3,400 ft. According to the *Fauna*, 1st edition, III, 165, nests were seen in Kashmir by Littledale, but unfortunately the locality is not given.

To what extent this bird is truly migratory we cannot say, but it is a widespread species covering most of Southern Europe, the mountains of North Africa, and South-West Asia, including the greater part of India wherein a definitely resident race, *bakeri*, is found in the south and also in Ceylon.

HABITS AND NESTING. Alpine Swifts are perhaps most often to be seen at the medium and higher elevations, sweeping over the upland meadows and side valleys of the Pir Panjal and main Himalayan ranges. It is at once evident, however, from their terrific speed and the manner in which they feed in one general direction, rather than backwards and forwards over a comparatively limited area as is the practice of the Eastern Swifts, that a normal day's run may quite easily take them from one end of Kashmir to the other. It is hardly surprising that with such wandering habits, few records of nesting places are extant. Take for instance our colony near Tithwal. Here in the gorge was a small cliff on the precipitous hillside. Some 40 or 50 ft up, the rocks were scored and split into crevices and narrow cracks; some horizontal, many almost vertical. About every quarter of an hour a screaming band appeared as if from nowhere, hurtling over a spur with the swishing noise of a passing shell. Whence they came and whither they went, it was quite impossible to tell, for their speed took them over the spurs in a matter of seconds. As the band swept past, one or two birds would turn into the cliff and without pause creep like bats into the narrow crevices. The vertical slits seemed just as popular as the horizontal clefts and their recesses were even more inaccessible. As these birds entered, adjacent holes would disgorge others to join the shrieking band, and during the intervals between visits faint twitterings could be heard issuing from within the recesses. Returning in the afternoon, not a bird appeared and not a sound could be heard. Perhaps the site was really deserted ; perhaps the entire band was merely on a foraging expedition a hundred miles away : we could not stay to find out.

Their great speed has been stressed, but we must admit that it cannot be compared with that of the White-throated Spinetail. The ease with which the latter overtakes an Alpine Swift has to be seen to be believed. We have spoken of shrieking bands, but in point of fact the note, if note it can be called, is far less of a scream than the shrill utterances of the Eastern and House Swifts. It is fuller and almost mellow in tone ; a pleasing rather than a discordant sound.

The Alpine Swift of India, *bakeri*, is said to nest from December to March. We have no records whatsoever on which to base an opinion as regards the type of nest or time of nesting of the typical race in the Himalayas, but to quote again from the *Fauna*, 1st edition : ' The nests have walls about an inch thick made of feathers, dry grass, etc., firmly cemented together by the saliva of the birds ; they are 4 or 5 inches in diameter, not lined. Several nests are often clustered together. The eggs are laid in Europe about May and June ; they are white, elongate, 3 or 4 in number, and measure about 1·2 by 0·75 inches ' [29·4 by 18·3 mm.].

EASTERN SWIFT

Micropus apus pekinensis (Swinhoe)
(See Coloured Plate III facing p. 136.)

BOOK REFERENCES. *F.B.I.*, IV, 326 ; *Handbook*, 311.
FIELD IDENTIFICATION. (2+) The general colour of this swift is an unrelieved ashy-brown. The dusky-white chin cannot be seen in flight and is nothing like so conspicuous as it appears to be from the plate. Usually to be seen in parties of considerable size in precipitous rocky country where they hawk flies high overhead, weaving in and out and backwards and forwards over a comparatively limited area of the sky. The narrow curved wings,

SWIFTS

which in flight are never closed to the sides, and the finely streamlined body impart very considerable speed to their movements.

DISTRIBUTION. Numerous and widespread in our area but not seen to any great extent over the main Vale. They are usually seen in bands high up the more precipitous and rocky mountainsides.

This species, so well known in the British Isles, contains a couple of migratory races which breed throughout the greater part of the Palæarctic region, wintering mainly in Africa. The Eastern form is the representative of the species from Cyprus and Asia Minor to Siberia, North-East China and Manchuria. In India it occurs only in the hills from Baluchistan to North Cachar. We have no information from our area about the date of its arrival and departure, but it reaches Baluchistan in April, though Swinhoe's and Hutton's first records for Kandahar were 6 and 20 February respectively.

HABITS AND NESTING. Osmaston puts down this bird as ' not uncommon between 9,000 and 12,000 ft '. Within these limits they are undoubtedly more numerous than at the lower elevations, but this in the main is due to their attachment to the more precipitous ground, where, in the many cliffs, they have their colonial nesting places. Swifts are rather powerless on the flat. The legs and feet are weak and the toes all turned forward : they are, therefore, unable to perch and do not easily rise from a flat surface. Consequent upon this, the Vale, karewas, and the immediately surrounding slopes are not patronized, for unlike our European birds they have not taken to using buildings, but wherever precipices and rocky gorges occur, there the Eastern Swifts are likely to be found, since it is here that they can creep into the fissures and launch themselves from the ledges. We have seen them flying over the lower parts of the Jhelum Valley and in the Kishenganga Valley have noted their presence at Pateka, Keran, and Khel, and indeed through to Gurais. Between Khel and Janwai they were undoubtedly using an immense vertical cliff. A well-known breeding-place in the Sind Valley is provided by the cliffs on the right bank of the river below Sonamarg village. Osmaston records yet another site in the crags near Tosha Maidan in the Pir Panjal at 10,500 ft. They must, of course, have many nesting places, for colonies are to be seen in every valley.

When feeding, the members of these bands have a curious habit of suddenly collecting into a close pack, all twittering and screaming together as if greatly excited. Soon the formation disintegrates and the birds resume their independent sweeping flights, swerving

PLATE 61

Pheasant-tailed Jaçana at nest

PLATE 62

European Little Ringed Plover with two eggs displaced by bird

Indian Whiskered Tern at nest

and wheeling back and forth high over the valley or their home cliffs. Towards dusk or in rainy weather, as the insect-supply sinks lower, they too come down with it and may then be observed at closer quarters.

Their nesting differs in no way from that of the European bird, except that as the valleys of Kashmir lack suitable buildings, their homes are exclusively in the narrow crevices in the crags and precipices and in their caves. They are usually quite inaccessible, often being hundreds of feet from the ground. Their nests, *vide* the *Fauna*, are massive structures of feathers and all sorts of wind-borne rubbish, matted together with the birds' saliva and apparently used for several years so that they become filthy and verminous. Thirty eggs sent to Stuart Baker from Yeneseisk and East Turkestan, and evidently laid in May and June, averaged 25·1 by 16·07 mm. We have no dates for Kashmir, but Marshall found nests in a cave in Baluchistan on 31 May, so the season is probably the same everywhere. Two to 3 eggs are laid which are white, very elongate, and have a fragile nearly glossless shell.

LONG-EARED OWL

Asio otus otus (Linnæus)
(See Coloured Plate V facing p. 328.)

BOOK REFERENCES. *F.B.I.*, IV, 393; *Nidification*, III, 500; *Handbook*, 342.

FIELD IDENTIFICATION. (4) A rather grey owl, in size that of the House Crow, possessing very prominent earlike tufts. The plumage is mottled and streaked in brownish-black, greyish-buff and buff. This dark coloration, in conjunction with a peculiar habit when alarmed of flattening the feathers so closely to the body that it gives the bird an appearance of inordinate length, renders it not unlike a dead stick amongst the branches in the midst of which it is perching.

DISTRIBUTION. The only records for Kashmir are of a nest taken in June 1895 in Gurais and of a single bird which was seen by Osmaston a few years ago in the Lolab Valley towards the end of April when it was being mobbed by a pair of Black-throated Jays. This latter bird may only have been on passage, as the Long-eared Owl is a by no means uncommon winter visitor to the Plains of a large part of Northern India, though Shelley, who found the nest

OWLS

in Gurais at about 9,000 ft, remarked that he also heard the birds lower than that elevation.

The Long-eared Owl breeds in Europe, North-West Africa, and Western Asia down to the extreme north-west Himalayas. In winter it penetrates into the Plains of the Punjab, Sind, and also reaches Kutch.

HABITS AND NESTING. Shelley's note appeared in Vol. X of the *Journal*, an extract from which reads : ' While after bear on the hills above Gurais in Kashmir on 4th June 1895, I found a nest of 4 eggs of this Owl in a Sycamore tree, at a height of about 12 ft from the ground, and I also shot one of the birds for identification. The nest, which was a mere platform of sticks, with no pretence of a lining, had doubtless, from its weather-beaten and dilapidated appearance, been originally occupied by some other bird, probably a Crow. The eggs were much incubated and would, I think, have hatched in a few days... On several occasions I heard owls hooting in the evening which I believe belonged to this species, so it is probably not an uncommon bird in Kashmir at suitable elevations.' The above accords well with the habits of this bird in Europe, which frequently converts an old nest to its own use, although it will also lay its eggs on the ground in a mere scrape amongst rank grass or other upland herbage. It usually affects tree growth of some sort, either scrub jungle or light forest, in which it spends the daylight hours resting up against some shaded tree-trunk. The mottled plumage then accords well with its surroundings, so that its presence is very apt to escape notice, especially with the plumage flattened so that its shape is not unlike a broken-ended stick.

Its voice is a dismal drawn-out moan, but like most owls it has a number of other calls for various occasions. Witherby syllabizes its usual call as *oo-oo-oo*, adding that when the nesting place is invaded, the birds cry continually, *oo-ack*, *oo-ack*, prefaced sometimes by a barking *woof woof*.

The eggs are pure white broad ovals measuring approximately 41·1 by 34·3 mm.

SCULLY'S WOOD OWL

Strix aluco biddulphi Scully
(See Plate 51 facing p. 204.)

KASHMIRI NAMES. *Rāta Mogul, Rāt Monglu.*
BOOK REFERENCES. *F.B.I.*, IV, 397; *Nidification*, III, 500; *Handbook*, 339.
FIELD IDENTIFICATION. (5—) A large brown owl without ear-tufts, matching in colour the tones of the thick fir forest in which it is usually found. Nocturnal in its habits, its presence will usually be first recognized by the loud call, *hoo*—interval—*hoo*—shorter interval—*hoo-ho-ho-hoo*, the final hoot being somewhat drawn-out. This call is, however, variable, some birds condensing it to *hoo-ho-ho*, or even *tu-hoo*.
DISTRIBUTION. A forest bird found almost everywhere outside the limits of the Vale, but particularly partial to the thick Blue Pine and fir forests between 7,000 and 9,000 ft. It has been recorded as high as 11,000 ft and the bird in the illustration was photographed at 5,500 ft. In winter it spreads farther afield, coming more freely into the main Vale, its hoots being heard commonly in and around Srinagar, but on the whole it is a resident species moving down somewhat under stress of winter conditions.

This race of the widespread European Tawny Owl, which also affects much of Western Asia, is found in the hills of North India, having been recorded as a resident in North Baluchistan, while Whitehead found it in the Kurram Valley. In the extreme north-western Himalayas it is common.

HABITS AND NESTING. In the daytime this fine form of the Tawny Owl remains concealed within the shady confines of its forest territory, but at night it will often come forth into more open country either to hunt or to alight high up on some isolated tree from which it can survey the ground around it or emit its eerie hoots. According to Osmaston a loud *quack* is also sometimes heard as well as a soft *coo* very similar to the call of the Ring-Dove. Each pair probably occupies the same territory for years. In early May we came upon a family party, the two adults with a couple of young ones whose feathers were still partially concealed by greyish down, a hundred yards up the wood immediately behind the Rampur-Rajpur Forest Hut. We had listened to a wood owl hooting in the same area on a previous visit seventeen years before.

In spite of its wide distribution the eggs seem to have been taken on but few occasions. In fact, Stuart Baker writes in the *Nidification*:

'The only two pairs of eggs in my collection undoubtedly referable to this species were taken by Ward near Sonamarg in May 1904. Both these clutches were laid on the ground on the banks of rocky streams running through forest. In neither case were they in regular caves, but were under large rocks overhanging shallows.' This is what one would have expected, but the steep slope in the forest, and not the actual proximity to a stream, was probably the chief inducement in the selection of these sites. Magrath actually met with them in the breeding season at between 11,000 and 12,000 ft where there were only a few scattered and stunted birch trees. He concluded that these particular birds were breeding in holes in cliffs. Scully's Wood Owl by no means confines its nesting to the ground. Whistler took two eggs at Baltal from a nest in a hole in a tree. Another pair to our certain knowledge had two young in a large cavity in an ancient pine behind the Forest Rest House at Rampur-Rajpur and yet another nest at Thiun in the Sind Valley was situated about 20 ft from the ground in a hole in a walnut tree. This nest contained the half-consumed body of a rat. Locals stated that this hole had been used by the owls for many years. We once came upon a young one in down upon an island in the Wardwan River which supported no possible tree site. Nevertheless, we are of the opinion that tree sites are used probably more often than any other type. Colonel Phillips has also taken the nest at Gulmarg in a hole in a tree.

The breeding season would appear to be from March to May, probably varying somewhat with the elevation, but the young are hatched usually by the advent of June. Two is evidently the number of eggs laid, which are the usual broad ellipses, white in colour and of fine texture. They measure approximately $50 \cdot 6$ by $43 \cdot 3$ mm., but they seem to vary considerably in size.

TURKESTAN GREAT HORNED OWL

Bubo bubo turcomanus (Eversmann)

(See Plate 51 facing p. 204.)

KASHMIRI NAME. *Rāta Mogul.*
BOOK REFERENCES. *F.B.I.*, IV, 413; *Nidification*, III, 510; *Handbook*, 342.
FIELD IDENTIFICATION. (5) A large solitary owl with well-developed ear-tufts which inhabits open rocky and scrub-covered

hillsides. Partly diurnal in its habits though shunning the direct sunlight. May be noted in the evenings perched on a high rock outlined against the darkening sky, every now and then giving vent to a dismal hoot. The call in question is a loud *hoo*, somewhat prolonged, but not unduly so, which is sometimes followed by a short cackle. It might also be described as *oo-oo*, a double note but the one running into the other. In flight the general appearance is of a warm yellowish-tawny bird heavily spotted and streaked with blackish. The wings are deep and rounded.

DISTRIBUTION. Distributed in moderate numbers about the barer rocky hillsides around the main Vale, where it is a permanent resident. As it occurs over a wide area in the Northern Indian hills, it will almost certainly be found in the Jhelum and lower Kishenganga Valleys, though we have traced no records.

As its name implies, this eagle-owl is the bird of Turkestan and Transcaspia and extends thence down to Persia, Afghanistan and Baluchistan, and to the Western Himalayas. Other races cover the rest of India.

This genus, comprising a number of species and sub-species, is practically cosmopolitan in its distribution.

HABITS AND NESTING. This fine eagle-owl is a bird of rock and scrub-covered hills, cliffs, old quarries, and even ancient buildings, but we can think of no better example of its normal habitat than the slopes of the Takht-i-Suleman at Srinagar where two pairs have lived for many years. There is no other owl in Kashmir which will be found in such surroundings and in its turn it is hardly likely to invade the precincts of others of its family. It is not, however, averse to entering clumps of trees in default of its favourite type of ground. According to Osmaston the chief food of these birds in Kashmir is rats, and this in all probability is their main diet everywhere, although no doubt they will descend upon other 'things that creep in the night' when occasion arises. The birds make no nest but lay their eggs in a bare hollow in a bank, under an overhanging rock on a steep hillside, or even in a shallow cave. The site shown in the illustration is a typical one and really requires no enlarging upon.

The number of eggs laid is up to 4, but 2 or 3 appear to be the normal clutch. They are large white ovals with no gloss, measuring 60·1 by 47·3 mm. In Kashmir the breeding season commences in March and extends to early May. This owl is, however, an early breeder and few nests will be found with eggs in the last-named month and then they will certainly be well incubated.

WESTERN COLLARED PIGMY OWLET

Glaucidium brodiei brodiei (Burton)
(See Coloured Plate III facing p. 136.)

BOOK REFERENCES. *F.B.I.*, IV, 450; *Nidification*, III, 532; *Handbook*, 346.

FIELD IDENTIFICATION. (2+) A minute hornless owl little bigger than a sparrow, but of course more bulky. The upper plumage and flanks are fairly closely barred in brown and rufous, the head appearing to be spotted with white. The underparts are white except for a narrow dark band separating the white chin and throat-patch from the breast. Its note is a ventriloquistic chirp or whistle reiterated for a considerable period rather after the manner of the well-known Coppersmith of the Plains.

DISTRIBUTION. Records show this tiny owl to be widely distributed from the lowest levels up to at least 9,000 ft. It is, in fact, comparatively common around the Vale and up all the side valleys, but being a quiet little bird and highly nocturnal in its habits, it is apt to escape notice. We have met with it at Domel (2,200 ft), in the Surphrar Nullah off the Sind Valley, at Ahlan, at Karabudurun (8,400 ft) in the Marbal Glen, and in the Ferozpur Nullah. Other records are from the Mahadeo ridge at 9,000 ft and from the Lolab Valley, so that our area is well covered with the exception of the Kishenganga and Wardwan Valleys where, however, there is no reason why it should not also occur.

This species is found throughout the Oriental Region from the Himalayas to China and Formosa, our race being found as far east as Western Nepal.

HABITS AND NESTING. Stuart Baker writes in the *Nidification* that this owl breeds in the North-West Himalayas in holes in trees standing either in forest or in well-wooded country. Our own experiences bear this out, for our first meeting with the species was in the dak bungalow compound at Domel, which may be described as well-wooded, but certainly not as forest. Hearing a lot of subdued squeaking, we flashed a torch into the trees near the fence overlooking the river. The beam fell upon a Pigmy Owlet seated on a horizontal branch about 20 ft up. It continued to utter, taking no notice whatsoever of the light, its behaviour being strongly reminiscent of the Crimson-breasted Barbet for it continually turned its head from side to side bobbing slightly with every squeak. The squeaks too, which we noted down at the time as being somewhere midway between the high-pitched squeak of a bat and that of a tree-rat, were spaced more or less evenly over a considerable period and not, as Osmaston stated of those he heard below Gulmarg, in series of four clear bell-like whistles. Our next encounter was opposite Ferozpur village where we obtained a nest in an old woodpecker boring about 12 ft from the ground. It was in the main trunk of a walnut tree just on the outskirts of the mixed forest surrounding a small meadow on the eastern bank of the Ferozpur stream. As late as 9 a.m. the owner of this nest seemed quite active. It seized a cicada from the ground close to our tent and proceeded to devour it on a nearby branch of a long-leafed pine, conveying it to its mouth in its foot in the same way that a parrot does. It then flew off to the nest, whence it gazed at us with its quaint face framed in the hole, looking for all the world like a little old man wearing Dundreary whiskers. At Karabudurun a family party used to spend its evenings in the thick branches of a fir under which our bearer had pitched his tent. When they first started up we thought he had a couple of young chickens in his tent. Their squeaks were certainly more melodious than our Domel birds', but being in chorus we could not say whether or not they had a four-noted call like Osmaston's birds. It is quite probable, as is the case, we find, with numbers of birds, that the voices of individuals vary to some considerable extent both in tone and cadence.

They seem to be rather partial to annexing the discarded borings of Pied Woodpeckers, and as those birds are so common in Kashmir such will probably turn out to be the sites chosen in nine cases out of ten. They do, however, when necessary, use natural hollows in trees. The main month for laying is May, but eggs have been taken both slightly earlier and later. We did not open up the nest in the Ferozpur Nullah, but from the behaviour of the owners it probably contained eggs, certainly not young ones. When we tried to look into the hole the bird within hissed and snapped its bill. The date, by the way, was 5 June. The eggs are typical, almost spherical and pure white : they average 29·7 by 24·1 mm. The number of eggs laid is usually 3 or 4, the limits being 2 to 5.

VULTURES

HIMALAYAN GRIFFON VULTURE

Gyps himalayensis Hume
(See Plate 52 facing p. 205.)

KASHMIRI NAME. *Grad.*
BOOK REFERENCES. *F.B.I.*, V, 13; *Nidification*, IV, 9; *Handbook*, 352.

FIELD IDENTIFICATION. (6+) A huge khaki vulture with short square tail and broad-ended wings whose turned-up primaries are in flight separated like the fingers of a hand. With a wingspread almost as large as the Lämmergeier's, it is easily distinguished from it even in the far distance by the above characters. Usually to be seen sweeping along hillsides on motionless wings or gathered in numbers on the ground around the scattered debris of a carcase.

DISTRIBUTION. This is the common vulture of the Western Himalayas and is to all intents and purposes the sole bird of its type to be found everywhere in our area from the lowest to the highest elevations. It is not often seen in the central plain of the Vale where, around the refuse heaps just below Srinagar, a few Long-billed Vultures and an occasional White-backed Vulture are usually to be found.

This fine vulture, a bird of the mountain ranges, is represented beyond the Himalayas from Turkestan and Afghanistan to Tibet. Bhutan is its eastern limit in the Himalayas.

HABITS AND NESTING. This grand bird is common, for grand it is apart from its repulsive habit of feeding exclusively upon carrion. Its powers of flight are remarkable and render it an object of wonder, since quite often from the time it comes into view a mile or so up the valley until it passes out of sight its wide-spreading wings remain outstretched and unmoving as if the bird were propelled by some invisible force. It instinctively makes the most of the air currents, and a vulture which one moment is a thousand feet below one, may in a short time be circling high overhead without having resorted to a single wing-beat. When it does require an expenditure of energy, the beats are slow and powerful.

The days are spent in searching the ground for dead or dying animals, and as these are mainly to be found during the summer months in the side valleys and on the upland pastures whither the Gujars drive their flocks of buffaloes and sheep, it is only to be expected that here too, rather than in the comparatively deserted main Valley, will these birds be seen sweeping along some grassy hillside or circling high above an open marg. It is seldom that

PLATE 63

Mekran Red-wattled Lapwing on nest

Nest and eggs of Mekran Red-wattled Lapwing

PLATE 64

more than one or two are in view at the same time, yet within a few minutes of the first bird's arrival a struggling, hissing, jibbering mass of 20 to 30 birds will be struggling for places at a carcase, eager to gorge themselves on the torn flesh of the departed beast. The birds obviously gather from a wide area, the first-comers dropping steeply from the skies immediately overhead with a loud windy noise of the air through the half-bent pinions. Later arrivals glide in at ever-decreasing angles till the last to appear come from still farther afield on steadily beating wings and probably have to remain perched on nearby trees or boulders owing to lack of room around the corpse.

We have never had the good fortune to inspect nests of these birds, for they choose inaccessible cliffs and precipices on the ledges of which they nest in company, though the nests are well separated and probably do not number more than 4 or 5 in the one locality, the lime-streaked faces of the rocks disclosing their resting-places. Osmaston succeeded in obtaining two nests with the aid of ropes. Even in the higher hills they prefer to nest in the cold-weather months, for these he took on 11 and 15 February at an elevation of 6,500 ft when snow still lay upon the ground. Ward, however, took nests in March and April. Each of Osmaston's nests contained a single unspotted white egg, the two measuring 96·0 by 69·0 and 88·4 by 69·7 mm. respectively. In tone the egg is rather a dull off-white and many are marked to a greater or lesser degree, a few heavily so, with streaks and blotches of reddish or pale brown.

As regards the character of the nest and the limits of breeding, we cannot do better than quote from Hume's and Oates's *Nests and Eggs of Indian Birds*. 'The nest, where there is one, is ordinarily a huge platform of sticks (at times the property, in past years, of some Eagle or Falcon, which the early nesting Vultures have seized upon long before the rightful owners have even begun to think of their annual matrimonial duties), placed, I believe invariably, on a rocky ledge of some bold precipice in the Himalayas, at least 3,000 ft above the sea. At times the whole nest consists of a few twigs or roots, or a little grass, and occasionally the egg reposes on the bare ground. I have never yet heard of their nesting on trees. Though generally gregarious in their breeding habits, large numbers rarely appear to breed together. Six is the greatest number of nests I have yet known of in one single locality.' As regards the time when eggs may be found, it is stated that they lay ' during the last week in December, the whole of January and February, and the first week of March. As a rule, however, by the end of February the nests contain young, huge, goslinglike creatures, thickly clad in long dingy yellow down '.

VULTURES
NEOPHRON, OR LARGE WHITE SCAVENGER VULTURE

Neophron percnopterus percnopterus (Linnæus)
(See Plate 53 facing p. 212.)

KASHMIRI NAME. *Patyāl*.
BOOK REFERENCES. *F.B.I.*, V, 22 ; *Nidification*, IV, 17 ; *Handbook*, 356.

FIELD IDENTIFICATION. (6) The plumage is largely white broken down by a certain amount of brown and black in the wings, but equally distinctive features are the naked yellow face and yellow bill. A loose ruff of spiky feathers round the neck gives it an untidy appearance. In flight it has much of the shape of the Lämmergeier, having the same down-drooping pointed wings and a wedge-shaped tail, but the smaller size and the great amount of white in the plumage readily distinguish it even at a distance.

DISTRIBUTION. Widely distributed, but not common, throughout our whole area. We have seen it at Domel, Garhi, Uri ; throughout the Vale and in the side valleys, including the Wardwan Valley, and about their margs up to 13,000 ft and more. Probably resident except at the higher elevations.

This, the typical race of the Egyptian Vulture, is found from South Europe and Africa to Western Asia, reaching Northern India from Baluchistan and Sind, through the North-West Frontier Province and the upper Punjab to the Western Himalayas. A smaller race covers the rest of India.

HABITS AND NESTING. Whistler in his inimitable *Handbook* calls this bird by its clumsy but less offensive pseudonym of Neophron, a name which befits the Kashmir bird rather than the more obvious but loathsome term of Scavenger Vulture. This latter name certainly describes the bedraggled bird seen strutting about Indian bazaars and the outskirts of insanitary villages, but in Kashmir the Neophron is more a bird of the open valleys and margs, of cliff faces and the gorges of the Jhelum and Kishenganga Valleys. We have never seen this bird within any Kashmir town, though it occurs on the Takht-i-Suleman ; nor have we yet remarked a nest of sticks and filthy rags on a temple roof or any other building. The Kashmir bird may have, and probably has, the same preference for a diet of offal and excreta as its Plains' relatives, since no one could ever deny the unsanitary conditions of the surroundings of any Kashmir habitation, but these unsavoury habits are not so much in evidence, for we have met with the Neophron in the wildest uplands, sailing steadily along with down-drooping wings and wedge-shaped tail like a smaller edition of the Lämmergeier, where none but a few hardy Gujars and *bakribans* with their herds of sheep, buffaloes, or goats were to be met with. Granted, the birds may live many miles from where seen, for their powers of flight are considerable. The nests we have been able to trace have, without exception, been placed on ledges of cliffs, sites worthy of eagles moving in the very highest circles ; but from what they have been built, with the exception of stout sticks which we have seen the birds carrying, we cannot say. The usual nest of this bird consists of a good foundation of sticks, with an admixture to form an untidy lining of leaves, wool, old rags and other rather dirty oddments. The same nests are obviously used over a period of years. There is one, inaccessibly placed we might add, under an overhanging cliff which overlooks the racing waters of the Jhelum near Uri. Three years running in the month of June we observed a bird seated upon it, while the fourth year, when passing in the car in May, we saw one of the birds land with a large stick. Two others, similarly placed on ledges of cliffs, can be observed between Ramban and Banihal, strictly speaking outside our area.

The egg—often only one is laid, though the full clutch is 2—is a large blunt oval with a thick pitted shell. The ground colour is dull white and this is closely stippled and blotched with red-brown. The markings are sometimes bold and well dispersed ; at times they practically cover up the ground colour. They average 66·4 by 51·8 mm. Osmaston states that fresh eggs are obtainable about the middle of April, but to this we consider May should be added.

HIMALAYAN BEARDED VULTURE, OR LÄMMERGEIER

Gypaëtus barbatus hemachalanus Hutton

(See Plate 52 facing p. 205 and Plate 53 facing p. 212.)

KASHMIRI NAME. *Patyāl*.
BOOK REFERENCES. *F.B.I.*, V, 26 ; *Nidification*, IV, 20 ; *Handbook*, 358.
FIELD IDENTIFICATION. (6+) The Bearded Vulture is readily distinguished in flight by its huge size, pointed wings and wedge-shaped tail. When sufficiently close the black, hairlike feathers forming the ' beard ' may be distinguished drooping over

VULTURES

either side of the bill. In old birds the head becomes nearly white. When by chance this wonderful exponent of soaring flight is seen at rest, additional points to note are its dark streaky appearance, again its great bulk, and the heavily feathered thighs.

DISTRIBUTION. Widely distributed throughout the wild uplands of Kashmir mainly above the tree line, but with its unique powers of flight it covers great distances and does not disdain to drop down to lower levels, so that it may be seen in the Jhelum Valley as well as over the more open spaces. We have seen it not infrequently from the motor road between Domel and Baramullah but seldom over the Vale, though its visits are more frequent in the winter, particularly in the Wular Lake area. A resident species except perhaps at the uppermost limits of its range, although in winter it may be seen amongst the lower hills almost down to the Plains, and it has, in fact, occasionally been noted as far afield and as low as Rawalpindi.

The lämmergeiers, of which there are a number of races, have a wide range in the mountains of Southern Europe, Africa, and Central Asia, including the hills bordering North-West India from Baluchistan to the Himalayas, in which it is represented as far east as Bhutan. It also perambulates the Salt range.

HABITS AND NESTING. The Bearded Vulture, or Lämmergeier, undoubtedly comes into its own in the open valleys and margs, the vast sweeps of grass-covered hillsides and the screes tumbling down from cliffs and snowy ridges above the tree line, where the eye can range for mile upon mile to a distance only limited by our poor sight or the windings of the precipitous walls of the valley cutting off the view. Here the Lämmergeier is a common bird, where it may often be seen quartering the hillsides, following the undulations of the valley bottom or the curves and projections of massive cliffs in ceaseless quest for the scattered bones of some defunct animal which has already lost its flesh to the almost equally huge Griffon Vultures. From afar the downward trend of the great bow-shaped wings with their upturned pointed ends marks them out; nearer at hand the long wedge-shaped tail and the black whiskers on either side of the cruel bill are equally distinctive features. Few such marvellous exponents of gliding flight exist, as for mile after mile the wings remain motionless, whether the flight be in a direct line, or the bird be rising in great spirals taking advantage of the rising air-currents. When the wings are required for more active work, the beats are slow and powerful. As a matter of interest, this bird was seen by a member of the Mount Everest expedition soaring at 24,000 ft.

The Lämmergeier does not stick only to the uplands, but may be seen anywhere in Kashmir, commonly down to about 6,000 ft and more sparingly down to Plains' level, be it ranging over the heavily afforested hillsides in such valleys as the Lidar about Pahlgam, or in more open wooded country such as the lower Sind. During a recent winter we were stationed where Lämmergeiers were comparatively common, and it was borne in upon us that birds of prey, like beasts of prey, may be creatures of habit, having their favourite beats which they perambulate unfailingly at fixed intervals. We came to look out daily for an old Lämmergeier by whom we could almost set the clock, for soon after midday he would sail across to the Fort from a high hill a couple of miles away, swing round the southern wall and so drift up the nullah past our quarters.

As already hinted at, their food consists almost exclusively of bones. They seldom if ever interfere with the vultures' repast, but wait until the flesh has been removed when they step in, as it were, to finish the task. Bones too big to swallow whole are often carried to a height and dropped so that they break into smaller fragments. The name Lämmergeier—Lamb Vulture—is a misnomer, as it does not attack living animals of any kind. In fact, in spite of its formidable appearance it is rather cowardly, giving way to any other vulture or bird of prey which wishes to dispossess it of a bone or even of its chosen nesting site. On the other hand, when sweeping along a hillside more often than not it quite disregards one's presence, sailing past so close that one can hear the air humming through its feathers.

The nest is an untidy affair of varying proportions of sticks, rags, wood, and dead leaves, usually placed on an inaccessible ledge of an overhanging cliff. The exception is said to prove the rule, and in one case Gujars actually caught the female and her nearly fledged chick at the nest and brought them down to an ardent egg-collector at the Atawat Rest House, while Dodsworth reports a nest where ' a wide ledge led from the top of the hill down some 40 or 50 feet, the grade being so gentle and the ledge so firm and wide that we could saunter down to the nest at the far end '. Recently, Captain Livesey informed us of an even more curious site near a cottage he was occupying in the neighbourhood of Almora, the nest being on top of a large bush not 12 ft from the ground. Mr A. E. Jones, who has examined many nests, informs us that the Lämmergeier usually has two or three alternative nesting sites; that for no apparent reason it often abandons one in favour of the others and yet will return to it in a subsequent year.

HOBBY

Up to 3 eggs are laid, but though 2 is the usual number and one only is often deposited, in any event it seems that never more than one young one is successfully reared. The eggs are variable in coloration, the ground colour ranging from a pale creamy yellow to a reddish-buff. At times this ground is almost obscured by close red stippling; at others the markings are much bolder but still cover the greater part of the egg. Their measurements average 85·0 by 67·4 mm.

The breeding season is early, about December to February or March, so that eggs are not likely to be found by the summer visitor to Kashmir. The young, however, remain for a prolonged period in the nest, not leaving it as a rule until late May or early June. During this period their food consists either of bones or carrion, e.g. dead rats or a crow. This is fed to them by both birds either by regurgitation or by being carried to the nest in the feet or in the bill. When the chick is small the food is torn up and fed to it piecemeal.

CENTRAL ASIAN HOBBY

Falco subbuteo centralasiæ Buturlin
(See Coloured Plate V facing p. 328.)

BOOK REFERENCES. *F.B.I.*, V, 43; *Nidification*, IV, 30.

FIELD IDENTIFICATION. (4—) The upper plumage is slate-grey, darker on the head. The white throat and sides of the neck throw up the dark moustachial streaks in bold relief, while the lack of the black tail-band easily differentiates this bird from the Kestrel. As opposed to the Kestrel's hovering and the Sparrow-Hawk's furtive hunting methods, the Hobby hawks flying insects and bats, twisting, wheeling and diving in dexterous pursuit, often high overhead.

DISTRIBUTION. This engaging little falcon is not quite so numerous as the Kestrel, but is found in all the better-wooded parts of the Vale as well as in and around the forests and cliffs of the side valleys up to the limits of the trees. Outside our area its breeding range takes it west to the Kurram Valley and east possibly as far as Assam. It also nests in Ladakh, Tibet and Central Asia. Other races increase the range from the British Isles to Japan and north-west Africa to South China. The Hobby nesting in peninsular India, Burma and Malaya to the Philippines belongs to a smaller species, *F. severus*.

HABITS AND NESTING. As we have remarked above, whereas the Kestrel wheels and hovers on vibrating wings over the open spaces and cultivation, and the Sparrow-Hawk flies silently low through the trees or close along the contours of a cliff face, the Hobby has a third method of hunting for it feeds largely on the wing. Nothing in the shape of the larger flying insects, such as locusts, beetles, and dragonflies, comes amiss. In the evenings, when the bats unfold their leathery wings, the Hobbies will be at work among them. The occasional small bird flying in the open is also deftly taken, though to have depicted the bird in the plate disembodying some large insect, rather than a bird, would have conveyed a more accurate impression of the Hobby's diet. In stormy weather the Hobbies seem to exercise their fine powers of flight in what can only be described as wild flying. Sweeping and wheeling, now close to the tree-tops, now many hundreds of feet in the air, they scream the while as if with excitement. Perhaps the buffeting of the wind elates their exuberant spirits, or is it that their winged prey is scattered and blown farther afield? Whatever the reason, it provides a thrilling display.

They can on the whole be described as forest hawks, but they eschew the thick pine woods without blanks and open patches, while they are common in country with plenty of scattered trees. A pair was noted in early July in the high trees of the Chenar Bagh at Srinagar, and within a mile of the Forest Rest House at Gogaldara, on the lower slopes of the Pir Panjal mountains, no fewer than three pairs were already on their nesting sites in June. They are, however, later breeders than the other falcons, and though on the 20th of the month one pair was chasing away every intruder from a discarded Jungle Crow's nest about 40 ft from the ground, on investigation it proved to be still empty. The other two nests were inaccessible without climbing-irons, being at the summits of colossal more or less isolated pines with no lower branches. A nest taken by Osmaston near Pahlgam was half way up a spruce fir on an island in the river and contained 3 fresh eggs on 28 June. He noticed another at Gulmarg in July, this time in a Silver Fir. Late June and July mark, in fact, the period within which they normally lay, and the site chosen is more often than not a Jungle Crow's old nest in a tall pine or fir. They are to be seen hanging around the faces of rocks and precipices, but we can trace no Kashmir records of cliff nests.

The eggs are like those of the Kestrel, but are generally duller and browner in tone and on the average a trifle larger, measuring 41·4 by 33·0 mm.

KESTREL

Falco tinnunculus Linnæus
(See Coloured Plate V facing p. 328 and Plate 79 facing p. 316.)

KASHMIRI NAME. *Bōher*.
BOOK REFERENCES. *F.B.I.*, V, 61, and see also under *interstinctus* on page 62 ; *Nidification*, IV, 41, under *interstinctus* ; *Handbook*, 385.
FIELD IDENTIFICATION. (4—) The Kestrel is the most easily recognized of Kashmir's falcons. Firstly the broad black terminal band in the rounded tail, which is often spread wide in flight, is distinctive ; but much more so is this bird's habit of facing up to the wind over some open space where it hangs motionless in the air on vibrating wings while it scans the ground below. The female lacks the blue-grey head of the male, while her back of rufous-brown is barred with black.
DISTRIBUTION. This bird breeds in Kashmir certainly as low as 3,000 ft. At this elevation a pair was noted occupying a site on a small cliff near Dhani in the Kishenganga Valley. They were seen frequently over all the more open parts of that valley through to Gurais. In fact the Kestrel is a widespread bird throughout our whole area from the lowest elevations to the margs and cliffs around at least 11,000 ft. The bird residing in Kashmir, and the north-west Himalayas generally and on to North Baluchistan, is probably the typical race of Europe and Northern Asia. In the winter there is an influx of Kestrels into the Plains of India, but birds are to be seen in the Vale of Kashmir throughout the year. It cannot definitely be stated, however, that Kashmir's Kestrels are resident as their places may be taken by winter immigrants from more northerly latitudes.

The Kestrel is represented by a number of races in Europe, Africa, and all Asia. Of these, besides the typical race breeding in the area given above and wintering throughout India, a darker race nests in the hills of peninsular India and Ceylon, while yet a third form, the heavily-barred race of China and Japan, comes in from the north-east in limited numbers in the cold-weather months.
HABITS AND NESTING. The Kestrel is found in Kashmir well within the forest's limits provided there are adjacent margs and open spaces over which it may range in search of the mice and insects upon which it preys. For the Kestrel is a beneficial species, paying attention mainly to the creeping things of the earth such as small rodents, lizards, grasshoppers and beetles, while its depredations amongst birds are confined to an occasional descent upon some luckless fledgling. Sailing out over the open spaces of the valley

PLATE 65

Nest and eggs of Common Sandpiper

Common Sandpiper on nest

PLATE 66

Newly-hatched Woodcock chicks in nest

Woodcock on nest

from its haven in a cliff face or amongst the tall pines on the flanking slopes, it steadies itself in mid-air and hangs on the wind, the wings vibrating rapidly when the air-currents are insufficient to hold it stationary. From a height of a hundred feet, or even treble that height on occasion, it searches its territory, sometimes circling to a lower level where it again hovers over the suspicious object. Having chosen its quarry it descends straight upon it from above. At times it will sit upon a rock or the summit of a tree and glide down to seize a beetle or other morsel rather after the manner of the Rufous-backed Shrike.

The Kestrel undoubtedly has a great preference for cliffs, upon the ledges of which a large proportion of its nests will be found, but where suitable precipices are unavailable it is not averse to nesting in trees. On such occasions it does not build its own nest but occupies the discarded structure of a Jungle Crow or other bird whose nest is sufficiently large. We have noted tree nests at Aru, Gogaldara on the Pir Panjal mountains, and in the Ferozpur Nullah not far from Tangmarg, while Osmaston speaks of two pairs nesting in tall trees at Gulmarg. Theobald records finding a nest in the Vale, which evidently held six eggs, in a hole in a serai wall near Shahabad. On the other hand few rocky cliffs are without their pair of Kestrels who scream to each other from the chosen cleft or ledge. For Kestrels, like so many hawks, are noisy and pugnacious birds when nesting, often keeping up a series of short sharp screams, *ki-ki-ki*, or a longer drawn-out *ki-kir*.

The cliff nests vary greatly in size, sometimes the eggs resting on nothing but a few twigs, bits of grass, and other rubbish, but again it may be a much more pretentious affair of some other bird, of stout sticks with a rough attempt at a lining of bits of wool, rags and other flotsam mainly round its rim. The eggs, usually fairly pointed, have no gloss. The ground colour is a shade of dull red, but it is almost covered up with small blotches, streaks, and smears of a darker shade of red-brown. The average size is $39 \cdot 3$ by $31 \cdot 6$ mm. The full clutch is said to range from 3 to 6, but we have seldom seen more than 4. The nest illustrated contained on 20 June two well-incubated Kestrel's eggs, the third light-coloured one being that of a Sparrow-Hawk which probably tried to seize the nest, but unsuccessfully as the Kestrels were still in possession. The eggs of the rightful owner were deposited on 3 and 5 June. Many eggs are laid much earlier than this. The Dhani nest was found on 22 April but rock plants prevented its contents from being seen. There is, however, a record of eggs being taken on 28 April. At the other extreme Ward took a nest of very hard-set eggs on 15 July.

EAGLES

BOOTED EAGLE

Hieraëtus pennatus Gmelin
(See Frontispiece.)

BOOK REFERENCES. *F.B.I.*, V, 79; *Nidification*, IV, 56.

FIELD IDENTIFICATION. (5) This small eagle is not unlike the Black-eared Kite but may be readily distinguished from that bird by its rounded tail. It is not common.

DISTRIBUTION. There is little information about the distribution in Kashmir of this widespread species. Osmaston says it is not uncommon and adds that at least two pairs breed in tall chenars in Srinagar. Davidson says he noticed it once or twice at Kangan in the Sind Valley and saw a pair on the Takht in June. Ward merely remarks 'the Booted Eagle was obtained in Kashmir in March', but he told Stuart Baker that his collectors also took a nest with eggs, he believes, near Ganderbal. A single bird paid us a visit at Astanmarg (11,500 ft) in July, alighting in the one and only tree close to our camp. It is certainly more numerous than would appear from the above, and Scully, writing in *Stray Feathers* on the birds of Gilgit, which is somewhat beyond our limits of course, states that it is a common summer visitor there from the middle of March to October and that it breeds at an elevation of 5,000 ft.

Stuart Baker, in the *Nidification*, gives the full distribution of this bird as extending over the greater part of Southern Europe, North Africa, and Western Asia to North-West India, though as a winter visitor the range includes Burma and even the Malay Peninsula.

HABITS AND NESTING. Despite the widespread nature of its range, Stuart Baker mentions only three Indian nests in addition to Osmaston's records; one from Salem, one from near Kohat, and the third from Murree. One of Osmaston's Srinagar nests contained a single fresh egg on 15 May. Scully procured a nestling captured on 12 July in the Gilgit district. The above gives the impression that the bird is indeed rare. This is probably not the case; some are undoubtedly overlooked. On 23 May 1938 at Sumbal we were shown a clutch of white eggs which at the time we did not recognize, but which probably belonged to this species. These had been taken on the way down from Srinagar and the collector said his shikari had got them from a nest near Srinagar. The bird's superficial resemblance to the kite is probably responsible for many of them escaping notice, through one not being on the lookout for them. Osmaston, in fact, is of the opinion that in most cases the Booted Eagle is the real culprit in the Vale where chicken-stealing is concerned, and not the much-maligned Black-eared Kite.

The nests so far taken have been built at the summits of very tall trees, one on a tree growing out of a cliff face. Except in one case where a lining of pine-needles is mentioned, the nest has been built exclusively of sticks and twigs with but a shallow depression for the eggs. Although Osmaston's egg taken in May was quite fresh, Rattray got two hard-set eggs at Murree on 20 March. The Indian-taken eggs have been white or white with a few pale flecks of reddish or grey, but in Europe some eggs are well marked or even blotched with reddish. Osmaston's egg measured 51·7 by 41·9 mm.

PALLAS' FISHING-EAGLE

Haliaëtus leucoryphus (Pallas)

(See Plate 54 facing p. 213.)

KASHMIRI NAMES. *Gāda Grad, Löz.*
BOOK REFERENCES. *F.B.I.*, V, 112; *Nidification*, IV, 79; *Handbook*, 367.
FIELD IDENTIFICATION. (6) A very large, dark brown eagle with lighter-coloured underparts and a conspicuous white band in the tail. The head and neck are a pale sandy colour. They are usually seen singly or in pairs about the rivers and jheels of the Vale, but they follow the larger side rivers for some distance into their valleys. The loud clanging call is most arresting, and may be heard even when the bird is circling as a mere speck in the sky.
DISTRIBUTION. The White-tailed, or Pallas', Fishing-Eagle is quite a feature of the main Kashmir Valley, being increasingly common in the lower half of the Vale as the Wular Lake is approached. It is restricted largely to the line of the Jhelum River and the big lakes below Shadipur, particularly the Wular Lake, around the western and southern boundaries of which it is almost numerous. We have seen it in the lower reaches of the Sind Valley as far up as Kangan. In fact with its fine powers of flight, it commonly makes excursions up all the side valleys for some miles, including the many small valleys at the head of the Vale.

This powerful eagle ranges from Southern Russia to Transbaikalia and south to Persia, Northern and Eastern India and much of Burma. It is common in the plains of the northern half of India down to about Bombay on the west and across to Bengal, where it is numerous in the Sundarbans. It keeps mainly to the larger rivers and jheels and stretches of fresh water, not patronizing the coasts to any extent. They are hardy birds equally at home in the grilling Plains or at high elevations in Ladakh and Tibet.

HABITS AND NESTING. This fine eagle draws attention to itself by its loud cries, which have been likened to the creakings of an unoiled cart-wheel. These cries are broadcast on the wing as the bird circles high in the air, but more often still from a conspicuous perch at the summit of some gaunt dead tree on the river bank or the margin of a jheel. When the water recedes as the season advances, they may be seen sitting on mudbanks and islands just as often as perching on trees. Although capable of catching really heavy fish, birds on the water, and even snakes, they also exist by piracy, stealing the catches of lesser birds of prey. Nor are they averse to feeding upon carrion thrown up on the river banks. On the other hand we once saw one at dusk stoop on a flying coot in a fashion creditable to any eagle. They have also a most disconcerting habit of robbing the sportsman, swooping down most unexpectedly to snatch up a fallen snipe or wounded duck, boldly risking a charge of shot which may sometimes be heard rattling off the stiff plumage with no other result than to cause a derisive shake of the tail.

The nest attains great size since the birds return to the same eyrie year after year. To quote from the *Nidification* : ' It is big even when first made, a new nest measuring anything up to 4 feet in diameter by a foot or more in depth but, as season succeeds season, it may be added to until it is nearly half as wide again and three or four times as deep. I have never seen any lining in the nest but sometimes it certainly adds grass, finer twigs, sticks and green leaves to the hollow in which the eggs lie. The sticks used for the body of the nest are often of considerable size, in some instances as much as 3 to 4 feet long and 2 or 3 inches thick, which must weigh several pounds.' As regards the lining we must add that practically every nest we have examined has been lined with grass or straw, this having been added to from time to time even after the hatching of the young. Rags, too, may be incorporated. This huge structure is generally placed on the summit of some giant chenar tree at no great distance from river or marsh. Betweenwhiles it may be appropriated by another bird of prey, but the intruders will quickly be turned out as soon as it is required by its rightful owners. On one occasion in Rajputana we noticed that a nest, which had only been taken over by a pair of vultures in October, was again in the eagles' possession in December.

The eggs are laid in the depths of winter or in early spring so that nidification is over before the arrival of the normal visitor to Kashmir. Up to 3 eggs, very occasionally 4, are laid, which are dull white in colour but which if held up to the light take on a dark greenish hue. They are quite smooth in texture but without gloss and they measure 69·7 by 55·1 mm. Both birds take part in incubation and in procuring and bringing food, mainly fish, to the nest, but the greater part of the task of feeding the young devolves upon the female. Incubation takes a month and the young are said to remain in the nest for about ten weeks. As many immature birds are to be seen in May, the majority of eggs must be laid in Kashmir about January, that is in the coldest month of the year when snow is often lying deep upon the ground. Meinertzhagen took a juvenile, the sole occupant, from a nest near Srinagar on 9 March. This bird was almost ready to fly, so on the above computation the egg must have been laid at the beginning of December.

BLACK-EARED KITE

Milvus migrans lineatus (Gray)

(See Frontispiece.)

KASHMIRI NAME. *Gānt*.
BOOK REFERENCES. *F.B.I.*, V, 124 ; *Nidification*, IV, 92 ; *Handbook*, 372.
FIELD IDENTIFICATION. (5) Wherever there are towns and villages, however humble, there too will the kite be seen either wheeling overhead or sweeping up refuse in its role of principal scavenger. The long wings with their whitish patches on the underside and the deeply forked tail, together with the peevish *cheeling* scream, mark out this bird. The more prominent white wing-patches differentiate it from the Pariah Kite of the Plains.
DISTRIBUTION. Common in and around the Vale and in the lower reaches of the side valleys. It nests in the Lidar as far up as Pahlgam and in the Sind certainly up to Gund. Just where in the Jhelum and Kishenganga Valleys the Black-eared Kite ousts *M. m. govinda* it is difficult to say, but we believe the Pariah Kite occurs in the lower reaches of both valleys. Only a good series

of skins will settle the point. This kite is resident in the Vale throughout the year. Although found mainly around the towns and villages, it will also be seen in the open country up to considerable elevations, but seldom if ever over the forests.

Races of the Black Kite cover a wide area of Europe, Africa, much of Asia and Australia, our race extending through the Western Himalayas to the higher ranges of Assam and Burma, and to North China, Japan, Mongolia and East Siberia.

HABITS AND NESTING. Kites are just as common in hill-stations as they are in Bombay and the Plains of India where the new arrival cannot help admiring their powers of flight. At the same time he notices, and perhaps dislikes, their peevish cries and positively loathes their unclean ways. The kite of Kashmir is a slightly larger bird and has well-defined whitish patches under the wings whereby to distinguish it from *Milvus migrans govinda* of the Plains. Its habits are the same, for it is an arrant scavenger, hanging around the towns and villages where it makes its living by sweeping up any offal it can find. It prefers to snatch up food in its talons without alighting and is amazingly dexterous and bold in so doing. We have seen a swooping kite in a narrow crowded street snatch the food from a man's hand as he left the shelter of the shop where he had purchased it, scattering the surplus in the roadway, and leaving the man gaping and empty-handed. Numbers, however, may be seen picking about garbage heaps and stumping around them with that deliberate gait of theirs, the body horizontal, the head held high. Carrion does not come amiss and one or two kites will generally be seen as hangers-on to the outskirts of a pack of loathsome vultures disposing of a carcase. Much chicken-stealing is laid at their door; indeed the cries of agitated womenfolk may often be heard driving them away from the newly-hatched chicks, but, according to Osmaston, the majority of chickens so taken are victims of the rapacity of the Booted Eagles. This may be so but the kites are by no means blameless. Perhaps we have painted their character in too dismal colours, for one must pay tribute, not only to their usefulness as scavengers, but also to their easy flight and the way they wheel tirelessly about the sky scanning the ground below. Sometimes a pair of birds will quarrel in mid-air over some disputed titbit and will come tumbling down, seemingly locked in mortal combat, squealing querulously, only to break away on the point of crashing, recovering their equilibrium with easy sweeps of the wide-spreading wings and much manœuvring of the long indented tail. The peevish cries of the kites are indulged in very freely and indeed mark their presence everywhere, whether on the wing, at the nest, or merely perched on some dead bough.

Giant chenar trees provide the Kites with lofty nesting sites and well over 90 per cent of the nests in Kashmir are built high up in these trees, many of them, judging by their size and considerable depth, eyries of some years' standing. A few nests may be seen in poplars. They are mostly inaccessible or nearly so, but one we reached in a village on the river bank near Sumbal was at least two feet in depth, massively built of sticks, and contained nooks and crannies into which noisy House Sparrows had stuffed their nests : the Kites evidently tolerated these interlopers.

The nest is made of dead sticks which are carried to the site either in the bill or the talons, and many are so large as to tax the bird's strength. It is no uncommon sight to see a building Kite circle its tree two or three times before gaining sufficient height to flop with its burden on to the half-finished structure. A Kite will incorporate all sorts of oddments in its nest, especially in the lining, such as tufts of sheep's wool, old rags, leaves, and a large feather or two, but there is no special lining in the real sense, the bed for the eggs being sticks somewhat finer than the rest.

The 2, or sometimes 3, eggs are very handsome. In shape they are broad ovals. The ground colour is a dull white on which are laid large and small blotches of red and red-brown. Sometimes these markings are very bold ; sometimes they are streaky and small, smeared over the surface. The average measurements are 57·9 by 45·3 mm. Breeding commences in March ; most clutches are complete by May, and only well-incubated eggs will be found by the end of that month. In fact on the 23rd of the month, besides the Sumbal nest which contained 2 eggs not far off hatching, another nest investigated contained large chicks.

LONG-LEGGED BUZZARD

Buteo rufinus rufinus (Cretzschmar)

(See Frontispiece.)

BOOK REFERENCES. *F.B.I.*, V, 137 ; *Nidification*, IV, 97 ; *Handbook*, 378.

FIELD IDENTIFICATION. (5) The buzzard is very variable in colour, some birds being a pale buffy-brown heavily marked with white ; the majority are much darker and, when circling monotonously overhead as is their common practice, not easily distinguishable from the Black-eared Kites. The rounded tail gives them away, and from below most birds show a crescent-shaped light patch at the bases of the outer flight-feathers. They possess a plaintive wailing or mewing cry quite different from the kite's shrill squeals.

DISTRIBUTION. Not uncommon from the level of the Vale to 13,000 ft or more and well distributed throughout our area. According to Whistler the Kashmir bird is the typical race which breeds from Greece, through Southern Russia and Asia Minor, to Western and Central Asia, including the mountain ranges which border the north-west corner of India. In winter it spreads over the northern half of the country, commencing to arrive in August and leaving the Plains again in February and March.

HABITS AND NESTING. Most accounts give the impression that this bird is seldom found in the breeding season below 9,000 ft. So far as Kashmir is concerned, this is not the case. In the Vale we have seen it circling over the Wular Lake, and a pair had a nest at the mouth of the Ferozpur Nullah at well under 7,000 ft. Davidson saw birds at Gund (6,800 ft) and Sonamarg in the Sind Valley, and to move but a little higher there was a nest in the forest by the Wurjwan Rest House opposite Inshan in the Wardwan Valley, elevation approximately 8,000 ft. At the higher levels nests were noted at 11,000 ft near Baltal, to which a bird was carrying sticks on 26 June ; 11,500 ft at Astanmarg ; and lastly, Osmaston records a nest found on 27 June above Gulmarg near the top of a big fir. All these nests were found in June, and of them, one had fully-fledged young which left the nest on 27 June ; Osmaston's nest, judging by the behaviour of the parents, was thought to contain young ; while our nest near Baltal was inaccessible. The bird was carrying sticks to it, but this does not imply that the eggs had not been laid. Indeed, it may have contained young ; the weather had been particularly bad and urgent repairs were probably in progress. These birds are, in fact, believed to nest from March to June.

Osmaston says they usually place the nest in precipitous rocky ground, while Whistler states that it is sited either on a tree or the ledge of a cliff. So far we have failed to find a cliff nest and all those mentioned above were stick structures placed high up in pines, the Astanmarg nest in a birch. Stuart Baker also writes, in the *Nidification*, that all the Kashmir nests of which he received notes were big structures of sticks built in trees, in some instances very high up. He adds that none of his correspondents made

PLATE 67

Fantail Snipe on nest

Ibis-bill chick in down

PLATE 68

Eastern Grey Heron at nest

mention of any lining. The eggs are similar to those of the kite, being a bluish-white with large red-brown blotches usually leaving considerable areas blank. They average 56·2 by 43·9 mm.

INDIAN SPARROW-HAWK

Accipiter nisus melanoschistos Hume

(See Coloured Plate V facing p. 328 and Plate 79 facing p. 316.)

KASHMIRI NAME. *Tsari Suh.*
BOOK REFERENCES. *F.B.I.*, V, 158; *Nidification*, IV, 110; *Handbook*, 380.

FIELD IDENTIFICATION. (4) This is the falcon to which so many of our better-known cuckoos bear such a striking resemblance. The upper parts are slate-grey, the breast lightly and closely barred with rust. With its rather long yellow legs, sharply-hooked bill, and piercing eye, not to mention its peevish screams, at close quarters it can at once be told from any cuckoo. Hunts by stealth and cunning, cruising rapidly and closely along the contours of a cliff face, wood or hedge, to pounce on any bird which it can take unawares.

DISTRIBUTION. We have found this hawk at the lower levels in both the Jhelum and Kishenganga Valleys; in the Vale, near the Wular Lake and in the Ferozpur Nullah; at higher elevations, near Pahlgam, Sonamarg, and in Gurais up to the topmost limits of the pine trees at about 11,000 ft. Though not particularly common it is, therefore, a widespread species.

This dark race of the well-known Sparrow-Hawk, so well represented in Europe, North Africa and Asia, breeds throughout the mountain ranges bordering India from North Baluchistan to the Himalayas and Eastern Assam. It also occurs in Tibet, Burma, Yunnan, and Szechwan. It is said to be resident, dropping to lower levels and adjacent plains under stress of weather, but it appears to have been found in many parts of Burma where its status is uncertain. This is also in all probability the form which nests in Eastern Afghanistan. The Central Asian race, *nisosimilis*, is found commonly throughout India and Burma in the winter.

HABITS AND NESTING. While staying in the Forest Rest House on the Bagtor marg, numbers of turtle-doves used to leave the shelter of the forest to feed about the newly-ploughed earth close to its rim. Time and again a Sparrow-Hawk, the possessor of a

SPARROW-HAWK

family higher up the slope, would glide swiftly by, craftily hugging the line of the outermost trees. Before the terrified doves could reach the safety of the tangled branches, an unsuspecting straggler would be seized on the ground, or, easily outpaced in a short chase, be ruthlessly struck down. Many heaps of feathers, often on and around a stump, testified to the numbers which had been slain, carefully plucked, and carried to the nest. Such is typical of the methods of this hawk which has truly gained its trivial name through its depredations amongst the smaller birds of the hedgerows. Magrath, in fact, summed up its character to perfection when he wrote of one he was watching : ' We eyed each other for a few seconds when he flew off following the line of the cliffs, and silently and swiftly gliding round each bend in the hope of surprising some unwary bird. This little Hawk might well be described as the panther of the bird kingdom, his flight is so silent and his manner of alighting in the middle of a tree or bush so stealthy.' It is found even in the thicker forests and likes the better-wooded areas with groves and patches of bushes amongst the cultivation to suit its hunting habits.

The Sparrow-Hawk breeds in the pine woods, choosing a lofty tree in preference to a ledge on a cliff, though such sites as the latter provide may at times be used. When they build their own home, which they often do, nearly always—according to Stuart Baker—using another bird's old nest as a foundation, the nest is a fairly large and loosely built structure of sticks with little in the way of a central depression. There is no lining, but as incubation proceeds, scraps of fluffy down are often found sticking to the twigs. At times the discarded nest of a crow or buzzard may be annexed and used without additions, and it appears possible from the following episode that attempts to seize occupied homesteads are not unknown. In June we found that a pair of Kestrels had laid two red-brown eggs in an old Jungle Crow's nest about 50 ft up a large pine at the mouth of the Ferozpur Nullah. Considering that the clutch might be incomplete, we revisited the nest on our return a fortnight later. Although the Kestrels were still in possession there were now three eggs, the new addition a dull white with large red blotches and clouds covering about half the shell. It proved to be fresh, whereas the Kestrel's eggs were well incubated and none too easy to blow. Before removing them we took the photograph reproduced on Plate 79.

This third egg was typical of that of the Sparrow-Hawk. The ground colour is generally a dull greyish or bluish-white, and the markings, of rich red-brown, are invariably bold smears and blotches with sometimes a few underlying clouds of purplish-brown. A considerable proportion of the ground colour is always visible, but there is no uniformity in the positioning of the markings; they may be well distributed but are just as likely to be collected about either end of the egg. The eggs average 39·1 by 32·6 mm. and 3 or 4 is probably the normal clutch, but Stuart Baker says clutches of 5 are common in the North-West Himalayas. Records of eggs taken in Kashmir are few and far between. Major Cock took a clutch of 3 eggs on 2 June from a nest in a pine tree at Sonamarg. We, also, have noted occupied nests in June whose contents we unfortunately failed to investigate. The Bagtor nest obviously had young by the fourth week in May, so the eggs must have been laid at the latest towards the end of April.

The birds are very noisy in the vicinity of the nest, and apt thereby to give away its position when incubation is in progress. In this connexion Major Cock wrote: 'While watching birds in a pine wood on a hillside near Sonamarg, Cashmire, I noticed some Hawk that now and again flew past the tops of the trees under which I was sitting: whenever this occurred I heard the shrill cry of another bird so I concluded a female was sitting somewhere near. After diligent search I found the nest three-fourths up a high pine.'

HUME'S BLUE ROCK-PIGEON

Columba livia neglecta Hume

(See Coloured Plate V facing p. 328.)

KASHMIRI NAME. *Wān Kotūr*.
BOOK REFERENCES. *F.B.I.*, V, 220; *Nidification*, IV, 148; *Handbook*, 393.
FIELD IDENTIFICATION. (4) The familiar Blue Rock-Pigeon is so common in its half-domesticated state throughout the villages and towns of India, especially in the vicinity of old wells, bridges and railway stations, that it hardly needs describing. The race found in Kashmir is a somewhat paler bird, having the rump a pale grey or even pure white. The general tone of the plumage is slate with dark bars in the wings and dark metallic reflections on the breast and hind neck.
DISTRIBUTION. Widely distributed throughout the Kashmir valleys from about 2,000 ft around Kohala and Domel up to

some 9,000 ft. Not uncommon in the Vale but on the whole it is not a village bird, preferring the lower rocky hills; in the side valleys, the gorges; and, along the rivers, banks and cliffs riddled with caves and deep ledges. As to its movements in winter there is little on record, but Ludlow informs us that he has seen numbers in winter at the mouth of the Sind Valley. It is probable, therefore, that there is a downward movement from the higher levels under stress of weather.

The Blue Rock-Pigeons are birds of wide distribution in Europe, Africa, Western and Central Asia down to India and Burma, with many races which grade imperceptibly the one into the other. Wild birds are much hybridized by contact with semi-domesticated varieties, which makes the determination of races and their distribution more difficult still. *Neglecta* is the form to which birds are assigned from North-East Persia, Turkestan and Afghanistan, North-West and North India, and Kashmir inclusive of Ladakh. Afghanistan birds are perhaps referable to the Persian race, *gaddi*.

HABITS AND NESTING. Nowhere in Kashmir have we seen this well-known pigeon in the large flocks in which the Plains' birds so often collect. For instance, we noted them along the entire length of the Kishenganga river up to Gurais, particularly in the rocky gorges such as that below Tithwal and again between Sharda and Reshna, but never more than a few pairs at a time. They have the usual habits of coming out in the mornings and evenings to feed on grain in the stubble and cultivation, moving about energetically in pairs or small flocks. The call of this pigeon is a wholly pleasant sound, a soft *kroo-kroo-karoo*.

The new nest is a loose collection of twigs and roots or straw woven together into a thin platform on a ledge of rock or in caves in cliff faces, more especially in the rocky vertical banks of the deeper gorges. The same sites are used for many years and this results in the older nests reaching considerable size and becoming much befouled, congealed masses of sticks, straw and feathers cemented together with droppings. A nest inspected at Thiun, Sind Valley, on 2 May had quite a pronounced egg-cavity. It was on a ledge varying from 2 to 4 ft wide on an enormous boulder on a steep hillside. Two eggs is the normal quota and these are laid from about May to August. They are pure white ovals, sometimes broad, sometimes elongated, and have a fair gloss. They measure about $38 \cdot 5$ by $28 \cdot 7$ mm.

NEPALESE SNOW-PIGEON

Columba leuconota leuconota Vigors
(See Coloured Plate V facing p. 328.)

KASHMIRI NAME. *Chut Kotūr.*
BOOK REFERENCES. *F.B.I.*, V, 224; *Nidification*, IV, 152; *Handbook*, 393.
FIELD IDENTIFICATION. (4) The Snow-Pigeon is coloured largely in white and soft greys to tone in with the snow-covered expanses and stony upland meadows among which it dwells. The plumage is extremely thick, rendering it a heavy rotund-looking pigeon which rises with considerable wing-clatter. It takes advantage of its obliterative colour pattern by remaining motionless on one's close approach, so that the sudden departure of the flock almost from under foot is at times the first intimation of its presence. Usually found in small flocks, pairs and odd birds being seldom met with.
DISTRIBUTION. A bird of the treeless uplands which, however, will descend the wider valleys to about 8,000 ft to feed in small flocks on open plough or waste land. Davidson, during his trip up the Sind Valley in 1896, noted flocks in early May as low down as Koolan at 7,000 ft. It is doubtful if even in the depths of winter it ever descends towards the Vale to a much lower elevation than that, though we should expect to see them at that season in the Kishenganga at such places as Sharda, Khel, and even Keran, and in the Wardwan at Inshan.

The typical race is a bird of high elevations from Chitral along the main Himalayan range to Sikkim. Another race carries the species through Tibet and extreme North Burma to Yunnan and Western China. Meinertzhagen calls it a bird of the transition zone, not extending much south of the great Himalayan watershed or far north into the desolate Tibetan Plateau.

HABITS AND NESTING. The Snow-Pigeon is to the Blue Rock what the Snow Leopard is to the panther of the Plains, a more softly and thickly befurred speciality built to withstand the bitter cold of an elevated habitat. But the simile may be carried further than that, for the soft greys and extensive whites of the Snow-Pigeon's plumage produce a pastel beauty which make the harder tones of the Blue Rock's garb appear almost crude and blatant. So it is with the pastel shades of the ounce's fur and the stronger colours of the panther, whose coat has to camouflage him against the deeper shadows and glaring sunlight of Continental India.

PIGEONS

As regards this question of camouflage, the Snow-Pigeon undoubtedly derives a full share of protection from his obliterative pattern, for it is extraordinary how inconspicuous a flock of Snow-Pigeons can be no matter over what type of ground they may be feeding. It does not require to be patchy with snow or broken up with stones; even on a vivid grassy slope the pattern appears scarcely less effective. That the birds themselves are aware of the fact is evident from the manner in which, rather than disclose their presence by immediate flight, they will often permit a close approach, stopping their search for grain and young shoots and squatting motionless on the bare ground, whence the whole flock will take to the air simultaneously with a considerable swishing and clattering of the wings, making off for all the world like great fluffy balls of sullied snow. Inhabiting as they do the open spaces and not entering the trees, a conspicuous plumage would render them altogether too easy a mark for the crouching ounce surveying his wide acres from some elevated view-point, and for the marauding eagle.

Even in the breeding season Snow-Pigeons will generally be met with in small flocks of up to a dozen or fifteen birds. Larger congregations are unusual, except that flocks may coalesce when feeding over the same ground. In the immediate neighbourhood of their breeding areas single birds and pairs may be met with. Seven birds shot by Meinertzhagen had the crops filled with green food consisting of crocus bulbs and other small roots, small hard seeds, and a large percentage of quartzite grit. The flight is swift, direct, and powerful, a flock sweeping along an open hillside being a truly lovely sight. The call seems to consist of an unmusical croak which Finn describes as being something in the nature of a hiccough.

We have said above that they do not enter the trees. This requires qualification, for, according to A. E. Osmaston, in winter in Garhwal they often sit for hours in some tree at the edge of the forest. No doubt this will be found true of Kashmir birds, as during the cold-weather months they undoubtedly descend the forest-flanked valleys to feeding grounds free of the snow.

Nesting is on a colonial basis, the nests being placed in caves, crevices, and on ledges of high cliffs and in rocky gorges, most often in inaccessible positions at elevations of from 9,000 ft upwards. A typical and definite nesting site is that provided by the steep cliffs close to Sonamarg village. Another, that deep and quite inaccessible narrow rocky gorge of the East Lidar River between Tanin and Burzilkot, elevation about 10,500 ft. The nests are like those of the Blue Rock-Pigeon, equally dirty and verminous and obviously used year after year. From the upbringing of successive generations, the sticks become much soiled and matted together with droppings and filth, straw and feathers, but they do not, it is said, attain such large dimensions. The only record giving a clue to the number of eggs laid appears to be Rattray's find of a small accessible colony on 10 June from which he took 2 eggs from one nest and 3 from another. We have no means of knowing whether 3 eggs should be considered an abnormal clutch. May and early June are probably the months in which eggs are deposited within our limits. Stuart Baker gives the average of forty eggs as 40·3 by 29·1 mm., adding that seven of them were exceptionally small, including Rattray's clutch of three.

HIMALAYAN RUFOUS TURTLE-DOVE

Streptopelia orientalis meena Sykes

(See Plate 55 facing p. 220.)

KASHMIRI NAMES. *Wān Kukil, Gūgi, Jangli Konkli, Kamīr, Kamru.*
BOOK REFERENCES. *F.B.I.*, V, 239; *Nidification*, IV, 160; *Handbook*, 394.
FIELD IDENTIFICATION. (4—) The bird in the photograph illustrates well the general characteristics of this dove. The feathers of the wing-coverts are blackish-brown, each one edged with rufous which produces the scalelike appearance of the upper plumage. The dark brown, almost black, tail is graduated and each feather boldly tipped with white. There are black and blue-grey markings in a band round the sides of the neck. The lower plumage is vinous-grey.
DISTRIBUTION. A common and widely-distributed bird throughout the well-wooded portions of Kashmir from about 4,000 ft in the Jhelum and Kishenganga valleys to 8,500 or even 9,000 ft. It is not to be found in the greater part of the Vale, where its place is taken by the next species, but it is met with the moment one reaches the Vale's rim. It is, however, to be seen in the thickly wooded portion of the deltalike country of the Sind River below Ganderbal, and we have more than once found the usual nest of two eggs there. From this area numbers sally into the fields bordering the Anchar Lake, and as many as half a dozen birds may be seen at a time

feeding in the stubble. In winter it migrates to the adjacent plains of the Punjab and North-West Frontier Province.

This species, closely allied to the turtle-doves of Europe, extends north and east from India, Ceylon and Burma to Turkestan and Siberia and Tibet to China and Japan. Our Himalayan bird breeds in South-West Siberia and Turkestan south to Afghanistan, the Sufed Koh and the Himalayas above 4,000 ft as far east as Western Nepal, where it meets the race *agricola*, a bird of somewhat lower elevations.

HABITS AND NESTING. The Himalayan Turtle-Doves are birds of the forests, but when not breeding they love to roost just within the outer fringes whence they can emerge in the early mornings and evenings to pick up their diet of seeds on the margs and newly ploughed fields. They progress easily on the ground, walking this way and that as they eagerly peck about in search of provender. They are strong fliers and provide sporting shots for a collector's gun and also a useful addition to the somewhat restricted fare on trek. At Bagtor in Gurais, when we were there in May, large numbers used to emerge on to the marg where ploughing was in progress, and it was noticeable that many fell victims to a murderous pair of Sparrow-Hawks which evidently had a ravenous family in the thick Silver Fir and spruce forest behind the Rest House. The cooing of this dove is a most pleasing sound, a mellow *kroo, kro-kro-kroo*.

When breeding they often choose rather thick parts of the forest, favouring especially close-growing spinneys of young firs, but nests may be found in the larger bushes and in the perrottias. The typical nest of sticks lined with roots is often well concealed amongst a mass of small twigs and branches, but at times it is conspicuously placed in bushes and small trees of various kinds. Though so obviously a dove's nest it often has a defined egg-cavity lined with a considerable quantity of roots and it may attain to some 10 inches across. Many nests, however, are skimpy affairs through which the eggs can be seen from below. Most of them will be found between 5 and 12 ft up, but there is one record of a nest at the base of a tree, and near Tsuidraman we found a nest on 22 May built actually on the ground, well concealed under a thick bush. The nest in the illustration, also in the Wardwan Valley, was built on a pile of fallen brushwood only about one foot off the ground.

Two eggs are laid, which are pure white glossy ovals measuring 32·3 by 24·0 mm. The nesting season is variously given by different authorities as May and June, June and July, and May to

PLATE 69

Indian Pond Heron sheltering young

Night Heron at nest

PLATE 70

Little Bittern on nest. Male on right

August. Doves have notoriously elastic breeding seasons, but April to July are probably the limits, though we ourselves have so far found no July nests. Our earliest date for a nest of two eggs, taken in the Chatragul Nullah, was 18 April.

INDIAN RING-DOVE

Streptopelia decaocto decaocto (Frivalszky)

KASHMIRI NAME. *Kukil*.
BOOK REFERENCES. *F.B.I.*, V, 248; *Nidification*, IV, 168; *Handbook*, 399.
FIELD IDENTIFICATION. (3+) A light grey bird with a semicircular black ring round the back of the neck. Flies up from the ground with considerable clattering and whinnying of the wings, showing a lot of white at the end of the tail.
DISTRIBUTION. Very common in the Vale, where in the summer it is the only dove to be seen, except below Ganderbal. Penetrates the side valleys for a short distance only. The Ring-Dove arrives from the Plains and lower foothills in April and leaves again in the autumn, a few stragglers lingering until November, chiefly at the eastern end of the Vale.

The two races of this species, found so commonly and so widely spread in India and Burma respectively, extend from the Balkan Peninsula throughout almost the whole of Asia to China, Japan, and down to the Indo-Chinese countries, the typical race being found westwards of Burma. Although found in Turkestan, Osmaston says that in Ladakh it is only seen sparingly on passage. Though it is a resident species over the greater part of its range, in the winter it tends to leave the higher hills and more inhospitable areas.

HABITS AND NESTING. This familiar dove, so common throughout the Plains, is just as tame and confiding in Kashmir as elsewhere, entering gardens and villages quite freely, where it will be seen feeding on fallen grass and other seeds. It walks rapidly here and there over the ground and only troubles to fly up noisily into some adjacent tree when one is too close to it. Its voice is that of other doves, but is trisyllabic, a quiet *cuck-uk-coo*. If disturbed suddenly, it often protests with a wheezy peevish *were* as it flies off to the cover of some bush or tree. It delights in showing off to its mate by ascending almost vertically into the air with much clapping of the wings, to plane down to earth again in a sharp curve, wheezing as it does so. It affects all types of open country well provided with large and small trees and hedges, but avoids forest.

The nest is the usual scant affair of thin sticks and roots loosely put together, seemingly without thought for the safety of the eggs, as it is often quite flat. It is surprising, however, what a lot of buffeting these flimsy nests will stand, and how the eggs remain on them in strong gales. We have, however, seen eggs tumbled out of the nest on the too-hurried departure of the sitting bird. The nest may be quite conspicuous, but more often it is fairly well hidden in some leafy tree. It is not as a rule placed very low down, seldom being within reach. At times it is 30 ft or more above ground. In Kashmir we have seen many nests in willow, mulberry or chenar trees, inside the village boundaries as well as far removed from them. In the chenars particularly the nests are frequently very well concealed amongst clusters of the great leaves where they are gathered about some horizontal limb.

The eggs are generally 2 in number, pure white glossy ovals measuring 30·2 by 23·4 mm. They may be found in any month from May to September, but chiefly in May and June.

KASHMIR KOKLAS

Ceriornis macrolophus biddulphi (Marshall)

(See Frontispiece.)

KASHMIRI NAME. *Plās*.

BOOK REFERENCES. *F.B.I.*, V, 312; *Nidification*, IV, 207; *Handbook*, 417.

FIELD IDENTIFICATION. (5—) In the Koklas the heavy black streaking on the silver-grey plumage of the back is diagnostic, as are also the elongate crest and long tufts growing from the sides of the crown and the bold white patches high up on the neck. The tail is of medium length and wedge-shaped. The female lacks the long crest and tufts and is more soberly coloured with buff or black streaks on most of her feathers. Both sexes are noticeably smaller than the Monal.

DISTRIBUTION. The Koklas is the commonest pheasant in Kashmir, being well spread out in the forest areas from some 6,000 ft up to the limits of the pines. The birds are perhaps most numerous above 7,000 ft, but do occur lower down. In the Kishenganga in early May we heard them crowing every morning of our stay at Sharda, which at river-level is barely 6,000 ft.

This Kashmir race of the Koklas is constricted in its range between the Common Koklas, the typical form found eastwards from Jammu, Kishtwar and on the outer ranges, and the more chestnut bird, *castanea*, immediately to the north-west. Stuart Baker says Gilgit birds and those from Ladakh north of the Indus are still *biddulphi*, but in Chitral and Afghanistan the bird is certainly *castanea*. We would not like to say offhand whether birds from the upper Wardwan belong to the Kashmir race or are typical *macrolophus*. Other races, whose distribution is imperfectly known, extend the range of the species to Yunnan and China.

HABITS AND NESTING. The Kashmir Koklas is a bird of the forests, starting in a zone somewhat lower than the Monal and being commoner and more evenly distributed than that bird. It is unnecessary to go far from Srinagar to come upon them, for L. S. White reported them in some numbers in the pine woods behind the Mahadeo ridge and even on the nearer slopes wherever there was sufficient cover. When Davidson went up the Sind Valley in 1896, he made the comment that around Gund in May it was common and to be heard crowing almost daily. This at once leads to the assumption that as with the other races the Kashmir Koklas breeds in May, and also June—Ward took a hardset clutch of six eggs on the 15th of July—for it is at nesting time that the male is so vociferous, calling particularly in the early mornings the loud *pok-pok-pokras* whence the species obviously gets its trivial name.

They are particularly fond of precipitous broken ground with good cover amongst the trees, either of rocks interspersed with scrub or of thicker bushes, and avoid the more open plain slopes and flats. They do, however, as indicated by Hume, affect ' some place in a gorge where a horizontal plateau is thrown out inside the gorge '. They afford excellent shooting, being very fast on the wing, for they have the habit of rising well above the cover and then, turning down the slope, they hurtle downhill on halfclosed wings. They are noisy birds and are said to respond with loud cries to any sudden clamour.

It is in the broken ground and thicker cover that the nest is placed under a rock, or more commonly in tangled undergrowth, or under a dense bush. It is seldom much of a nest, just a collection of the surrounding grasses, leaves, sticks and odds and ends with a scratched-out hollow for the eggs. These, up to 6 in number, possibly more, are usually a fairly rich buff in colour stippled all over with small specks of reddish-brown : occasionally they have larger blotches. The shell is stout and in shape they are nearly true ovals. They average 50·9 by 37·8 mm. This pheasant appears to be monogamous and both cock and hen remain with the brood.

IMPEYAN PHEASANT, OR MONAL

Lophophorus impejanus (Latham)

(See Frontispiece and Plate 55 facing p. 220.)

KASHMIRI NAMES. Male, *Sunāl, Suna Mūrg* ; Female, *Hām.*
BOOK REFERENCES. *F.B.I.*, V, 335; *Nidification*, IV, 217; *Handbook*, 418.
FIELD IDENTIFICATION. (5) Found on precipitous wooded slopes from 7,000 to 11,000 ft, particularly where the ground is much broken and grassy spaces occur. The white rump of the male is conspicuous in flight. The female is a typical, but short-tailed, 'hen pheasant', dark brown with broad conspicuous streaks of buff.
DISTRIBUTION. Is well distributed amongst the woods and crags throughout our area from about 7,000 ft, lower in the Kishenganga Valley, to at least 11,500 ft. It is not at all numerous except in the Wardwan and Kishenganga valleys where it is more often to be met with, while it is scarcest on the Pir Panjal range. It is a resident bird, dropping but little in the winter.

The Monal, with no races, is a purely Himalayan pheasant, that is if the Sufed Koh can strictly be included in the Himalayas, since it ranges from those mountains to Chitral and thence along the Himalayas to Bhutan. Sclater's Monal, a different species, is found farther east in Assam, South-East Tibet and extreme North Burma to Yunnan.
HABITS AND NESTING. This grand pheasant is perhaps the most beautiful game-bird in the Himalayas, most certainly in Kashmir. It is not so often seen, for it is shy, and haunts difficult terrain in the higher mixed and pine forests where precipitous nullahs with scattered bush and rocky declivities render it hard to move about, diverting the searcher to easier paths. With the exception of the odd bird which gets up with a whirr of chestnut wings and a flash of the broad white rump, the woods often seem

untenanted, but should the watcher care to ensconce himself in May and June in the dawn hours near some open grassy slope adjacent to the broken cover, he may have the chance of witnessing the courtship display. The brilliant metallic bronzes and greens of the upper plumage of the cock are offset by drooping chestnut wings and the wide fanned-out tail being jerked high over the back. As he struts about the dewy grass, the smaller dark brown and buff hen often shows little concern in his posturings. When disturbed the Monal hurls itself downhill emitting a volley of shrill whistles and often flies a considerable distance over the nullah before sweeping to rest.

The nest is generally on steep ground in the shelter of a grass-embedded rock, at the foot of a tree, or near a sheltering bush. It is often well concealed, as was the case of a nest in some precipitous rocks on the edge of a log-shoot near Pahlgam. This was placed in a grass-overgrown cleft between two stones overlooking a 50-ft drop with the twigs of a baby fir draped over it. Had it not been for this foliage we should have gained the hen's photograph, for she sat quite motionless while the camera was being erected and the grass pushed aside. It was only when an effort was made to remove the last obstructing twigs close to her face that she went off with a great commotion. As against this, an open nest was found on 26 June low down amongst the Blue Pines at 8,000 ft or so in the Wardwan Valley where the gradient eased off and undergrowth was particularly scanty. It was embedded in the pine-needles between the trunks of two trees and contained three eggs when our shikari unwittingly flushed the sitting bird. The dried remnants of two more eggs were later found in a depression some feet down the bank whither they had rolled some time previously. Unfortunately a dastardly Jungle Crow had been trailing the shikari, and on arrival upon the scene only twenty minutes later we found the eggs had been smashed and sucked. Another Wardwan Valley nest, found at an elevation of 10,000 ft on 24 May when it contained three eggs almost ready to hatch, was on an open hillside strewn with boulders, tumbled-down birches, and some scattered tree growth. Consisting of a few grasses in a hollow under a rock, it was imperfectly hidden by a couple of fallen trees. The nest illustrated was also ill-concealed ; the date of its discovery was 7 May and the six eggs were quite fresh. It will be noticed that here the eggs rest on a bed of leaves. There is often a small quantity of moss and other matter, but it may also consist of nothing but a hollow scratched out in the pine-needles or earth, containing only the debris which falls into it.

PARTRIDGES

In the past, the number of eggs laid by this pheasant has been variously estimated at from 2 to 9, and Stuart Baker in his second edition of the *Fauna* lists our Pahlgam nest as containing 8 eggs. True, it did, but the nest had been 'doctored' by an avaricious local, and we consider that it can be taken as certain that whenever more than 6 eggs are found in a nest, human agency has been at work; the feathers of a cock Monal are no doubt of considerable use to the makers of trout flies, and it is easier to keep an eye on them if the contents of two nests are put under the one hen!

The eggs are long ovals, pale stone or buffy-white to a rich reddish-buff in colour. They are evenly stippled with small reddish-brown spots, a few of which may coalesce into small blotches. The shell is stout and faintly glossy. They average about 63·5 by 44·9 mm. They are laid mainly in May and the beginning of June, early May being the time when the majority are deposited.

CHUKOR

Alectoris græca chukar (Griffith & Pidgeon)

(See Plate 56 facing p. 221.)

KASHMIRI NAMES. *Kāk, Kāku, Kākov.*
BOOK REFERENCES. *F.B.I.*, V, 402; *Nidification*, IV, 253; *Handbook*, 428.
FIELD IDENTIFICATION. (4) A plump ashen-grey partridge with a conspicuous black ring encircling the face and throat, and a double row of black, chestnut, and grey bars on the flanks. Bill and legs red. It, in fact, closely resembles its near relative of Europe, the Red-legged Partridge, or 'Frenchman', excepting that it lacks the fringe of black spots outside the necklace. Flies strongly in series of rapid noisy wing-beats followed by a lengthy glide, but prefers to get out of the way by running, usually uphill, rather than by taking to flight.
DISTRIBUTION. Particularly common on the barer rocky hillsides around the northern and western rims of the main Vale. We also found it plentiful on the scrub- and stone-covered northern flank of the lower Kishenganga. It extends up to considerable heights on the drier, more open hillsides over our whole area, to at least 10,000 ft, and is resident except perhaps in the highest parts of its range. Unwin writes: 'Pairing commences early

in March, when many birds return up hill for the season, and during summer, and even in October, Chukar may be seen up to over 9,000 ft elevation. Many, however, seem to breed near their usual winter haunts, and these birds form the stock which supplies sport in early September.' By *usual winter haunts*, he presumably refers to the 'Chukor ground' around the rim of the Vale.

The Red-legged or Greek Partridges are widespread in Europe and in Western and Central Asia. Two races occur commonly on the northern frontiers of India and also in the Salt range, the more easterly form extending along the Himalayas from our area to Nepal.

HABITS AND NESTING. The Chukor is a hardy bird which thrives on the barest and hottest of hillsides at low elevations, yet will also be found in varied types of ground, except in thick forest, right up to the open uplands at 11,000 or 12,000 ft, even where snow is still lying in extensive patches. The stony slopes around the Vale, where small boulders lie amongst tufts of coarse grass and low prickly or scrubby bushes, and where water is conspicuously hard to come by, are their most favoured haunts, but they will be found on the better-clothed hillsides as well, and within limits even in the forest blanks.

We had better explain what is meant by the last remark. The little Rampur-Rajpur Valley, perched some 2,000 ft above the Wular Lake, is entirely surrounded by forest. On these wooded hillsides there are a number of narrow strips forming clearings of sorts, often not more than 50 yards wide and varying in length from a hundred to two or three hundred yards. Although forest encloses them on all sides and they consist of a mixture of grass, boulders, fallen logs, bushes, pine and fir saplings and an occasional large tree, and are constantly grazed through by herds of cattle, we found that even the smallest held its Chukor population.

Chukors, like all partridges, run strongly and it is surprising how quickly their rather waddling gait will take them up a hill. They possess a varied repertoire of harsh cackling notes, which they utter freely. The nest is a mere scraping in some sort of a hollow, protected by the roots of a bush, a stone, a fallen log or even a coarse patch of grass, and is not as a rule easy to find, for the sitting bird is hard to flush. It is scantily lined with grass, a few dead leaves, or merely scraps of rubbish. At times there is a more substantial lining of grass or roots to which a few of the birds' own feathers are added as incubation proceeds.

Every authority we have consulted gives the number of eggs laid as normally from 8 to 12 or 14. Hume is stated to have found never more than 12, but in his day Kashmir was comparatively unworked. We have a feeling that the number 14 is a well-worn repetition of an early estimate, and as far as Kashmir is concerned, the word ' normal ' appears of doubtful use in describing the number of eggs to be found in Chukors' nests. Every number from 8 to 21 has been recorded, with the middle figures from about 11 to 15 cropping up perhaps more often than any others : yet, of four nests found by us in quick succession, one in the Erin Nullah and the other three a few days later at Rampur-Rajpur, the eggs numbered 17, 15, 15, and 19 respectively. It is often suggested that these large clutches are the products of two hens. Surely it is unlikely that birds in a purely feral state will habitually favour joint nests ? Unfortunately eggs from a second nest are not infrequently added by human agency the more easily to keep both clutches under observation, so that to distinguish genuine large clutches from those which have been tampered with is no easy matter.

The eggs are a pale stone colour to a light buff and are freckled more or less evenly over the entire surface with reddish-brown. They measure 42·1 by 30·7 mm. Breeding certainly commences early. We saw a covey of chicks at Keran at the end of April. Great numbers of eggs are to be found in May and June, while at the higher limits of their range they may lay on into early July.

INDIAN BLACK PARTRIDGE

Francolinus francolinus asiæ Bonaparte
(See Coloured Plate V facing p. 328.)

BOOK REFERENCES. *F.B.I.*, V, 408; *Nidification*, IV, 257; *Handbook*, 430.

FIELD IDENTIFICATION. (4) Confined in our area to the slopes along the Jhelum Valley road up to about Uri, where the harsh call *kuk-kiya, kiya-kuk*, may be heard emanating from the rough fields, and grass- and scrub-covered hillsides bordering the river. A little smaller than the Chukor, and the dark plumage at once differentiates it. The female lacks the white cheek-patch and the conspicuous chestnut collar, while her underparts are narrowly lined with black rather than being boldly white-spotted on a deep black ground. The artist has depicted a bird with an abnormally wide collar. The skin used as a model was produced for our

PLATE 71

Little Bittern's chick seizing parent's bill to induce regurgitation

Protective camouflage: Little Bittern with neck fully extended

PLATE 72

Mallard on nest

inspection and had this feature even more strongly developed than it has been painted.

DISTRIBUTION. We have noticed this bird only in our peregrinations along the motor road from Domel to the vicinity of Uri. We did not come across it in the lower Kishenganga where it might confidently be expected to occur.

There are a number of races of this favourite game-bird ranging from Asia Minor to Assam and Manipur, of which three are resident in parts of the northern half of India. The race under review occupies Northern India, exclusive of Baluchistan and Sind, to Western Nepal and Bihar, and southwards to a line running approximately through Gwalior and the Chilka Lake in Orissa. It is everywhere resident and occurs in the Himalayas in places up to 8,000 ft, not only in grasslands and scrub but even, as around Simla, within open forest well provided with undergrowth and blanks.

HABITS AND NESTING. The call of this partridge is too distinctive to be missed. Its very harshness makes it difficult to believe that it belongs to this beautifully patterned bird. At close quarters a high-pitched *chuck* can be heard preceding the grating notes which in the Punjab are popularly transcribed into ' Subhan, teri kudrat (O Omnipotent, thy power) '. We must confess that it is solely on the strength of this call that we admit the Black Partridge to our list of breeding birds, for we have no nesting records to fall back upon. During their breeding season, which in the Plains extends from April to October, though most eggs are found in June, the male delights in mounting some eminence, such as a mound or wall situated in the vicinity of his sitting spouse, to give vent to his feelings, and it is in the above months that one so frequently hears him during the journey to and from the Vale. The usual nest is but a hollow in the ground lined with leaves or grass in varying quantity and concealed amongst rough grass or standing crops.

The eggs generally number from 6 to 9, but variations on either side of those figures are frequent. Stuart Baker mentions a clutch of 17, perhaps the product of two hens. They are very broad eggs, glossy with a hard shell, and are often well pointed at the smaller end. They range in tone from a pale stone, or olive, to a deep olive-brown. In size they average 37·8 by 31·3 mm.

HIMALAYAN SNOW-COCK

Tetraogallus himalayensis himalayensis Gray

(See Frontispiece.)

KASHMIRI NAMES. *Kabuk, Gourkāku, Rām Chukār.*
BOOK REFERENCES. *F.B.I.*, V, 426; *Nidification*, IV, 271; *Handbook*, 428.

FIELD IDENTIFICATION. (5) Adults are usually darker than the bird figured, being a richer buff above, inclining at times almost to chestnut, although in the old birds the plumage does vary considerably, some being much greyer than others. The Himalayan Snow-Cock is large, males running up to six and even eight pounds and having a wing span of $3\frac{1}{2}$ feet. It will only be met with, usually in coveys or small flocks of more than one family, in bare stony ground at high elevations well above the forest limits. The call is a loud whistle, and when walking they carry the tail like a domestic hen.

DISTRIBUTION. Snow-Cock are not uncommon in summer between 12,000 ft and the snow-line, particularly on the slopes of the main Himalayan range. They occur, but probably in smaller numbers, on the Pir Panjal mountains, Osmaston recording having seen four fully grown birds above Gulmarg at 12,000 ft and a larger covey of about ten north-east of Tosha Maidan. They will doubtless be found generally distributed along that range all round the Valley. They are resident birds, moving down to somewhat lower elevations in the winter but never entering forest. The full range of this species is in the area enclosed by the Tien Shan and the Hindu Kush through the Pamirs and Eastern Afghanistan, the Sufed Koh and the Western Himalayas to Garhwal.

HABITS AND NESTING. These fine birds are not to be seen as often as their generally wide distribution would imply. This is easily accounted for by the fact that in the summer they live only at high elevations in bare stony country comparatively seldom visited. In the area with which we are concerned, the trekker generally takes the better known routes, and these, including the passes from one valley to another, usually run through country of a more verdant nature than that in which Snow-Cock elect to pass their time. Bare stony ridges with scant bush and short grass, so orientated that one face is sheltered from the prevailing wind, are this bird's preference; country with little to interfere with wide observation and not much in the way of cover except stones and boulders. They are not on the whole shy birds and can at times be driven up the hillsides almost like a flock of geese, for, like so many of these hill partridges, they prefer to walk uphill rather than to fly, and so make their way to the crest of the ridge before taking to wing. The flight is strong with rapid wing-beats, and once under way they do not as a rule stop before putting a considerable distance between themselves and the disturbing element. Wilson, incidentally, states that the Golden Eagle is an inveterate annoyer of these birds, but as they are always on the alert and fly off before the eagle comes to sufficiently close quarters, this monarch of the air does not often strike one down. We once saw such an escape near the head of the Wardwan Valley when an agitated Snow-Cock whirred over our heads whistling like a runaway engine.

Descriptions of the call are full and interesting. According to A. E. Osmaston the breeding call is a prolonged whistle uttered with great persistence. B. B. Osmaston states that the call of the cock bird is of two kinds, ' (a) A clear whistle repeated three or four times at intervals of about a second, the notes rising in scale so that the fourth and last note is an octave above the first ; (b) a cry beginning *kuk kuk* repeated about a dozen times, gradually quickening and rising in scale until the last high-pitched notes follow each other so rapidly as to remind one of the acceleration of a ping-pong ball '. Wilson adds that when they get on the wing the whistles are shriller and very rapid and that no matter how far they fly the whistles are continued till the birds alight and for a few seconds afterwards. Lastly, when the female is sitting, the male is generally perched not far off on the top of a boulder or other eminence whence he can warn her by a loud whistle.

The female constructs no nest but lays her eggs in an unlined hollow scratched out in the shelter of a stone or boulder on open ground, usually near the crest of a ridge at anything between 12,000 and 15,000 ft, sometimes much higher. She sits closely, according to Stuart Baker often until she is almost trodden upon, especially if the person approaching is moving up against the wind. The eggs often number 5, but Osmaston says he received a clutch of 7 slightly incubated eggs taken at about 14,000 ft, while Ward obtained an incubated clutch of 3. The ground colour varies from a *café-au-lait* through pale buff to a dull greenish, and this is freckled all over with pinkish-brown or chocolate spots and specks. They are coarse-grained eggs sometimes having a fair gloss. They average 67·2 by 45·7 mm.

These partridges commence laying about the end of April or early in May, but Meinertzhagen obtained a chick in down near

Leh (Ladakh) at the beginning of July. Snow-Cock are gregarious to some considerable extent and a number of coveys often join up soon after the chicks are hatched. A. E. Ward, in a letter to Stuart Baker (see *Game Birds of India*, IV), mentioned a flock of three or four families all sunning themselves on the side of an almost bare rocky hill. Chicks and old birds alike were lying on their sides with their limbs stretched out so as to get as big a share as possible of the warmth of the sun. Colonel Meinertzhagen (*Ibis*, 1927) gives some interesting notes on courtship and habits. They read: 'Their courtship is simple. The neck-ruff and tail are extended, the white undertail-coverts forming a conspicuous feature. The male then walks slowly round the female with lowered head. At intervals he makes a dash for a small rock or mound which he ascends, throws his head well back and gives his wild five-noted whistle.' And again: 'When walking up-hill, the tail is held erect, the white undertail-coverts hanging down a bit and showing conspicuously. The wings are held low and in line with the axis of the body, the neck feathers being very slightly ruffed out.'

TURKESTAN WATER-RAIL

Rallus aquaticus korejewi (Zarudny)

(See Plate 57 facing p. 228.)

BOOK REFERENCES. *F.B.I.*, VI, 6; *Nidification*, IV, 289.

FIELD IDENTIFICATION. (3+) A dark bird of brown and black streaked plumage with a very short tail and rather long legs, which sneaks like a rat through the thickest cover in the larger secluded marshes. The flanks, lower abdomen and under the tail are barred black and white. The foreparts are noticeably grey and much of the bill is red.

DISTRIBUTION. The Water-Rail, being a particularly shy bird, appears to confine itself to the larger and more secluded marshes of the Vale, especially to the State *rukhs*. It is rare, if not quite absent, from the numerous small jheels and those in which it is likely to be disturbed by the activities of the Kashmiri boatmen and visitors.

This species nests from the British Isles across Europe and Asia to Japan. Two races occur in India, but one only remains to breed, and this, *korejewi*, nests from Transcaspia, Persia and

Turkestan to Kashmir. Stuart Baker includes Ladakh where, he says, Ward's collectors obtained its eggs, but Ward himself in his list of birds of Kashmir and Jammu does not mention Ladakh in his remarks on this bird.

HABITS AND NESTING. This and the next five birds all belong to the *Rallidæ*, or Rail family. They are skulkers and one and all inhabitants of marsh and bogland. At the two extremes come the Common Coot, which has rather the habits of a diving duck and loves to cruise in herds about reed-fringed open waters; and this bird, the Turkestan Water-Rail, which is the most arrant skulker of them all, creeping about unseen in the densest cover of the large marshes which have a surrounding jungle of tangled grass and thick reed-beds standing in the shallows. It is not uncommon in Kashmir, but owing to these habits is seldom noticed; in fact, it is more easy to come by the eggs than to obtain glimpses of the bird. The other Kashmir rails which are to be found in the grassy cover of the marshes, Baillon's and the Ruddy Crake, may both be flushed off the nest, the former indeed often is, but the Water-Rail prefers to sneak silently off its eggs before the intruder approaches too close. When not hurried it struts about with the tail intermittently flicking up at an angle and the head jerking backwards and forwards. It is an excessively wary bird, so avoids disturbance by nesting almost exclusively on those large jheels which are little frequented, such as Hokra and in and near the *rukh* portion of the Anchar Lake and marshes of like character farther down the Vale; indeed it is obvious that the State *rukhs* provide its main habitat. In spite of its wariness it is somewhat noisy.

Osmaston writes of it that it calls chiefly in the evenings, a curious cry reminding one of the squeal of a small pig. In point of fact the Water-Rail possesses a considerable repertoire of curious calls, from this subdued squeal, which we prefer to describe as a sheep wheezing, through the Moorhen's *ker-ick* to an almost inaudible piping. When distressed the bird also makes a drumming sound.

The nests are always well hidden, usually in some kind of rank growth such as long grass, tall water-weeds, or broken-down reeds in a few inches of water. They may be on the banks of ditches or on flat wet ground near the water's edge, or again well out in the matted reeds. In the last case the nest sometimes attains considerable proportions both as to width and height. It is a loose collection of rushes, weeds and a little grass, usually about 6 inches across and quite thin, but large nests may attain to a foot across and half that in height.

The eggs are fairly broad ovals, often with considerable gloss,

RAILS

a pale cream to a pinkish-buff in ground colour not very thickly spotted and blotched with reddish or purplish-brown with underlying markings of lavender or grey. They average 36·3 by 25·6 mm. Full clutches are said to run from 5 to 8, being obtainable during the latter half of May, in June, July and the first half of August. Five happens to be the largest clutch we have seen. Both birds incubate and both help in the construction of the nest.

EASTERN BAILLON'S CRAKE

Porzana pusilla pusilla (Pallas)

(See Plate 58 facing p. 229.)

KASHMIRI NAMES. *Grettu, Gralu.*

BOOK REFERENCES. *F.B.I.*, VI, 14; *Nidification*, IV, 293; *Handbook*, 437.

FIELD IDENTIFICATION. (3—) The smallest of our rails. A skulker in the marshes which is seldom seen, but at times pops up from under foot soon to drop into the cover again after only a short flight. Its rotund little form and short wings remind one strongly of a quail, but it is easily recognizable by the long legs hanging down behind and the slower wing-beats. The plumage is mainly rufous-brown streaked with black, while the hinder parts beneath have alternate bars of brown and white. A point to notice at close quarters are the peculiar flecks of white on the upper plumage, like smears of paint, which show up well in the illustration.

DISTRIBUTION. Breeds profusely in and around all the swamps in the Vale, but we have no records of its occurrence in other parts of the State. Although Baillon's Crake is resident on the jheels of Northern India and the United Provinces, the birds of these areas are almost certainly augmented by a winter influx from the Himalayas, but we have no information as to whether any crakes remain in Kashmir for the winter.

A widespread species in Europe, Asia, and Africa, Baillon's Crake has occurred and nested in England. The Eastern, typical, race occurs from Afghanistan to China, the Indo-Chinese countries and Malaya. In India it is known to nest commonly only in Kashmir, but its eggs have been taken as far apart as Deesa and Etawah. In winter it is common in the Northern and Eastern Indian jheels, and over a great part of Burma.

HABITS AND NESTING. Although a common breeding bird in

Kashmir this little rail is not often seen for it remains constant to thick cover, be it the soft grasses around the marshes' rim, dwarf rushes and derelict rice-patches, or matted reeds and water-plants. It is not an inhabitant of the tall reed-beds which stand in the water and have little or no cover between their stems, no matter how close-growing they may be. As Osmaston truly says, they are far more often heard than seen, preferring to run like rats through the concealing grass and tangled cover of the bogs rather than take to flight. They can also swim quite well, jerking the tail with each stroke in the usual rail manner. Not infrequently they remain upon their nests until they are almost in danger of being stepped upon. It is then that they shoot up at one's feet after the manner of a quail—which they are by no means unlike at first sight when flying directly away low over the cover. They do not of course get up with the startling whirr of that delectable little bird and their wing-beats are far slower. Besides, the long legs and prominent feet trail behind drooping at an angle from the body. The call is a high-pitched *crake*, first a single note, followed after an interval by another which is then repeated at decreasing intervals until the notes coalesce.

The nest varies both in bulk and materials according as to where it is placed, but it is invariably tucked away out of sight covered over by enclosing rushes, grass or other vegetation. When sited amongst the soft marsh grass around the shores of such lakes as Hokra Jheel or the marshes around Sumbal, favourite haunts incidentally, it is often but a pad of grass mixed with a few leaves of the surrounding weeds with the concealing grass arched well over it. When on wetter ground, in short rushes or broken-down rice, the base is apt to be more substantial and built of these same materials with a number of the rushes or rice-stalks bent over into a sort of protective roof.

The eggs are very large for such a small bird and when one sees the area taken up by a nest of 8 eggs it seems extraordinary that such a tiny creature should be able to incubate them successfully. In colour they have a yellow-brown, olive or pale brown ground streaked and freckled profusely and boldly with reddish-brown. They usually have a certain amount of gloss, which is sometimes well pronounced, and they average $28 \cdot 4$ by $20 \cdot 6$ mm. Captain Livesey records a clutch of 9; we have a couple of 8's, but 6 and as few as 5 are often incubated. They lay from early in May, great numbers breeding in Kashmir in the months of June and July, while some lay as late as August. We have taken a nest of 6 slightly incubated eggs on the 11th of that month.

RAILS

NORTHERN RUDDY CRAKE

Amaurornis fuscus bakeri (Hartert)

(See Plate 57 facing p. 228.)

KASHMIRI NAME. *Gillu.*
BOOK REFERENCES. *F.B.I.*, VI, 21; *Nidification*, IV, 300; *Handbook*, 437.
FIELD IDENTIFICATION. (3) A slender warm-brown rail with a blood-red bloom about the face and lower plumage. Never takes to the wing unless forced, but sneaks through the cover at the edges of rice-fields, weedy ditches, and the rushes about the rims of the marshes. Jerks the head and flicks the tail after the manner of a Moorhen. In fact its habits are those of the Moorhen on land than which, however, it is considerably shyer, smaller, and more lightly built.
DISTRIBUTION. Widely and fairly commonly distributed about the wetter portions of the Vale, being particularly fond of weedy ditches, rank herbage bordering rice-fields, and swampy patches on the outskirts of the larger marshes. In the side valleys they probably occur up to about 6,000 ft, that is up to the limits of the irrigated rice-fields. We saw one at about that elevation in the Lidar Valley.

This species has a wide range in the Oriental and Chinese regions from India to China, Japan, the Philippines, and throughout the Indo-Chinese countries and Malaysia. The Northern race is found from the Afghan Frontier and Kashmir to Assam; also in Sind, Bengal, and eastwards through a great part of Burma to the Shan States.
HABITS AND NESTING. This rail is apt to be met with when looking for Baillon's Crake, but whereas that bird is most numerous in soft cover encircling the large jheels, the Ruddy Crake is not so much at home there as out in the rice-fields where it haunts the irrigation ditches and weedy channels between them. It is not at all fond of the extensive reed-beds which surround the large jheels, but is common amongst the swampy patches and rank marsh grass and vegetation in their vicinity. It very seldom uses its wings, much preferring to sneak off through the grass, running easily and quickly on its longish legs, just like a lightly built edition of the Waterhen. When swimming, at which it is adept though seldom indulging in that pastime, it again moves like the Moorhen, the head oscillating back and forth with each stroke, the short

PLATE 73

Nest and eggs of White-eyed Pochard

White-eyed Pochard on nest

PLATE 74

Coot on nest

Indian Little Grebe on nest

tail being periodically flicked up to show the white-edged undertail-coverts. In the daytime it remains in thick cover, but during the night comes out more into the open swampy meadows, never, however, venturing far from cover into which it can slip at the first signs of danger. The call indulged in late in the evening or in the early morning hours is a metallic *tewk* repeated every 2 or 3 seconds, usually followed by a bubbling note, much like that of the Dabchick but louder.

Nests are to be found in a variety of situations. We have seen them anchored a few inches above the water in dwarf rushes and amongst the rice-stalks, and also on the banks between the rice-fields where they rested on dry ground protected by some thick vegetation, the roots of a small bush or an overhanging stump, while sometimes they are hidden in the grass and weeds on flat ground. An exceptional nest was placed three feet above the water in a bush growing in a marsh. On one occasion we were shown three nests in rice all within 20 yards of the main Srinagar-Baramullah road near Shaltin. The nest built over water is supported on a few bent-down reeds or rice-stalks and is more or less concealed by more stalks being bent over the nest into a loose lattice-work roof. The nest itself, which is never bulky, consists of grasses and the dry leaves of aquatic plants.

The eggs are large for the size of the bird, averaging 32·6 by 23·5 mm. We were once presented with a female with an egg in the oviduct which could not only be felt but so distended the abdomen that its contours could be plainly seen. She appeared uninjured but was hardly able to fly, and when released at shoulder height fluttered down unsteadily on to the water of a nearby ditch and swam rapidly away hugging the thin line of reeds at its edge. The ground colour of the egg is a very light cream or sometimes pale buff fairly boldly spotted more or less all over with red-brown and purplish-grey. Six to 9 are laid, chiefly in late June, July, and August, but a few full clutches may be found even in May. Both birds incubate, but it is uncertain whether or not the male helps with the construction of the nest.

MOORHEN

INDIAN MOORHEN

Gallinula chloropus indicus Blyth
(See Frontispiece and Plate 59 facing p. 236.)

KASHMIRI NAME. *Tech.*
BOOK REFERENCES. *F.B.I.*, VI, 28; *Nidification*, IV, 308; *Handbook*, 438.
FIELD IDENTIFICATION. (4+) A black ducklike bird seen swimming about the margins of reedy ponds and reed-fringed canals. The Moorhen, or Waterhen as it is often called, swims with the stern well out of the water and the short tail almost erect, every few moments jerking it still farther upwards to expose thereby its white marginal feathers and undertail-coverts. This in conjunction with the yellow-tipped red bill, red shield on the forehead, and otherwise dark plumage except for a white wing-patch, renders it quite unmistakable.
DISTRIBUTION. A common bird throughout the marshes, reed-fringed canals, and those portions of the lakes where bulrushes or osier-beds exist around their shores. We have not met with it outside the Vale, though an investigation of the reedy pond between Domel and Garhi would probably disclose its quota of these birds. It appears to leave the Vale in the winter.

The Moorhen is almost cosmopolitan in its distribution, being absent from Malaya and Australasia. Of its many races one only occurs throughout all India and Ceylon extending eastwards without its limits through Burma to Indo-China and China.

HABITS AND NESTING. In the Vale of Kashmir we have always found the Indian Moorhen to be a shy and cunning bird, for it is much persecuted by the Kashmiri boatmen who on every possible occasion take its eggs for food. In consequence it is most cautious in its approach to the nest and deserts its possessions on the least provocation. It seems useless to put up the hiding-tent close to a nest with a newly-hatched chick in it, or with eggs on the point of hatching, or yet to try a gradual approach to a sitting bird by advancing the hide a yard or two daily to give the birds an opportunity to get used to it. In the former case, the parents will remain out of sight in the reeds, clucking to the chick until it eventually clambers out of the nest to join them. Thereupon the unhatched eggs are deserted. In the latter case, the appearance of the tent, and the lengthy period it remains *in situ*, usually lead to the discovery of the nest and the filching of its contents by some rapacious local.

Although the Waterhen will at all times return to cover on one's near approach, it is certainly less allergic to being watched than most members of the rail family, so a bird or two may always be seen swimming across the canals or hugging the nearby reeds as the shikara skims the waterways. They will never, however, be seen on the open waters of the large lakes. When swimming the tail is held high, every now and then being given an upward flick while the head jerks rhythmically backwards and forwards. On land the Waterhen moves freely, stalking about with upright carriage and the same flirting of the tail, and may be flushed from thickets some distance from the bank whence it will run or fly back to the protection of the reed-beds. It is no mean flyer, second only to the Coot, taking to the air more easily than the other rails, but as usual flying with the legs and feet trailing down behind and soon dropping into cover.

The nest, which consists of a pad of rush leaves, is usually supported on a number of bent-down reeds from a few inches to as much as a couple of feet above the water. It measures about 8 inches to 10 inches across. It is generally in the dense reeds but often borders a more open space within them, so that when pushing through a thick bed, on coming to an open patch of water, however small, it is profitable to make a search for Waterhens' nests immediately around it. Occasionally a nest may be found as much as 3 feet up on a bent-down willow or even in a tangle of undergrowth on a bank, but the reed-bed nests are vastly in the majority. The chick, which is covered with a grey-black down, is capable of leaving the nest within a couple of hours of hatching; and unborn chicks, even before breaking the shell, understand the calls of their parents, ceasing their subdued chirping the moment they hear the danger call. A perturbed parent will swim around the nest in the cover of the surrounding reeds uttering clucking noises and occasionally slapping the water with the wings in an endeavour to draw away the intruder.

The nesting season is prolonged. We have found their eggs in all months from May to August. This is due as much to the many robbings which occur as to the probability of their being double-brooded. The eggs are often very handsome. They are said not to differ from those of the European bird, but in our view the ground colour is on the whole paler and the markings richer. The ground is usually a very pale pinkish stone colour with the markings bold blotches of deep red and reddish-brown. Sometimes, however, they are smaller and more numerous. We have seen as few as 5 incubated eggs in a nest, but up to 8 or even 9 is usual, while as many as 12 have been recorded. In size they average 41·4 by 29·6 mm.

INDIAN PURPLE COOT

Porphyrio poliocephalus poliocephalus (Latham)
(See Frontispiece.)

KASHMIRI NAME. *Wān Tech.*
BOOK REFERENCES. *F.B.I.*, VI, 32; *Nidification*, IV, 312; *Handbook*, 440.
FIELD IDENTIFICATION. (5—) An ungainly-looking bird of dense reed-beds, which nevertheless clambers up and down the stems with considerable agility thanks to the long legs and inordinately long toes. Usually seen in small parties which at times come into the open to feed. No other large bird has this distinctive plumage of smoky-purple with conspicuous red bill and forehead. In spite of the heaviness of the bill, the head, when the rather slender neck is stretched out, appears small in relation to the size of the body so that at times the bird rather reminds one of the domestic guinea-fowl.

DISTRIBUTION. Uncommon but increasing in the Vale of Kashmir, where it is to be met with in small colonies in a few of the secluded marshes with dense reed-beds, such as the *rukh* portion of the Anchar Lake, and those near Sumbal, Pampur, and Sopor. It is usually a resident bird, moving about locally after breeding, and may be more common in Kashmir than it appears to be in the summer. In winter it leaves its secluded quarters for more public waters. Colonel Phillips now writes, 'It is quite a sight to see numbers of them on the duck jhils in winter', while in a further note he states they are abundant on Haigam and at Pampur.

The species ranges from Asia Minor and Transcaspia to Siam and Malaya. The typical race is found in suitable areas throughout the greater part of India and Burma, birds from farther east and in Malaya belonging to a different race.

HABITS AND NESTING. The habits of the grotesque Purple Gallinule are quite different from those of the Common Coot, being in a love of seclusion more on a par with those of the typical rails. On the other hand, like the Common Coot, it is a sociable bird, which, usually in small colonies during the nesting season, does at times gather into herds of considerable size in the winter. We have seen, in the United Provinces for instance, a couple of hundred of these birds feeding together on open wet ground bordering the Udainagla Jheel. The Purple Coot is not, however, a swimmer, and, when not paddling about marshy ground and swamped fields at the edge of a jheel, prefers to clamber about in the heavier patches of bulrushes, the longish legs and attenuated toes being specialized for this particular purpose. We have unfortunately little information in regard to Kashmir of this bird's movements in winter, except for Colonel Phillips's remark alluded to above, but in the Plains of India it is a local migrant only, tending to spread from its breeding haunts after nesting is over.

In Kashmir—and possibly elsewhere—the Purple Coot would appear to be double-brooded, for Colonel Phillips reported finding a colony nesting in April, the members of which were again laying in July. In the South of India some birds lay as late as September, so it may be a universal habit. This bird takes to wing only when forced to do so, the greater part of its time being spent in clambering in slow motion through the marsh grasses. When shooting we have frequently come across parties of these birds in dense reeds bordering open water, some standing close to the surface, others on flattened reeds on the fringe, while more have been climbing slowly up the stouter rushes above them, so that the whole has conveyed the impression of a well-planned museum group. Purple Gallinules are dainty feeders. When confronted with a morsel considered too large to swallow whole or requiring dissection, they have the quaint but effective habit of transferring the food to the foot, using the hind toe as a thumb to wedge it between the bases of this and the inner toe while pecking at it. They are said to be noisy birds, particularly in the breeding season, harsh groans, clucks, and much cackling issuing from their quarters in the thick cover.

The nest, placed amongst close-growing reeds, consists of a considerable collection of rush leaves and grass blades built up on bent-down reed stems. It is often of considerable depth so that it stands a foot or two above the water. The eggs are not unlike handsomely marked Waterhen's eggs, much larger of course and more elongated, but the ground colour has a pink tinge about it and is sometimes quite a deep salmon. The markings are large and small blotches of red with secondary spots of lavender or stone colour distributed more or less all over the egg. From 3 up to 7 are laid, though 5 seems to be a popular number. They are large blunt ovals averaging $50 \cdot 5$ by $35 \cdot 7$ mm.

COOT

Fulica atra atra Linnæus

(See Plate 74 facing p. 293.)

KASHMIRI NAMES. *Kolūr, Kāvput, Kolru.*
BOOK REFERENCES. *F.B.I.*, VI, 34; *Nidification*, IV, 314; *Handbook*, 441.

FIELD IDENTIFICATION. (4) A bird with the entire plumage a blackish-grey, the head and neck somewhat darker, the underparts greyer, which rides like a duck on the open waters of the larger reed-encircled jheels. The bill is white, the upper mandible projected backwards into a conspicuous shield which covers the forehead, providing a quite unmistakable identification mark.

DISTRIBUTION. During the breeding season the Coot is not particularly common in Kashmir, but it is found on a few of the larger jheels which have plenty of open water with adjoining reed-beds and patches of water-weeds. We have never seen one during the summer on the Dal lakes, Manasbal, or the open parts of the Wular, but they nest on Hokra and the Anchar Lake and one or two other of the less frequented jheels of like type. In the winter they spread farther afield or are augmented from outside, and may then be seen in herds on the Wular Lake and adjacent lagoons.

The Coot, a familiar bird in the British Isles, is found throughout most of Europe, Asia, and Northern Africa with allied forms in the Australian region. So far as the Indian region is concerned it is absent only from Ceylon. It is a resident bird wherever suitable conditions persist throughout the year, but after breeding deserts areas which tend to dry up, often moving great distances in the process. The result is a great cold-weather influx of birds into Indian and Burmese limits.

HABITS AND NESTING. In the cold weather Coots collect in droves of hundreds to cruise about the open waters of large lakes as well as collecting in small parties on the most restricted ponds. During the breeding season, however, these packs break up into pairs, but the individuals still retain a hint of their sociability in that they defend only a limited territory, most of the open water constituting neutral ground, the occupants of a jheel going wheresoever they please, mixing freely and sometimes nesting quite close to one another. Coots dive freely, obtaining their vegetable food, supplemented by a goodly quantity of fish, insects and molluscs, as much below the surface as upon it. They have no particular love of flight, but when urged to do so leave the water with considerable splashing and spattering over its surface. Still, they fly well when once on the wing, although the legs trail behind and their action looks somewhat laboured.

The nest is either on the fringe of the rushes or floating on aquatic vegetation of any type capable of supporting it. When resting upon beaten-down reeds, or well supported by other means such as the branches of half-submerged osiers, it is a flat thin affair of rushes, but when more or less floating amongst the reeds or in shallow water with its foundations on the bottom, it often attains large proportions, projecting some inches above water like the nest in the illustration. The upper portion is then composed of dry brittle reed stems sometimes with a few rush leaves as lining. It is often unconcealed in any way, so that the bleached stems catch the eye at some distance. The eggs vary little, having a strong glossless shell of a pale stone or stony-buff colour, marked with small spots and tiny blotches of black. These markings are pretty well distributed about the whole surface. They are elongated eggs, somewhat pointed at the small end, and average 53·1 by 35·6 mm. The number in a clutch is elastic, but round about 8 seems normal, with limits between 5 and a dozen. Osmaston took a clutch of 8 on the Anchar Lake on 20 May, while we came upon newly-hatched young in a nest on the 27th of that month. We have seen a good number of nests in June, but Stuart Baker says of Kashmir that they lay on into July and even August.

There is no doubt that at times two broods are reared. We once watched a pair on two consecutive days which were feeding an almost fully-fledged quartet of very demanding youngsters. To our surprise one or other of the parents would periodically make off to a patch of weeds where for a space it would settle down to incubation on a conspicuously tall nest. Investigation showed it to contain three fresh eggs.

When the eggs are on the point of hatching or a young one is still in the nest, the old birds are very demonstrative, barking freely and slapping the water with their wing-tips in efforts to draw away the intruder. The down plumage of the chick is rather striking, being so very different from what one would expect of such drab parents. The body is black both above and below as are also the legs. The wings are a fleshy red surrounded by orange-coloured down. The crown is red with a complete ruff of orange but above the eye is purple. The forehead, frontal shield, and part of the upper mandible are scarlet and only rather more than the tip of the bill is at this time white.

JAÇANA

PHEASANT-TAILED JAÇANA

Hydrophasianus chirurgus (Scopoli)
(See Plate 61 facing p. 244.)

KASHMIRI NAMES. *Gūnd Kāv, Gair Kov.*
BOOK REFERENCES. *F.B.I.*, VI, 42; *Nidification*, IV, 322; *Handbook*, 457.

FIELD IDENTIFICATION. (4—) No one can fail to recognize this peculiar bird at a glance for there is none other in Kashmir, or indeed in India, anything like it. Lily Trotter, to give it one of its sobriquets, describes it well, for it will always be seen moving about the lily leaves and aquatic weeds growing on the water's surface, and never on dry land. The contrasting black and white plumage relieved by the yellow patch on the neck, the drooping pheasantlike tail, and in flight the abnormally long toes, which trail behind like a bundle of twigs, are quite distinctive; to say nothing of the mewing cry.

DISTRIBUTION. Confined to the marshes and lakes of the Vale of Kashmir, being common wherever lotus and weeds reach the water's surface. In the nesting season we have never met with it outside the Vale, but it is said at times to straggle to high elevations. F. A. Betterton observed a specimen on the Vishan Sar tarn which is situated at 12,050 ft amongst the high ridges between the Sind Valley and Gurais.

The Pheasant-tailed Jaçana inhabits practically the whole Oriental region, wherein it may be classed both as a resident and as a local migrant. It appears to leave Kashmir altogether in the winter, probably retiring to suitable jheels in the somewhat warmer adjacent Indian Plains. Being a bird which often nests on small impermanent patches of water, after the autumn moult it packs into small flocks which move about locally from one pond to another.

HABITS AND NESTING. Every visitor to Kashmir is bound to take notice of this curious yet beautiful bird. A trip on the Dal lakes in a shikara is sufficient to introduce it to the most unobservant so that no bird-lover can possibly miss it. Its mewing cries, ability to progress freely over the water where but the scantiest of weeds have reached the surface, and its striking colour-pattern—the long black tail, white wings and yellow and white head—make it one of the most extraordinary of India's birds. Its calls carry far over the water and are as distinctive as the bird itself. The chief cry is a loud musical *me-e-ou, me-e-ou* or *me-onp*, often taken up

PLATE 75

Nest of Pale Bush-Warbler

Brown Hill-Warbler

PLATE 76

Among the chir pines in the lower Kishenganga Valley, elevation

The distribution in Kashmir of the Slaty-headed Parakeet is

by its mate or other bird in the vicinity, but many shortened variations are made use of, perhaps the commonest being a loud open *kloo*. Its liking for the lotus leaves and the weed-covered shallows is catered for by the immensely elongated toes and claws which are plainly visible when the bird is on the wing, for they trail beneath the loosely waving tail-feathers like a bundle of twigs.

The nesting is equally noteworthy. The eggs, 4 in number, rarely 3 only, are bronze in colour, often with a greenish or olive tinge, and have a high gloss so that they blend remarkably well with the bright green hues of the aquatic plants which surround them. The nest is a sketchy pad of a few hastily collected floating weed stalks or lotus stems. It hardly rises above the surface and is barely sufficient to accommodate the eggs and to support the sitting bird. To ensure that the eggs will not be lost, they are very broad, flattened at the base and so sharply tapered to a narrow point that, placed point inwards, they cannot possibly roll apart.

Lately a further safeguard has come to light. In a recent exhibition Colonel Boyle showed a remarkable photograph. This portrayed a Jaçana removing an egg from its nest by wedging it between the bill and its breast and then retreating backwards from the nest to the selected place of safety. Some years ago we found a new nest to have been prepared in the space of one hour, and the eggs removed 12 feet to it by a pair which objected to the erection of the hiding-tent, and we have since heard of similar instances. The habit has not of course been evolved purely to avoid the attentions of importunate bird photographers. Jaçanas build largely around the margins of tanks and jheels whose areas are apt to shrink rapidly, particularly in the Plains where this beautiful bird is equally common. Numbers of nests would, therefore, be left high and dry by the receding waters, and the eggs be open to destruction, were it not for the adoption of this useful practice.

Both the birds are said to build the nest, but the female, the larger of the two incidentally, alone incubates. In Kashmir nesting appears to commence about the second week of May, and to continue throughout June and into July. The eggs average 37·2 by 27·8 mm.

PAINTED SNIPE

Rostratula benghalensis benghalensis (Linnæus)
(See Plate 60 facing p. 237.)

KASHMIRI NAMES. *Gug, Gūgū.*
BOOK REFERENCES. *F.B.I.*, VI, 45; *Nidification*, IV, 323; *Handbook*, 478.

FIELD IDENTIFICATION. (3—) Slightly larger than, but in some ways resembling the Fantail Snipe. The bill, however, is shorter, somewhat bulbous and bent downwards at the tip, the flight direct and rather slow, and the legs trail behind. The spotted and ' painted ' plumage, the wide band round the eye, and the white gorget rimming the breast are very distinctive.

DISTRIBUTION. The distribution of this so-called snipe is much the same as that of the Fantail Snipe, but it will often be found in coarser reeds and on wetter ground. It is a certain find on the Krahom swamp with its large area of semi-reclaimed land. But it is common, and said to be resident, in any of the larger swamps of the Vale and may also be seen in the rice-fields. Ludlow, however, says he has not met with it in the winter months.

The typical race of the Painted Snipe is well distributed in Africa and Asia with another race in Australia and Tasmania. It occurs in the mountain ranges of India up to 5,000 ft as well as throughout the Indian Plains, and is mainly resident throughout its range though prone to local movement.

HABITS AND NESTING. The Painted Snipe is looked upon with disfavour by the sportsman, who considers its poor powers of flight and consequent ease with which it can be shot as not worthy of his attention. He is in the main right, for the ' Painter ' is not a true snipe and though superficially resembling it in many respects it in fact bears much closer affinities to the rails. Like the rails it flies with the legs dangling ; nor does it leave the ground with the speed and suddenness of the snipe, but rather does it flutter into the air and fly off in a straight line to drop again into cover at no great distance.

It is, on the other hand, a bird of unusual interest to the naturalist. Not only is its plumage very handsome, but its ways of life are striking, for it is one of those freaks of nature in which many of the functions of the sexes are reversed. The female is polyandrous, and after laying her eggs she leaves their incubation and the bringing up of the young ones to the male while she goes off in search of another husband. As might, therefore, be expected, the male is more soberly attired than she is in order to render him less conspicuous on the nest and in attendance upon his young. He lacks that rich chestnut hue on the neck and breast, the white rims round the eye are not so conspicuous, and the remainder of his plumage is greyer than in the female. A characteristic of the 'Painter' is the large luminous eye which points to nocturnal habits and we think it is a fact that this bird feeds in the twilight and at night, lying up by day in the thicker cover. The sounds emitted by the Painted Snipe consist usually of low grunting notes; no others have been recorded except a hissing sound during the display. Nor has it any artificial aids to communication such as the drumming of the true snipes. When cornered, however, these birds adopt a protective and menacing attitude, raising one or both wings and bringing them forward so that they reach beyond the tip of the bill. The tail is expanded and the wing-feathers spread out to show off the buff markings. At the same time the hissing note is emitted. We have noted young birds behaving in the same manner even down to emitting the hissing note and not the low plaintive whistle as described by Herbert. When flushed from the nest, the male does not as a rule resort to flight, but sneaks off in a typical rail-like manner to hide in the adjacent reeds or other herbage. The bird in the illustration showed no inclination whatsoever to take to wing, merely squatting by a grass tuft a few yards from the nest, moving off in a crouching attitude when approached and at once starting to creep back to the nest as soon as we turned our backs.

The nest is merely a pad of grass just sufficiently large to take the eggs, though when slightly raised from the ground, as is sometimes the case, it may be more substantial. The eggs are far smaller than true snipes' eggs, are rather pointed and have considerable gloss. The ground colour is yellowish-buff with bold well-distributed brownish-black markings. They average $35 \cdot 4$ by $25 \cdot 3$ mm.

INDIAN WHISKERED TERN

Chlidonias hybrida indica (Stephens)

(See Plate 62 facing p. 245.)

KASHMIRI NAMES. *Krind, Krew*.
BOOK REFERENCES. *F.B.I.*, VI, 111; *Nidification*, IV, 359; *Handbook*, 482.

FIELD IDENTIFICATION. (4—) A graceful light-grey tern to be seen flying up and down the reed- and lotus-covered marshes, and over newly planted rice-fields. The top of the head and the abdomen are black, the bill and feet red. Nests in colonies, making a floating raft amongst water-plants. It also draws attention to itself by a harsh oft-repeated *kreek*, from which its Kashmiri names are probably derived. The tail is not noticeably forked.

DISTRIBUTION. Found in colonies throughout the marshes and lakes of the Kashmir Valley excepting those lakes with open shores devoid of reeds and floating vegetation; Lake Manasbal for instance. Even here the odd visitor may appear, but as a general rule they frequent only the reed-fringed, water-lily covered waters from which they also make regular excursions over the rice-fields. They are summer visitors to the Vale, arriving from the Plains in April. Meinertzhagen states that they were still present during the first week of September but had nearly all gone by the 21st.

The Whiskered Terns inhabit Central and Southern Europe, Africa, and Southern Asia to Australia. There are a number of races of which two nest in Indian limits. Our race nests in Kashmir, the United Provinces and Bihar, and probably in the Punjab Salt range, while *javanica* breeds in Assam and farther East. The Indian bird migrates in winter farther into India and Ceylon and possibly to Burma.

HABITS AND NESTING. The Whiskered Tern is Kashmir's most graceful marshland bird. Who could help but admire the grace of its easy flight as it twists and wheels back and forth over some reed-fringed lagoon in its search for aquatic insects and small fry, every now and then dipping momentarily to kiss the placid surface? The wide sweep of the tapering wings lifts a streamlined dusky body with every stroke, imparting an almost butterfly quality to the flight, while the birds call to one another with a scratchy note which somehow or other is part and parcel of their delightful personalities, since it proclaims their presence from afar. For this tern often fishes in conjunction with other members of its colony, systematically working a stretch of water from end to end.

At first sight it would appear that in the last couple of decades the Whiskered Tern has progressively dwindled in numbers, for, where previously colonies of large size were common, even on the Dal lakes, small parties of as few as five or six pairs are now the rule on these more public waters. The decrease, however, is probably only an apparent one, for the invasion of even the most out-of-the-way corners of the Dal, and other lakes open to the ever-increasing number of visitors, has undoubtedly led them to seek out more secluded waters. Indeed, our last tour of the Dal lakes disclosed only one small colony nesting near the Nishat Bagh and another amongst some floating gardens, the latter owned by a pundit who, it is pleasing to note, brooks no interference with the birds. In consequence there were some six or seven nests containing their full quota of 3 eggs. It is usually difficult to find nests with above a single egg in them for the boatmen raid the colonies systematically, leaving one egg here and there to ensure that the birds will continue their efforts to breed. A certain proportion of their spoils is hawked round the houseboats to be sold as plovers' eggs, but the majority are filched for local consumption. They are well able to compete with that delicacy both in appearance and flavour. Where diminution in numbers has, in fact, taken place, it is this systematic pilfering which is mainly responsible, and while it continues the Whiskered Tern will only hold its own so long as there are State *rukhs*, game reserves, where they can only be interfered with by the occasional poacher.

There is no question of a colony's nesting activities remaining unnoticed ; it is far too obvious a hive of industry. The area is at once marked out by the birds wheeling over it and dropping into the reeds or landing on the water-lily leaves ; it is usually close to a reed-bed as the nests require anchorage of some sort, either the occasional reed or other aquatic herbage which reaches the surface. In addition the harsh skreeky notes are freely indulged in and carry some distance over the water. While nest-building is in progress, birds will be seen collecting reeds and water-lily stems of most inordinate length, for the nest is a crisscross straggling collection of vegetable matter floating on the water, projecting in the middle an inch or so above the surface. The stems composing it may easily measure a couple of feet or even a yard in length, and the birds are, alas, not at all averse to filching material from their neighbours' nests. Indeed Colonel Phillips records a case where such a thief tipped a completed clutch of eggs into the water. The thinner decomposed strands are coiled up in the centre and the firmer wands are merely laid across each other with the ends floating out at all angles like the spokes of a wheel. The middle is sufficiently large to hold the maximum clutch of 3 eggs, which are of course placed point inwards. For these eggs are tapered to prevent them rolling from the flattened nest into the water and to ensure the clutch taking up as little room as possible. They are large eggs for the size of the bird, since to ensure their safety the chicks must be able to leave the nest immediately on hatching. In their exposed nests they would otherwise quickly fall prey to the omnivorous Booted Eagles, kites and crows, not to mention the less common Marsh Harriers. They can swim from birth and are clothed in a mottled down of black and buff so that lying motionless and half submerged in the shelter of an upturned lotus leaf they are well-nigh invisible. In the water they are just like corks, and if pressed down bob up to the surface without a single drop of water adhering to their protective down ; and when disturbed they keep their little eyes closed and move not a muscle till they think all danger is past.

Thanks to persecution the nesting season is prolonged. An odd egg was noted on 14 May, while we have seen birds still building new nests at the very end of July, although at the same time numbers of fledglings were noted, drawn up in parties in adjacent fields awaiting visits by the parent birds, who often transferred food without ever alighting amongst them. The ground colour of the egg varies considerably, from a pale grey to greenish and on to some shade of yellowish-brown. The markings are equally variable in size. Often collected mainly around the larger end, they consist of blotches and spots of deep chocolate or purplish-brown. They have no gloss and average 37·0 by 27·4 mm. Both birds take part in building the nest and incubating the eggs. While photographing one particular pair, nest-relief took place every 30 to 40 minutes.

EUROPEAN LITTLE RINGED PLOVER

Charadrius dubius curonicus Gmelin

(See Plate 62 facing p. 245.)

KASHMIRI NAME. *Kola Katij*.
BOOK REFERENCES. *F.B.I.*, VI, 171; *Nidification*, IV, 391; *Handbook*, 463.
FIELD IDENTIFICATION. (3—) This small plover with the large head is in Kashmir mainly to be found on the islands and shingle

beds of the lower reaches of the large side rivers. Here, except when moving in characteristic short little runs, the black collar around the white neck and the band over the top of the head so break up its outline that it is extraordinarily hard to pick out from its surroundings. The yellow eye-rim and yellow legs are also diagnostic. The Common Sandpiper inhabits the same localities but is larger, has a slender neck and an attenuated bill.

DISTRIBUTION. Shingle beds on the banks and round the larger islands of the big side rivers where they reach the Vale provide the main habitats of this interesting little plover. The zone they occupy is narrow and in it they are comparatively common, but they cease to be found as soon as the river-beds steepen and the valleys begin to close in. For instance, along the Madmatti they occur on shingle beds which commence opposite the little village of Dachhgam and extend up it to within a mile or so of Sonarwain, a distance of not more than four miles in all. They are also to be found sparingly on a few sandbanks along the main river between Baramullah and the Wular Lake. We have seen them on the Tulamal Jheel, a large area of which, thanks to its many drainage channels, is firm enough for ponies to graze upon, and we have taken its nest there. A footnote in the *Nidification* informs us that Ludlow obtained it on the Bringh and Duncan on the Wardwan River, where we too saw its nest near Inshan at an altitude of some 8,400 ft. It winters in India. In fact this race has been found as far afield as Ceylon and Burma. It returns to Kashmir in April and early May.

The Little Ringed Plovers have a far and wide nesting range from South and Central Europe, North Africa, and a great part of Asia south to India and eastwards to Japan. Those nesting from the foot of the Himalayas throughout the Indian Plains are considered a separate race, *jerdoni*, on account of their slightly smaller size. If this separation be valid, then we consider that specimens are needed from the numbers of Ringed Plovers nesting at the mouths of the side rivers debouching into the lower half of the Vale, and from those nesting in the Vale itself, to determine to which form they belong. The birds from across the main Himalayan range and the isolated birds in the Wardwan, as well as Ludlow's Bringh specimen, will no doubt be *curonicus* but we have doubts as to the others which, as pointed out by Stuart Baker, lay an egg agreeing in size with the Indian bird, and which we feel might prove to be invaders from the Plains which find their way up the Jhelum gorge. There seem to be few skins from these areas in the museums, and all Kashmir specimens from that State's wide territories on

PLOVERS

both sides of the main Himalayas appear to have been considered together.

HABITS AND NESTING. The Little Ringed Plovers are essentially birds of shingle and sandbanks. Standing or sitting motionless, the colour-pattern is such that in this habitat they are almost invisible at a distance of a few yards. They are not particularly numerous in Kashmir and may easily escape notice, but are concentrated in certain well-defined areas. The best procedure to discover their presence, and the whereabouts of their nests, is to walk boldly over a shingle bed to disturb whatever birds may be on it, and then to take up a hidden point of vantage some distance off. They return to the nest in short diagonal runs, standing motionless in between each little advance, whereupon they immediately blend in with their surroundings to a remarkable degree. When a bird appears to have settled down, it is necessary to mark its position with the utmost precision, for the eggs are even harder to distinguish amongst the shingle than the birds. It is quite possible to have to re-institute a search for a nest you have been standing over only a moment previously.

They are silent little birds on the whole, only giving vent to a plaintive piping when disturbed at the nest. This consists of little more than a circular depression in the pebbles scraped clear and then hollowed out by the bird pivoting upon its breast. It may be lined with a few water-worn scraps of wood, chips of bark, etc., picked up at the water's edge, or consist merely of a bed of the finer pebbles. At times it is marked out or protected by the presence of a large stone or two, but just as often there appears to have been no particular reason for the choice of position. We found one nest a hundred yards or more from the water's edge and only six inches from a frequented path which was used not only by all and sundry from a neighbouring village, but by numbers of cattle as well. The eggs in it were none the less successfully hatched off.

As is usual with most plovers, 4 eggs are laid, but we have found incubated clutches of 3 on more than one occasion. Whether the clutch be 4 or 3, the eggs are invariably placed symmetrically with the pointed ends inwards. They are drab or stone-coloured eggs, quite without gloss, and are stippled somewhat closely with small black markings so that they match a background of sand or shingle to perfection. Osmaston gives the measurements of eight eggs as averaging $28 \cdot 9$ by $21 \cdot 5$ mm. The breeding season is comparatively early, fresh eggs being hard to come by after the third week in May. Numbers in fact must be laid in April, while

Taobat, lower Gurais, in the Blue Pine and fir zone which stretches up to nearly 11,000 ft

A corner of the Dal lakes, typical of the marshes and village areas of the Vale

PLATE 78

Above the tree line except for scattered birches, the haunt of Ruby-throats, Redstarts, and many upland species

The Naubug Valley, characteristic of the lower reaches of the side valleys

very few indeed will be found in June. The chicks are equally difficult to spot, as when danger threatens they have the usual habit of the family of lying motionless with eyes closed and heads flat on the ground.

MEKRAN RED-WATTLED LAPWING

Lobivanellus indicus aigneri Laubmann
(See Plate 63 facing p. 252.)

KASHMIRI NAMES. *Hatatūt, Hatatertu.*
BOOK REFERENCES. *F.B.I.*, VI, 186; *Nidification*, IV, 397; *Handbook*, 460.

FIELD IDENTIFICATION. (4) Draws attention to itself by wheeling anxiously around the intruder, calling out plaintively *Did-he-do-it, did, did-he-do-it.* The black and white wings—brown when folded—black face, and deep black breast contrasting strongly with the white of the remainder of the lower plumage, and the long yellow legs are distinctive features. The red wattles in front of the eyes are not easily seen except at the closest range. Found on the shingle-strewn islands and on waste ground along the banks of the larger rivers and on the margins of extensive jheels.

DISTRIBUTION. The Red-wattled Lapwing is not uncommon along the Jhelum Valley from Domel upwards wherever the valley widens sufficiently for fields to border the river. It is also to be found in the Vale near the banks of the main river and wherever there is extensive waste ground. It occurs too where marshes have to a certain extent been reclaimed, as for example the Tulamal Jheel, but it is undoubtedly commonest amongst the islands in the lower reaches of the large side rivers. It is apparently resident wherever found.

The Red-wattled Lapwings are found from Iraq to Indo-China, the Malay peninsula and Sumatra. Our race is spread over the greater part of India down to Madras, where it meets the typical race of Southern India and Ceylon.

HABITS AND NESTING. To find these plovers in any numbers it is best to repair to the lower reaches of the larger side rivers, in particular the rivers Lidar, Sind, and Madmatti. Where their valleys widen out and the fall begins to ease off, rough stony or grassy flats of waste land border the banks; and islands, some of them of considerable size and partially cultivated, appear in their beds. Such is the ground sought after by the Red-wattled Lapwing and here the birds are not uncommon, almost every small island having its pair and the bigger ones perhaps a larger population. This bird is, however, strongly territorial and resents fiercely any intrusion across its frontiers by outsiders of its own species and in fact by any bird, animal or man abnormal to the accepted population of its area. For this reason there is never any question of finding Red-wattled Lapwings in small flocks and parties as is so often the case with other plovers, so that pairs are well spaced out over the suitable areas.

Who has not experienced what almost amounts to a mobbing by these excitable birds, which suddenly rise from the ground, perhaps even a couple of hundred yards away, and fly towards the intruder wheeling about him, sometimes swooping to quite close quarters, keeping up *ad nauseam* loud plaintive protests of *Did-he-do-it, did-he-do-it, did did did-he-do-it*? If one of the pair is sitting at the nest when its watchful mate gives the first alarm, usually a subdued *pit* uttered on the ground before starting its uneasy flying, it will sneak away to some distance before joining in the mobbing.

We do not consider the plumage of this lapwing to be of a particularly protective nature. True, with the wings closed the upper plumage becomes an unobtrusive brown matching its surroundings well enough, but the deep black of the face, foreneck and breast contrasts strongly with the white bordering it and makes it easy to pick out from the front or side a motionless sitting bird even at some distance. When sitting on its nest amongst stones the size of one's fist or larger, it does, however, blend sufficiently well with its environment.

The nest is interesting even though it is but a scrape in the bare ground, for the birds have a habit of lining and surrounding the slight hollow with small pebbles or little bits of hardened earth and other matter found in the neighbourhood. Sometimes this collection of odds and ends is almost absent, at others it forms a bed for the eggs as well as a considerable rampart around them. No doubt it has been evolved to combat encroaching sand, as is the case, for instance, with some of the desert-nesting larks which surround their nests with a rampart of stones of such a size that their strength must be sorely taxed in collecting them. Large numbers of sheep are driven up the Sind and Lidar valleys in the spring and early summer just when the Red-wattled Lapwings are preparing their nests and we have found eggs on the islands in the lower Sind completely surrounded by sheep droppings in addition to the usual little pebbles or bits of earth. In one case the numbers of such dried-up pieces of dung ran well into hundreds in addition to bits of earth, dried horse-dung, bark, and tiny pieces of driftwood, the flotsam of the islands. In this instance the space covered by the nest was well over 8 inches across.

As a rule the site chosen is in the open where the owners have a good all-round view, and in addition to this it is often placed in an area slightly higher than the surrounding ground further to facilitate observation and to provide good drainage. This is, of course, not invariably the case and we have found nests in hollows from which the sitting bird could be flushed at close quarters. We use the term ' sitting bird ' advisedly since both incubate. In accordance with the general rule amongst the plovers the full clutch is 4, the sharply pointed eggs fitting well and compactly together, lying as they invariably do with the small ends inwards. They are well camouflaged eggs, always being hard to pick out from their surroundings, for they seldom have any gloss, are generally a fairly dull buff, often with a tinge of ochre about them, and are well marked with large and small irregular black blotches. Sometimes the blotches are so big as to be almost continuous, especially around the larger end ; sometimes they are smaller and well separated. They average 42·9 by 31·1 mm. The birds appear to lay mainly in May and the first half of June.

BLACK-WINGED STILT

Himantopus himantopus himantopus (Linnæus)

(See Plate 64 facing p. 253.)

KASHMIRI NAME. *Longzeyet.*
BOOK REFERENCES. *F.B.I.*, VI, 193 ; *Nidification*, IV, 403 ; *Handbook*, 464.
FIELD IDENTIFICATION. (4—) Nearly always seen wading in clear water a short distance from the shore. The most striking feature about the Black-winged Stilt is the inordinate length of its thin red legs in comparison with the slender body. As its name implies the wings are black, but they have a green gloss, while in the breeding season the top of the head of the male assumes an increased amount of mixed black and white. The remainder of the plumage appears almost entirely white, although again in the case of the male it takes on a rosy tinge about the neck and

STILT

breast. In keeping with the rest of this remarkable bird, the bill is long and straight. In the female a sullied brown replaces the male's black, and the crown and nape are often mixed grey and white.

DISTRIBUTION. Confined almost exclusively to the shallower marshes and lakes of the lower part of the Vale, but Osmaston records having seen it in swampy ground near the mouth of the Lidar River, the only time he met with it. It is never common and is very capricious in its appearance and distribution in Kashmir but is seldom entirely absent in summer as a breeding species. W. T. Loke came upon a bird in July at Vishan Sar (11,500 ft) which, of course, was on passage.

In addition to Kashmir this distinctive bird breeds also in the Punjab, Sind, round the Sambhar Lake, and in Baluchistan, nowhere in great numbers. In the winter there is a great influx, extending to the whole of India, from its wider range outside for it is almost cosmopolitan in its distribution. The typical race occurs and nests in South Europe, Africa, and much of mid and South Asia, while other forms are present in America, Australia and New Zealand. The Ceylon bird has also been separated.

HABITS AND NESTING. Black-winged Stilts breed before and during the rains in many parts of India, particularly in the northern Plains, which are intensely hot in summer. One would hardly expect, therefore, to find them nesting in the comparative coolth of May and June in the Kashmir Valley. Nevertheless, they are at times not rare around the eastern and southern shores of the Wular Lake and about the extensive jheels on both sides of the river below Shadipur and Sumbal. They do not appear to have quite made up their minds about the advantages of a sojourn in Kashmir, as quite often hardly a bird is to be seen, and why they should be more numerous in some years than in others is by no means clear. It does, however, appear to us that their numbers may be dependent upon the state of the water at the usual time of passage, that is about early May. In a cold late year, resulting in a low water-level in the early months, we noted them in the first week of June in comparatively large numbers feeding around exposed mud flats in the Wular Lake. A couple of years later, when in the same month the water-level was much higher, we failed to trace a single bird, although we had two men searching the marshes. They are in any case uncertain in their movements and prone to seek new haunts for no apparent reason.

As they are never numerous in Kashmir, it is easy to overlook their presence in the Vale. Food-supply is probably responsible for these local movements, as they are dependent on water of a depth enabling them with their long legs to pick up seeds, aquatic insects and small molluscæ which other shore birds cannot touch. Their movements in the water appear somewhat grotesque, as they withdraw the partially-webbed foot clear of the surface before each forward step to avoid resistance. As one would expect, they can swim quite well but are not particularly given to that habit.

Unfortunately we did not visit the Wular Lake mud flats on the occasion referred to to see whether the birds were breeding thereon, and so have records of nests only on the bunds of rice-fields and round the margins of the larger marshes, where they often choose ground from which the water has receded leaving the surface in a somewhat caked state. Here they collect a few aquatic weeds or grass and at times a little mud into a mound which projects an inch or two above the surface. The eggs are laid in a slight depression on a scanty lining of grass which is usually soaked through by the soggy weeds and mud on which it is spread. They are noisy, demonstrative birds when disturbed at the nest-site, wheeling about overhead with the outlandish legs trailing awkwardly behind. They also have the habit when agitated of beating the water with the wing-tips. The usual number of eggs laid is 3 or 4. These are by no means unlike Red-wattled Lapwings' eggs, but the texture is finer and they usually have a fair gloss. The ground colour is a clearer brown or buff and the markings fairly numerous spots and blotches of black, blackish-brown and rich brown, sometimes including a few streaks and being perhaps more numerous towards the larger end. They average 44·0 by 31·0 mm. There is little published data upon which to fix the extent of their breeding season in Kashmir, but in June they were nesting around shallow jheels between Shadipur and Dangarpur on the Jhelum's left bank, where the subject of the illustration was found on the 14th. They commence to lay in that month, certainly not earlier, and promptly lose the majority of their first layings which go the way of so many water-birds' eggs in Kashmir. That is, they find their way into the bazaars, or are hawked round the houseboats as plovers' eggs. They lay again but even then many of the second clutches are filched. Both male and female incubate. The nest-relief is carried out very frequently and it is an amusing sight to see the relieving bird shake any mud and water off the feet before mounting the nest.

IBIS-BILL

Ibidorhyncha struthersii Gould
(See Coloured Plate V facing p. 328, Plate 67 facing p. 268 and
Plate 79 facing p. 316.)

BOOK REFERENCES. *F.B.I.*, VI, 196 ; *Nidification*, IV, 406.

FIELD IDENTIFICATION. (4) A dove-grey bird of the larger rivers, which in build and habits is a mixture between the larger sandpipers and the plovers. It affects shingly islands in certain of the larger side rivers, and flies from island to island well above the water's surface, at times uttering a loud musical *klew klew*. When picking up its food it has that habit so noticeable in many of the plovers of pivoting or tilting its whole body forward.

DISTRIBUTION. The Ibis-bill is nowhere common. Odd pairs have been noted on the islands of the Sind and Lidar rivers at 6,000 to 8,000 ft, and Osmaston records having seen a pair on the east Lidar River at Praslun. We found it more numerous on the Wardwan River between Inshan and Tsuidraman (approximately 8,500 ft) where there are many islands of considerable size with stony margins. Ludlow also found it about here and remarks that he saw several specimens of this strange bird in and near the islands just below Inshan in July 1919. It is undoubtedly a sure find in that neighbourhood in larger numbers than elsewhere in our territory. In the winter it descends the rivers into the foothills, being found a few miles into the adjacent plains. The downward movement probably starts in September ; in that month we saw a bird well below Ganderbal.

The Ibis-bill occurs at medium to high elevations from the Pamirs to Kashmir and Ladakh, Tibet, and along the Himalayas to North Burma and North-West China. The downward movement in winter along the rivers to the foot of the hills is general throughout its range.

HABITS AND NESTING. The Ibis-bill is a unique bird, as well as rather a rare one. It has, of course, no affinities with the ibises, to which it bears little resemblance except in its bill. In fact it brings to mind three very different types of birds : stone-plovers, curlews, and sandpipers. It is comparatively large, with a loud disyllabic whistling note which it utters on the wing. Its flight from island to island is usually well above the water, not close to its surface like that of the sandpiper. The wing-beats are slow and in this it recalls the Red-wattled Lapwing. Its cry is not unlike that of the Greenshank, but consists only of a double clear whistle, a loud rather mournful *klew klew*. Magrath made some interesting observations in the *Journal* after his first encounter with this bird at Pahlgam in July 1911. ' In habits ', he wrote, ' the Ibis-bill differs little from the Sandpipers and allied waders, but in some respects it approaches the Plovers. In the evenings they were often to be seen feeding and probing for worms on the short grass by the river during which they would run for a few steps and then stop much as the Plovers do. When watching them one morning I also noticed a rather characteristic and almost " cincline " habit of wading breast high into fast flowing water and dipping the whole head and neck under water picking up food of sorts from the bottom. The Sandpipers act similarly, but not in such a dipperlike manner and in such heavy water.'

Further light is thrown on the cincline habits of this bird and on the reason for its peculiar structure in some notes on one of the Everest expeditions written by Major R. W. G. Hingston. ' The reason ', he explains, ' why this bird has a bill with so peculiar a curve is disclosed by observing its behaviour. It runs about on the water-worn boulders, sometimes wading into the torrent up to its breast, thrusting its long bill under the round stones in the hopes of finding insects beneath them. Sometimes it curves its bill around the front of the stone, sometimes it inserts it from one side. The bill is an excellent instrument for this purpose. Were it straight it would not suit the roundness of the pebbles. Its peculiar curve is a necessary part of it, and is excellently adapted to the habits of the bird. For it is curved in such a way that it fits neatly around the boulders when the bird is probing for food.'

In the above quotation great stress has been laid upon the peculiarity of the bill's curvature. Examination of the illustration facing page 328, and of Colonel Phillips' excellent photographs in the *Journal of the Bombay Natural History Society* (45, 347), shows that in the living bird the curve is fairly even and but little more exaggerated than in that of a curlew or ibis. It is certainly not bent downwards so abruptly as in the sketch on page 197 of Vol. VI of the *Fauna*. We have, however, examined a number of skins of this bird and in them the curvature of the bill varies to an appreciable extent and does show a decided increase over the last third. May not this be distortion during the process of drying, so that the shape of the bill in the museum specimen soon differs from that of the living bird ?

With the completion of its motor road Pahlgam has changed immeasurably since Magrath's day, with the result that the Ibis-bills have long since been driven away from the islands in its immediate vicinity. They certainly prefer more secluded stretches, where they are to be found in pairs on islands well provided with stony and wide shingly margins. They are not averse to running about the interiors of the islands, threading their way between the scrub from one open patch to another and hiccoughing between each little advance like a sandpiper, but they delight chiefly in stony shores and those islands provided with stony ridges and open areas of shingle above the normal flood level. It is here that the nest is generally to be found in a slight depression hollowed out amongst the shingle and lined with small stones.

In this connexion Ludlow's and Bailey's remarks are most interesting and well worth quoting. The former writes : ' The nest seems always to be placed on high ridges on the island and it is no use looking for it on low ground where mud on the shingle shows it liable to be flooded. The eggs, 4 in number, are laid in a shallow depression amongst very small pebbles which the parent birds evidently take pains to collect.' Bailey's nest, also of 4 eggs, ' was made of small stones about half an inch in diameter, forming a smooth and perfectly flat surface '. Judging by an unfinished scrape, the ground is first cleared of all stones over a patch measuring about a foot in diameter and the small stones are then carefully relaid after the manner of crazy paving. Although 4 is the normal number of eggs laid, Whymper in Garhwal took some incubated clutches of 3, while Osmaston, on 14 June, took a full clutch of 2. This, however, he considered was a second laying. Colonel Phillips records clutches of 3, 4 and 5. The last must be quite abnormal. The usual nesting time in Kashmir is undoubtedly April, so that the young may be hatched off before the snows begin to melt, but where early floods have destroyed the first nests, second layings may on occasion be found in May and even, as Osmaston shows, in June. In Kashmir proper, however, the islands are invaded by Gujars and their flocks, with the result that second layings have little chance of survival.

Stuart Baker states that the colour of the legs and feet of breeding adults is blood red, but Colonel Phillips, who photographed this bird in colour at its nest on an island in one of the Kashmir rivers, states positively that on the two occasions on which he has seen these birds at the nest, their feet were a livid and greenish-grey colour. Stuart Baker describes the eggs as follows. The ground colour is a very pale grey tinted with greenish, yellowish or buff, the tint always very faint and indeterminate. The markings consist of pale or moderately dark reddish blotches, generally rather small, fairly numerous at the larger end, but sparse elsewhere.

PLATE 79

Crow's nest, occupied by Kestrel, also holds fresh egg of Sparrow-Hawk

Ibis-bill islands near Inshan in the Wardwan Valley

PLATE 80

Delta at Basman, Wardwan Valley, which supports a great concentration of Hodgson's Rosefinches

Weed-covered areas near Suknes, Wardwan Valley, elevation 9,500 ft, in which the Paddy-field Warbler nests in some numbers

The secondary markings, of the same character and equally sparse, are of pale grey. In shape the eggs are rather broad ovals, not as broad as those of the Woodcock and also less pointed. The texture is fine and close while the surface has a fair gloss, but the shell is fragile in proportion to the egg's size. They average 51·0 by 36·9 mm. A chick in down picked up at Inshan on 22 May, and judged to be from five to seven days old, had the down grey above paling to almost white below. The crown and sides of the head were a shade darker. The bill, about an inch long with a tendency to bend downwards towards the tip, was black. About the wings, back, rump, and sides of the body and near the eye the grey down was stippled with light brown markings. The feet were a pinky-brown and the claws black.

COMMON SANDPIPER

Actitis hypoleucos (Linnæus)
(See Plate 65 facing p. 260.)

KASHMIRI NAMES. *Tōnt Kon, Kola Kavin, Bularu.*
BOOK REFERENCES. *F.B.I.*, VI, 217; *Nidification*, IV, 408; *Handbook*, 466.
FIELD IDENTIFICATION. (3) Grey above, mainly sullied white below, the Sandpiper is a small wader which shares with the Little Ringed Plover the shingle beds and islands from the level of the Vale to about 6,000 ft. Above this elevation the Ringed Plover fades out, leaving the Sandpiper in sole possession of the larger rivers up to at least 10,000 ft. Its main characteristics are the long straight bill and somewhat spindly legs, on which it persistently waves the body up and down while intermittently bobbing the head on its slender neck. It usually flies low across the water with a rapid jerky flight, the wing-tips bent downwards so that they almost flick the surface. A whitish band bordering the trailing edge of the wing is very prominent in flight. The call-note is a piping whistle, while a more querulous chittering is connected with the display.
DISTRIBUTION. Common is its name and common it is, being a feature of the bird life of all the larger rivers of Kashmir up to an altitude of at least 10,000 ft. It does not penetrate the side streams and mountain torrents to any extent, preferring the islands of the larger rivers and to a lesser degree their open banks. It is common

SANDPIPER

in the Wardwan Valley, and in Gurais, but in May we did not meet with it on the Kishenganga in any numbers below Taobat. It will not be seen around the margins of the Kashmir lakes and only seldom along the main river from Baramullah to beyond Islamabad, where the currents are sluggish. In the *Nidification* we are credited with finding it numerous around the Dal lakes, but this statement must be due to a slip of the pen for we have never recorded it from there. It, of course, occurs on the Pir Panjal side of the Vale, and Phillips found a nest at 9,000 ft at Gulmarg. In the winter it is a great wanderer, soon spreading throughout the whole of India. Numbers start to move down as soon as the rains are well established in the Plains, many not returning to the hills again until late April, though Meinertzhagen states they were already abundant in the Sind Valley between 6,000 and 8,000 ft in late March.

The Common Sandpiper, a species with no races, nests throughout almost the whole of Europe and across Asia to Japan, in winter migrating into Africa and to India, Burma, and the East Indies to Australia and Tasmania. In India it nests for certain only in the extreme Western Himalayas to Lahoul, though a few may nest in North Baluchistan.

HABITS AND NESTING. In the breeding season the chittering whistles of the Sandpiper, *tisitit-tisitit*, draw attention to it as surely as does the peculiar bobbing walk or the low rapid flight close over the water with down-drooping pointed wings. They cannot progress over the stones and boulders at the water's edge without interminably waving the rear half of the body up and down, at the same time intermittently jerking the head. They are not particularly shy and often allow of quite a near approach before flying off to the opposite side of the river, piping as they go. When disturbed on the nest they generally sneak quietly to the water's brink before taking to wing. They then stand afar off protesting plaintively until the danger is past.

We have taken a great many nests in Kashmir, and it is obvious that sandpipers prefer to nest upon the scrub-covered islands, but nests will also be seen along the river banks. Territory appears to be ill-defined, and nests will often be found quite close to one another; we once discovered four on one small island. They are usually well concealed in the roots of a bush, such as lonicera, tamarisk scrub or wild indigo, or even in amongst docks and other weeds. Sometimes a nest will be under a large stone or overhanging stump and very occasionally it may be in the open, but this is quite exceptional. Nests are seldom more than three or four yards from the water but this rule is not invariable. The bird first of all scrapes out a circular hollow which it rounds off by shaping it with its breast. Inside this it collects a scant nest of odds and ends, such as small scraps of wood, dead leaves, and often a certain amount of grass.

The eggs are always 4 in number; we have never seen more. Buchanan, however, had three clutches of 6 and a number of 5's which he found around Pahlgam, but we cannot help wondering whether these nests had been tampered with. On more than one occasion we have been shown nests of different birds to which eggs had been added in the hopes of increased reward for something out of the ordinary. Sandpiper's eggs are rather handsome pyriform ovals. They are placed with the pointed ends inwards and are rather glossy. The markings are small and numerous spots and blotches of reddish-brown. Boldly marked eggs are to be found, but in the majority the markings are small with equally small secondary marks of pale grey. Their ground colour is usually pale creamy buff, a noteworthy fact being that, taken as a whole, they are markedly paler than Sandpiper's eggs taken in the British Isles. We find this applies to numbers of other species breeding in Kashmir; for example, to the Woodcock, Water-Rail, and Moorhen. In size they average 35·5 by 26·0 mm. The normal breeding season is May and June. It is exceptional to take fresh eggs later than June, albeit the eggs in Phillips' Gulmarg nest were hatched off in July. Few eggs will be found before 15 May. Both birds take part in incubation.

WOODCOCK

Scolopax rusticola Linnæus

(See Plate 66 facing p. 261.)

KASHMIRI NAME. *Zar Batchi.*
BOOK REFERENCES. *F.B.I.*, VI, 252; *Nidification*, IV, 413; *Handbook*, 472.
FIELD IDENTIFICATION. (3+) The diagnostic characteristics of the Woodcock are firstly its large size in comparison with the Fantail Snipe—it weighs up to 12 ounces to the Snipe's 4 or 4½—secondly the comparatively shorter bill and warm buff plumage with dark brown transverse bands across the head—in the Snipe the bands run longitudinally—and the finely barred underparts. It is a bird of the forest glades with a peculiar owl-like flight, flitting

easily this way and that between the tree-trunks as it flies silently away.

DISTRIBUTION. Not uncommon on the northern slopes of the Pir Panjal range from about 7,000 ft. We have flushed it on the slopes of the Kazinag mountains, and above the Wular Lake, while two clutches of eggs we saw recently came from the Bandipur Nullah. We also observed it in the Surphrar Nullah where it was seen on several occasions rôding at dusk. According to Ward it breeds both in Kashmir and in Kishtwar. It is said to be not uncommon in the Kishenganga Valley. In winter it has a peculiar distribution, dropping lower down into the foothills, but migrants penetrate to the hills south of the Brahmaputra and to the ranges of Southern India, a few even reaching Ceylon. There are, however, practically no records of movements in between these hill ranges, and there is nothing to connect these South Indian birds, which may come from much farther afield, with our Kashmir birds which may only seek somewhat lower elevations in the Himalayas ; for throughout the winter a certain number are still to be found in their summer haunts, probably remaining near the warmer springs in the forests.

The Woodcock is another bird which is spread far and wide through North Europe and Asia and Japan, extending farther south in winter into North-West Africa and into the southern countries of the Asiatic mainland. So far as India is concerned, its breeding haunts are confined to the Himalayas from Chitral and Hazara to Bhutan.

HABITS AND NESTING. The ' Cock ' is an inhabitant of the mixed forests containing lavish undergrowth and bordered by open damp patches and little streams with muddy margins. To these glades the Woodcocks repair at dusk to feed throughout the night on their almost exclusive diet of worms. These feeding-grounds may be at considerable distances from the haunts where they breed, as the nest is often amongst the dead leaves on some sloping dry hillside in the forest, usually where there is plenty of undergrowth but at times right in the open.

The so-called nuptial flights, usually attributed to the males, and known as rôding, have been described many times, but we cannot do better than quote the description of this interesting performance given in the *Nidification*. ' He told me that just before dusk the birds came out of the cover and sailed backwards and forwards in front of it. At first the flights were high but gradually the birds got lower and lower until, reaching the level of the scrub, they disappeared into it. Each flight was said to be in the form of a long arc, the highest point being reached at its end and commencement, while in length they were anything from fifty to two hundred yards or more. The bird was described as flying slowly with plumage puffed out, head thrown back and bill pointed somewhat upwards. I unfortunately did not inquire how long these nuptial flights lasted but the impression I gained was that the flights were numerous and lasted a long time.'

The nest is merely a circular depression scraped out in the dead leaves. A flat layer of leaves forms the lining, occasionally having bracken incorporated in it, if there happens to be any in the vicinity. It is usually placed amongst the scrub but we have seen a nest quite in the open. The number of eggs is usually 4, but 3 only may be laid; Captain Livesey took a nest of 3 in the Ferozpur Nullah which showed distinct signs of incubation. The eggs are large broad ovals, generally somewhat pointed at the small end, and are very handsome. Those Kashmir eggs which we have seen have all had a very pale creamy background, much lighter in tone than any series of English-taken eggs. This evidently does not hold good for the Himalayas as a whole, Anderson (*Stray Feathers*) describing a clutch taken by him in Kumaon as ' far darker and redder than the usual run of Woodcocks' eggs '. The markings consist of blotches of varying size, sometimes smudged rather thickly round the large end. In colour they are reddish or reddish-brown with underlying markings of lavender or grey. They measure 40·9 by 33·8 mm. According to Stuart Baker the breeding season in Kashmir is from late in May to July. Captain Livesey's incubated eggs were, however, taken about halfway through May. Osmaston took a nest with three rather hard-set eggs from amongst skimmia on 20 July. This was at 9,000 ft in the fir forest east of Gulmarg. This may have been a second brood. Although there is little on record, and nothing as regards Kashmir, what evidence there is goes to show that the Woodcock is double-brooded, at any rate in some countries. Stuart Baker says both sexes incubate but it is generally accepted that the female alone incubates, both sexes tending the young. These when hatched take some time to dry off, but can leave the nest after a few hours.

The Woodcock's appetite is so voracious that it would undoubtedly be difficult for the parents to keep the young supplied at the nest. It is alleged, therefore, that this has led to the evolution of the parents' habit of transporting their chicks one by one over considerable distances, thereby getting them in a short space of time from the dry surroundings of the nest to the boggy feeding-grounds in which the adults' sensitive attenuated bill can probe deeply for the succulent worms beneath. Whatever the reason, it is now an established fact that Woodcock do carry their young from place to place, but there still appears to be a measure of doubt as to the method of transportation. Some hold that the chick is grasped in the claws, others that it is pressed up against the breast between the feet, or yet again held between the thighs. There is nothing improbable in all of these methods being employed, although it is generally considered that the Woodcock's claws are too weak for this purpose. Other reasons for the practice would appear to be to get the young over obstacles which in their flightless state they are unable to negotiate, and to remove them from immediate danger. On a sudden threat, it is unlikely that there would be time for an agitated parent to pick up its young always in the same manner.

Unfortunately, the one occasion upon which we thought we were to witness this interesting spectacle was undoubtedly a case analogous to 'injury-feigning', for to quote from the *Ibis* (January 1946, 13), 'When a Woodcock in company with its brood is disturbed by the close approach of a human being or a dog, it commonly behaves in a peculiar and characteristic fashion. It flies slowly and heavily, with its legs hanging down and its tail much depressed, usually not rising more than a few feet from the ground, and very frequently settling again after a comparatively short flight. ... the impression gained by the human observer who has not previously seen this performance is that the bird is carrying something fairly heavy with which it is only just able to fly. The tail depressed between the legs completes the illusion by suggesting a largish object carried in this position. One or two of our correspondents have referred to this behaviour as the " decoy-flight ".' To this we would like to add that the birds which so successfully diverted our attention from their brood also made a great deal of noise when first disturbed by our dog, squealing and groaning like a bird or animal in pain, and subsequently maintaining a froglike croaking until they pitched into cover at the bottom of the nullah.

FANTAIL SNIPE

Capella gallinago gallinago (Linnæus)
(See Plate 67 facing p. 268.)

KASHMIRI NAME. *Chāh*.
BOOK REFERENCES. *F.B.I.*, VI, 261; *Nidification*, IV, 419; *Handbook*, 475.
FIELD IDENTIFICATION. (3—) A marsh bird which rises at one's feet with extreme suddenness, giving an impression of pointed wings, a very long straight bill, and a brown body heavily streaked and lined with black. The flight is swift and jinking and on landing the wings are momentarily held extended high over the back. In the breeding season they have a habit of circling over the nesting area at a considerable height, every now and then diving steeply, during which dive the well-known drumming sound or 'bleat' is produced.
DISTRIBUTION. From the large numbers of passage migrants which pass through in the autumn and spring, a limited number remain to nest on boggy ground and marshland grass around a few of the more secluded jheels in the Vale. The extensive rim of soft grasses and wet ground around Hokra is perhaps their main stronghold, but the *rukh* portion of the Anchar Lake, the Krahom swamp which has been partly drained, and some of the marshes near Sumbal have their quota. A few Jack Snipe are sometimes to be seen amongst them, probably pricked birds for there are no records of this latter species ever having nested within Indian limits. Fairly large numbers of Snipe arrive in Kashmir from the north in the autumn, a goodly proportion staying throughout the winter to provide the sportsman with the trickiest of shooting. Still larger numbers pass on to India and beyond, retracing their steps in the spring. Kashmir is on the southernmost fringe of the Snipe's breeding range so only a meagre proportion of the winter's population remains on throughout the summer.

The Snipe's main nesting area lies farther north, from Great Britain and Scandinavia to the Pyrenees and across to Northern and Central Asia. There are also races in America. Excepting Kashmir, the only other reported Indian breeding area is in the Santhal Parganas. In the winter, hordes of Snipe migrate south, not only into India and Burma, but from Europe into North Africa and in Asia down to Malaya.
HABITS AND NESTING. The Fantail Snipe is, of course, well known to everyone keen on shooting. It must be admitted that

SNIPE

snipe-shooting in India is one of the most satisfying of sports, not perhaps calling for that skill required to lay low the wild birds of an Irish bog, but providing larger numbers and consequently plenty of sport. But whether one is walking up the Snipe to shoot it or to find its nest matters little, for its ways are the same in either case and it will rise off or from near its nest in the same manner as it does before the sportsman ; a sudden springing to life with a harsh croak, that well-known nasal *pench*, the dark wings showing sharply pointed and the long straight bill thrust slightly downwards. Soon it starts jinking to left and right, flying for some distance not many feet from the ground. It then rises high into the wind to fly rapidly around its area in wide circles until vouchsafed a chance of dropping abruptly to earth with bent-back wings, only checking its fall close to the grass with a few fluttering beats. On landing it melts into its surroundings and becomes once again well-nigh invisible to even the best-trained eye.

When feeding, the Snipe quite obviously has a rooted objection to wetting its breast, so it is no use expecting to flush it from ground where it cannot either walk over the vegetation or wade without constantly dipping its feathers in the water. What it prefers is wet or muddy ground sufficiently damp for its attenuated bill to probe beneath the surface, and for the sensitive ends of the mandibles, which are plentifully supplied with nerves and muscles, to be able to feel for and open up in order to seize the worms and grubs come upon in the slime.

The commencement of nesting is marked by the drumming flights of both sexes, later, when the female is sitting, carried on by the male alone. Mounting high in the heavens, frequently uttering a sharp high-pitched double note, *chick-chack*, *chick-chack*, he commences to circle his territory. Suddenly he dives steeply and, as he describes a wide arc on rapidly moving wings, a vibrant deep-toned whinny reaches the ear. This he achieves by fanning out and depressing the tail. On gathering speed the stiff outer feathers vibrate rapidly in the rush of air. The sound may be reproduced by binding these feathers at an angle to a cork and twirling the contraption round the head on the end of a string.

The Snipe's nest is a scanty pad of dry grass, rush blades, or an occasional leaf, usually well concealed in a tussock of grass or thin tuft of reeds, but the bird may also rely on the protective coloration both of itself and of its eggs, for the latter even when laid open to view in short grass are not easily spotted. They are a pale buff or greenish-buff, smooth but with no gloss, and the greater part of this ground colour is covered over with large and small smears and blotches of blackish-brown and grey. They are strongly pointed, large for the size of the bird, and average 37·1 by 26·8 mm. The Snipe lays early, late April and May being the months for eggs. We have failed to find them in June.

EASTERN GREY HERON

Ardea cinerea rectirostris Gould
(See Plate 68 facing p. 269.)

KASHMIRI NAME. *Brag*.
BOOK REFERENCES. *F.B.I.*, VI, 340; *Nidification*, IV, 457; *Handbook*, 507.
FIELD IDENTIFICATION. (6) A dignified ash-grey bird standing about three feet in height. The long-legged storklike build; the thin neck with interrupted blackish lines down its front; a sharply-pointed straight yellow bill springing from a whitish head with flattened black crest in which are intermingled a few straggling white feathers; usually seen alone standing motionless in the shallows of a river, on the margins of the marshes, or flying steadily overhead on wide rounded wings with the head drawn back but the legs pushed out straight behind. These are the main points which mark out the Grey Heron. A solitary bird when feeding, but nests in colonies on tall trees around the lakes.
DISTRIBUTION. Grey Herons are widely spread and comparatively common in the Vale, particularly in the lower half with its large lakes and many marshland areas where there are large breeding colonies. They are not confined to the Vale, for they penetrate the side rivers up to some 7,000 or 8,000 ft, but probably only on fishing expeditions. We have seen them in the middle Kishenganga on a tree-covered island near Danudra where there might possibly be a heronry, in the Wangat Nullah, as far up the Lidar River as Aru, and along the many trout streams prolonging the Bringh, such as those of the Ahlan Nullah and the Naubug Valley. They are resident in the Vale of Kashmir.

The Grey Herons occupy nearly the whole of Europe, Africa and Asia, the Eastern race being a common bird throughout India and extending from Iraq across to the Philippines. Throughout this range they are mainly resident but move locally and collect during the nesting season into heronries, sometimes of considerable size, which they often share with other species. The typical race straggles into Baluchistan in some numbers and also into Sind.

HABITS AND NESTING. There can be few people unfamiliar with the appearance of the Heron. The bluish-grey plumage shows up well as it stands in the river's shallows or fronting the tall reeds rimming some Kashmir marsh. The long legs and the straight daggerlike bill on a thin drawn-out neck, poised to strike rapidly at any fish or frog which ventures in reach, are very distinctive features. For the Heron hunts mainly by the simple method of standing motionless till fish or other fry are rash enough to venture within range, whereupon a swift dab effects their capture. Occasionally the bird wades a few paces to another pitch, moving forward slowly and deliberately with the neck slightly curved into a graceful S : when the water is deep the body becomes almost horizontal and the curves of the neck more emphasized. Only very occasionally does the Heron swim, but we have twice seen it do so, floating high in the water with its neck stretched upwards to its limits. White also records seeing one settle on deep open water. The flight of the Heron is very stately, though it leaves the ground in a rather ungainly manner. But the neck is soon folded so that the head is just above the shoulders with the sharp bill pointing straight to the front while the legs are thrust out stiffly behind. With measured continuous flapping of the wide rounded wings the bird moves majestically by, mounting at times to a considerable height when intent on covering some distance and emitting at wide intervals a loud *frank*.

On these fishing expeditions the bird is invariably alone. Even in the nesting season it is unusual to see both birds of a pair fishing together, despite the fact that they nest colonially in what have become known as 'heronries' in the British Empire and 'rookeries' in the Americas. This applies no matter whether the collections of nests are those of herons, egrets, cormorants, and other birds such as spoonbills and ibises, alone or severally mixed together. In Kashmir the Grey Heron and Night Heron sometimes choose to nest in company, but many heronries are composed of the larger bird only. The trees selected are nearly always the largest chenars in the vicinity of water or not far from a lake's shore, so it is not surprising that the magnificent trees of the Shalimar and Nishat baghs hold heronries of long standing. A couple of huge chenars in Bandipur have also been patronized for many years.

Amongst the shady topmost branches of a single tree many nests —Meinertzhagen counted as many as 47—can be built at a height guaranteeing freedom from interference, save from the thieving crows. The nests are shallow collections of sticks, except where a protected site has favoured the survival of a nest from year to year, in which case it may attain considerable dimensions. There is in general a scanty lining of grass or leaves. The eggs, which are more often than not 3 in number, are long true ovals, although pointed eggs may sometimes be found. In colour they are a pale sea-green which soon fades to a lighter shade on being blown. They average 58·6 by 43·5 mm. Nesting commences very early in the year, all activities in the heronry being over by the beginning of June; Meinertzhagen, in fact, remarks that many young in a heronry he investigated had hatched out by the first week in March. Osmaston recalls that at the commencement of breeding many fresh eggs may be found lying below the trees and suggests that these eggs may be ejected in the struggle for nesting sites. Is it not possible that this is due to the depredations of crows, whose destructive habits in regard to the eggs and young in unguarded nests of any birds make these black bullies a disgrace to their order? The young are fed by regurgitation. In this connexion we would draw attention to our description of the feeding habits of the Little Bittern. This habit on the part of the chick of seizing the parent's bill in its own is also followed by the young of this and of the Pond Heron and, in fact, appears to be universally practised by the young of all the herons and bitterns.

INDIAN POND HERON

Ardeola grayii (Sykes)

(See Plate 69 facing p. 276.)

KASHMIRI NAME. *Broku.*
BOOK REFERENCES. *F.B.I.*, VI, 354; *Nidification*, IV, 470; *Handbook*, 512.
FIELD IDENTIFICATION. (4) In breeding plumage, with which we are chiefly concerned here, the dumpy little Pond Heron is really quite a handsome fellow. The normal drabness is doffed for brighter though still sober hues which, however, retain for him that ability to remain quite unnoticed whilst standing still in the shallows of some wayside pool until a sudden unfolding of white wings and a protesting croak reveal his presence. A long crest of three or four narrow white feathers springs from the head.

The back and the long scapulars, which help with the buff inner secondaries to conceal the folded wings, are rich maroon, the head and neck yellowish-brown, while the long pointed feathers of the breast and upper flanks are ashy-brown with pale yellowish streaks. The remainder of the plumage is white.

DISTRIBUTION. The Pond Heron cannot be termed a particularly common bird in the Kashmir Valley. In fact of late years we have failed to discover it anywhere except near the villages around the Anchar Lake, in the Sind River delta between Ganderbal and Shalabug, and at a heronry on the shores of the Wular west of Sopor. In the Jhelum Valley it is not uncommon between Domel and Garhi, while we have seen it from the car as far up the Valley as Uri.

This species occurs west to the Persian Gulf and east throughout India and Burma to Siam and Malaya. It ascends most of the hill ranges of India to about 4,000 ft. According to Osmaston the Kashmir birds are residents, and in fact, like the last bird, it is in the main a resident everywhere, moving about only locally.

HABITS AND NESTING. Everyone in India is familiar with the ways of the Paddy Bird, that squat little heron which is content to stand motionless for minutes at a time with its head drawn back between its shoulders as it waits patiently at the edge of any patch of water, no matter how dirty nor where it may be, for some wandering frog, fish or insect to provide it with a meal. Although appearing sluggish in its movements, often moving step by step with extreme slowness and deliberation, when need arises it strikes with great quickness, revealing that after all it possesses a long neck like others of its family. Any type of smooth water attracts it, and it may be seen by a filthy roadside puddle, on a bund between rice-fields, or at the edge of a clear pool.

In the Kashmir Valley they seem for some reason to be confined practically to the margins of the Anchar and Wular lakes, particularly to the waterways and reaches bordering the few villages on their shores, while extensive colonies also inhabit the bushy delta of the Sind River between Ganderbal and Shalabug. Near the villages they nest in small colonies, usually in willows, the nests being at no great height from the water, often in fact at no more than 6 or 7 ft up. The half-dozen or so nests in these colonies are usually spread about in different trees and are seldom very close together. In the thickly wooded delta, colonies are often larger; we have seen some which held quite a hundred pairs. Here in the inundated bushes the nests are built low down, many within 5 feet of the water's surface. Five miles above Domel there is a

PLATE V

1. Indian Black Partridge (adult male). 2. Kestrel (adult male). 3. Eastern Red-billed Chough. 4. Hume's Blue Rock-Pigeon. 5. Indian Sparrow-Hawk. 6. Nepalese Snow-Pigeon. 7. Long-eared Owl, 8, Ibis-bill, 9. Central Asian Hobby

long island with a considerable volume of the Jhelum's turbulent waters swirling past on either side. The lower end of it is covered with close-growing bushes and low trees. It is here that the majority, if not all, of the Pond Herons of the Jhelum Valley stretch obviously collect to breed, and as one passes up or down the road in May, June, and July, birds can always be seen winging their way to and from the island, standing on the low trees, or flapping their way into the interiors of the large bushes where lie their nests.

In form the nests remind one of nothing so much as over-sized doves' nests, being loose flat collections of fairly small and thin dead and half dead sticks torn from the trees by the birds themselves and interlaced with ill-defined depressions for the eggs. They do, however, appear to be slightly more substantial affairs than those built by their sisters in the Plains. The eggs, 4 to 6 in number, are rather long ovals, pale blue in colour but rather more blue than those of the other herons. They measure 38·0 by 27·9 mm. We noted building operations on the 9th of May which had obviously been in progress for some days. Eggs may certainly be taken from mid-May and throughout the whole of June, while we visited a colony near the village of Batsapur on 11 July which still had incomplete clutches of fresh eggs. This extended season may be due in part to the depredations of the House Crows which succeed in smashing and sucking the contents of many nests. Both sexes incubate and bring material for the nest, but the female does all the building. When it comes to looking after the young, the male again takes his full share. As is the case with the Little Bittern and the Grey Heron, the chicks have the habit of demanding food by seizing the parent's bill; he or she then disgorges the food by depressing the bill towards the breast so that the food-remains fall to the bed of the nest.

NIGHT HERON

Nycticorax nycticorax nycticorax (Linnæus)

(See Plate 69 facing p. 276.)

KASHMIRI NAME. *Bōr.*
BOOK REFERENCES. *F.B.I.*, VI, 359; *Nidification*, IV, 475; *Handbook*, 514.
FIELD IDENTIFICATION. (5—) The Night Heron, with its short legs, stout deep bill and large head hunched between the shoulders

on thick bull-neck, is the very antithesis of the Grey Heron. Its sole resemblance to that much-lauded bird lies in its loose black crest containing long white feathers which are sufficiently soft and hairlike to be waved about by the wind. The crown, neck and back are black glossed with metallic green, and the remainder of the plumage pale blue-grey with the cheeks and underparts white. Young birds are brown, much spotted with pale rufous. A nocturnal and secretive bird, it rests by day in parties in the upper foliage of large trees, flighting at dusk in ones and twos or small flocks to its feeding grounds in the ponds and marshes. While passing overhead it frequently emits a loud *quark*. Nests colonially, sometimes in company with the Grey Herons.

DISTRIBUTION. The Night Heron is fairly plentiful around the larger jheels of the main Vale from the vicinity of Srinagar down to the Wular Lake. It is a resident species.

The Night Heron has an immense range in South and Central Europe, almost throughout the African continent and the greater part of Central and Southern Asia east to Japan and the Philippines and south to Malaya. Another species of the genus occupies much of North and South America.

HABITS AND NESTING. This squat heron is in some ways as regards build and appearance not unlike a larger edition of the male Little Bittern when that bird is seated at the edge of a reed bed with head drawn in between the shoulders. It is seen only occasionally by day, standing morosely on a bent-down reed at the edge of the open water. It is nocturnal in its habits and appears to be a shy and furtive bird, but as it quite often roosts in the trees hard by a village this impression may be conveyed by the deliberately evasive manner with which it creeps out of sight into the upper foliage, looking over its shoulder like a thief with a guilty conscience. The shape of this bird when at rest is distinctive, since the head is large and its contours merge into those of the well-groomed body as if it were neckless.

Throughout the daylight hours Night Herons sleep amongst the upper foliage of the tall trees, often in the huge chenars of some lakeside village, but L. S. White has some interesting remarks to make about the large heronry he located in a chenar in the Nishat Bagh. During the day numbers of the Night Herons roosted in withy beds near this tree, many of them carrying sticks for building purposes at all hours of the day. Occasionally in dull weather a few were to be seen feeding in daylight. Those that were roosting in the withies would start carrying sticks to their nests with great energy as soon as they were disturbed, and it was through this curious habit that he first discovered the colony. Periodically they seem to change their quarters, for one year we noticed many close to Hazratbal Post Office, whereas the next season there were none to be seen. They often attach themselves to the heronries, by which is meant the breeding places of the Eastern Grey Herons such as those near the Shalimar Bagh on the shores of the Dal lakes and at Bandipur on the Wular. Another colony was noted in tall willows in a village on the western shore of the Wular Lake not many miles from Sopor.

At dusk they come to life and may then be seen flying off at intervals to satisfy their hunger by fishing for frogs and fry in the marshes. They do not disperse in a body but leave the roosting place in ones and twos or small parties of up to a dozen, and as they flap their way straight and deliberately with slow but continuous wing-beats, one or other of the party gives vent to a loud *quark* at almost fixed intervals. The flight has often, and rightly so, been likened to that of the Flying Fox.

The nesting season in Kashmir starts comparatively early, eggs being found in April and May. The nests are shallow platforms of sticks loosely put together so that at times the eggs can be seen from below as through open lattice-work; lining there is none, and the useful portion of this untidy structure is often but 8 inches across. In marshes and flooded areas in continental India, we have seen Night Herons' nests within 2 ft of the water's surface, but where, as in Kashmir, the nests are built in trees standing on dry land, for safety's sake they are invariably amongst the upper branches of lofty trees, usually in the shady chenars. Brooks stated that in Kashmir they sometimes nest in the reeds but we have so far never come upon such colonies. The eggs run from 3 to 5. Four is the most usual number, while Stuart Baker writes of occasional 6's. They are typical herons' eggs, elongated ovals with one end slightly more compressed, but the blue-green of the glossless shell is often so pale as to be nearly white. They average 49·5 by 34·8 mm. Both sexes incubate and help with the construction of the nest.

LITTLE BITTERN

Ixobrychus minutus minutus (Linnæus)
(See Plate 70 facing p. 277 and Plate 71 facing p. 284.)

KASHMIRI NAMES. *Goi, Kurgoiu.*
BOOK REFERENCES. *F.B.I.*, VI, 364; *Nidification*, IV, 479; *Handbook*, 516.

FIELD IDENTIFICATION. (4) The Little Bittern is that awkward bird so often seen squatting motionless on the outskirts of a reed bed. When approached the neck is elongated and the bill thrust upwards so that, thanks to the brownish stripes down the neck, much intensified in the female, the bird becomes part and parcel of the reeds amongst which it is standing. The blue-black back and wings of the male are in the female brown, each feather with a darker shaft stripe. As with the Pond Heron, in flight the head is drawn back between the shoulders and the beats of the rounded wings are rather slow and heavy.

DISTRIBUTION. Widely distributed throughout the Vale's lakes, waterways and marshes wherever there are reed beds to provide it with the necessary cover.

Besides Kashmir the Little Bittern breeds in suitable localities along the outer Himalayas, according to Stuart Baker as far east as Nepal. It nests in some numbers in Sind and Stuart Baker says it is also resident in the United Provinces. It is a winter visitor to the Punjab plains.

The species has an immense range elsewhere than in India, the typical race nesting in South, Central and Eastern Europe, North Africa and Asia Minor to the Caucasus, and a great part of Western Asia to Turkestan and Afghanistan. There are other races in tropical and South Africa and also in Australia and New Zealand.

HABITS AND NESTING. There are certain birds which are almost symbolic of their country. For instance, what bird-lover could ever think of Kashmir without recalling the little blue and chestnut kingfishers, the brilliant orioles, the noisy reed-warblers, and the shimmering Paradise Flycatchers? To these must be added the Little Bitterns, for it is impossible to pass even the most public reed beds of the Dal lakes and their waterways without seeing these squat little herons which always look as if they suffer from curvature of the spine. They love to fish along the margins of the reed beds, half the time standing forlornly in their characteristic hunched-up attitude, periodically coming to life to prod at an insect on a nearby stem or to lunge with unexpected speed at a passing sprat. Occasionally a froglike *wuk* will be heard from a patch of reeds followed by the bird flapping across the waterway on blunted wings with head drawn well back into its shoulders, to flop down again into the cover on the opposite side. Should one approach rather too close, the bird will probably freeze, extending the neck to incredible lengths so that the feathers flatten against its sides, at the same time thrusting the beak stiffly upwards. Standing immobile in this attitude against a background of reeds the females become almost invisible, and even the more brightly coloured males must find it useful.

Their nesting habits are equally noteworthy. The nest is a pad of rushes a few inches to as much as 3 ft above the water's surface, built in part by bending down a number of the reeds, and then adding other pieces of reed and their leaves to form a shallow platform. These nests are usually found in those reed beds where the water is shallow, that is from a few inches to knee depth. In fact it is generally a waste of time to search elsewhere, although occasionally we have seen nests where the water was so deep that wading was impossible. The majority are certainly built quite close to the water's surface, and, where this is not so, in four cases out of five the water has receded, the nest having been commenced when the water was above its normal level. The density of the reeds appears to count little, provided there are clumps sufficiently close-growing to provide support for the nest, but the thicker beds where visibility is restricted to a very few yards seem to contain the most nests. In the Anchar and Shalabug area are many willow plantations. These are mostly inundated and some have a reedy growth amongst them. Here, too, many Little Bitterns' nests are to be found which have their foundations resting on the small willow branches up to 2 or even 3 feet above the water's surface. These nests are often made entirely of fine twigs and are provided with a well-marked depression for the eggs.

Both sexes take an equal share in building, maintaining the nest in repair, and in incubating the eggs. When engaged in the latter occupation they sometimes sit so closely as to allow of their being caught. On being approached up goes the head with beak thrust into the air as high as it will go, so that the bird, thanks to the brown streaks running down the neck from the throat to the breast melts into its reedy background. If one attempts to circle behind it, gradually and imperceptibly the bird turns with you so that the same reedlike aspect remains presented. This protective upthrust is not, however, carried out in one. A preliminary quick thrust extends the neck to about half its limits after which the head is raised almost imperceptibly in a series of short jerks in accordance with the extent of the danger, as if the bird were aware that continuous movement might give its presence away. It is our considered opinion that this initial upward thrust, in which the bill is raised at an angle of about $45°$, is a defensive rather than a purely passive movement. In this position the bill is a formidable weapon held like a soldier's bayonet at the 'On guard', and the low siting of the Bittern's eyes enables it to watch its antagonist's every move without lowering its 'point'. Nevertheless, Little Bitterns still suffer from the depredations of their enemies since House Crows destroy numbers of their eggs, mainly when the nest is momentarily unattended.

Up to 7 eggs are laid, 5 being a common number, and although they are deposited at two-day intervals, incubation usually starts with the laying of the first one. This results in great disparity in the size and development of the chicks, and it would appear that the puny last-born can have little chance of survival in competition with its already well-developed elder brethren. This seems to be compensated for in part by an amazingly aggressive spirit possessed from birth. As with incubation both parents tend their young, feeding them, and sheltering them from the sun's rays in the heat of the day. The nest-relief is carried out frequently and usually in silence, the birds moving on and off the nest in a ludicrously deliberate and detached manner. The chicks, clothed in a loose growth of down of a beautiful shade of cinnamon, do not remain many days in the nest, but soon take to the surrounding reeds, clambering about with great facility with the help of their long toes which develop disproportionately quickly. They return to the nest, however, when called by their parents and probably collect there at night. At this stage, with down sticking out from their crowns like the bristles of a bottle-brush, they are the quaintest little creatures. They are fed in the nest by regurgitation, one of the chicks seizing the parent's bill at its base crosswise in its own and following the food down to its tip. The regurgitated food, however, often falls to the bed of the nest where it is seized by whichever chick can grab it first. The close co-operation of the parents at all stages is really rather remarkable, for the male does not possess the same degree of protective coloration as the female, who can render herself to all intents and purposes part of the rushes. Yet we know that in the case of the European Bittern, in which both sexes are equally protected by their similar colour pattern and whose choice of nesting site is very similar to that of the Little Bittern, the male takes no part in his home affairs.

The nesting season of the Little Bittern is in the main rather late, as nest-building cannot begin in the reed beds until they have attained sufficient height and strength. Eggs, therefore, are generally obtainable after the first week in June and in July, and birds which have had their nests destroyed by floods may even have eggs in August. We have, however, found a nest under construction on 6 May and have seen a good number of nests with eggs in the osier beds from the middle of that month. The eggs are rather elongated ovals, white in colour and without gloss, and their measurements average 34·5 by 25·5 mm.

MALLARD, OR WILD DUCK

Anas platyrhyncha platyrhyncha Linnæus

(See Plate 72 facing p. 285.)

KASHMIRI NAMES. Drake, *Nilij* ; Duck, *Thuj*.

BOOK REFERENCES. *F.B.I.*, VI, 419; *Nidification*, IV, 506; *Handbook*, 526.

FIELD IDENTIFICATION. (5) The Mallard scarcely needs to be described since many of our domestic ducks, being its direct descendants, resemble it sufficiently closely. The drake can be recognized quite easily by the white collar round the deep green neck, while, when examining it in the hand, one must not forget the nicely turned-up quiff of stiff feathers in the tail, which as Mr Sálim Ali humorously puts it in his *Birds of Kutch*, is ' meant for sticking in sportsmen's hats ! ' The duck, of course, is not so colourful a bird, but both sexes are much bigger than the other breeding duck of Kashmir, the darker-hued White-eyed Pochard.

DISTRIBUTION. Confined to the secluded marshes of the Vale. The Wild Duck can no longer be deemed prolific in the summer months, the great majority migrating beyond the confines of India to areas far afield. In the State *rukhs* they nest annually in small numbers but even on such well-protected waters as Hokra Jheel it is doubtful if more than half a dozen pairs breed in the season. An occasional nest may be found elsewhere in the Vale, for example in the Anchar, but few of these nests can ever escape pilferage.

Within Indian limits the Mallard almost certainly nests only in Kashmir territory. It appears probable, however, that it breeds in Southern Tibet, which will mark its southernmost breeding limit. The Mallard is almost cosmopolitan in its distribution, the typical race breeding throughout Europe and Asia across to Japan, south to the Sahara in Africa and the Himalayas in Asia. It also breeds in North America. In winter it spreads farther south, to Abyssinia in Africa, Borneo in Asia, and Panama in America. So far as India is concerned, though large numbers will be found in winter in Kashmir and the northern provinces of India, its numbers rapidly decrease as one goes southwards, so that in the Central Provinces and Orissa few will fall to the sportsman's gun.

HABITS AND NESTING. It seems well established that in Kashmir, of the vast hordes of duck and teal which pass through the Vale or remain to spend the winter on its lakes and marshes, limited numbers of only two species remain to breed. Of these two, White-eyed Pochard remain in greater numbers than Mallard, and, except in the reserves such as Hokra mentioned above and the *rukh* portion of the Anchar Lake, the latter are now positively scarce. Stuart Baker is wrong in saying that in Kashmir they breed in vast numbers (see *Nidification*). It is hard to believe the stories that at one time boat-loads of both Mallards' and Pochards' eggs were sold in the bazaars of Srinagar, but this is what Major Cock wrote in Hume's *Nests and Eggs of Indian Birds*, published in 1890 : ' This species breeds in large numbers on the Anchar, Dall and other lakes in Cashmere during the months of May and June. Boat-loads of their eggs are brought to the Sirinugger bazaars for sale, together with the eggs of the Coot and White-eyed Duck.' Another quotation from the same page reads : ' There is quite a trade in the eggs of this species and *Fuligula nyroca* at Sirinugger, and my man went out daily almost for a month in one of the egging boats. The boatman told him that they *had* found as many as 16 eggs in one Mallard's nest ! ' These depredations must have had disastrous repercussions, for only six years later Unwin writes in Sir Walter Lawrence's *Valley of Kashmir* : '. . . arrives in late October and leaves in March. I once shot a female in May near Sumbal and it is possible that a few pairs remain to breed here, but the majority certainly migrate and I can find no evidence to support the statement in the *Game Birds of India* that this bird breeds extensively in Kashmir and that the eggs are sold in large numbers in Srinagar.'

There is an amazing divergence in these two descriptions written comparatively close together. Unwin has undoubtedly underestimated the numbers remaining to breed, but, although they must formerly have been more common, we suggest that even in Cock's time a careful analysis of the 'boat-loads of eggs' would have disclosed a great preponderance of White-eyed Pochards' eggs over those of the Mallard, and a very mixed bag, vastly outnumbering both ducks' eggs, of Moorhens', Coots', Little Grebes', and probably Terns' eggs, as well as those of the Painted Snipe and Jackdaw. Even on closely-guarded Hokra Jheel, during a visit on 18 June 1921, we succeeded in locating only one nest. During a subsequent visit three years later on 24 May, the head watcher stated he had seen but two nests that season, both of which he showed to us. These three nests were built in particularly thick tangled masses of reeds. The many ducklings the 1921 nest contained succeeded in escaping into the surrounding reeds before we could reach them, leaving but two of their number and a third which was struggling to detach itself from the broken shell of its late prison. Of the 1924 nests, one contained an addled egg—the ducklings had left two days previously—the other, eight much incubated eggs. We have at times come upon empty or water-logged nests in the neighbourhood of Shalabug. These were built against or on cut-down willow stumps growing in the reeds; in fact, more often than not Mallards' nests rest on a solid foundation. We found one on almost bare ground at the foot of a poplar near the water's edge, but it is a noticeable fact that Mallards' nests are usually built more in the heart of the reeds than are those of the next bird. Nests built in the river-fed marshes are unfortunately very prone to destruction through sudden changes in the water-level. Late April, May, and possibly early June are, we believe, the laying months in Kashmir.

The nest consists of a quantity of dry grass and rush leaves, while some of the bird's own down is added as incubation proceeds. They are at times bulky affairs rimmed with a thick well-felted layer of the down, which at other times may be almost absent. Any number from 6 up to 12 eggs appears to be laid. These have a smooth faintly glossy greenish-white shell but they become browner and discoloured as incubation proceeds. They average 56·6 by 40·3 mm.

WHITE-EYED POCHARD

Aythya rufa rufa (Linnæus)

(See Plate 73 facing p. 292.)

KASHMIRI NAME. *Harawŏt*.

BOOK REFERENCES. *F.B.I.*, VI, 453; *Nidification*, IV, 514; *Handbook*, 538.

FIELD IDENTIFICATION. (4+) Smaller and much darker than the Mallard. The general tone of the plumage is dark brown, the breast a rufous-brown. There is a white patch in each wing and the underparts from below the breast to the vent are white and show particularly when the bird is flying. When swimming, however, only the white undertail-coverts show up. The white iris of the drake is conspicuous at close quarters.

DISTRIBUTION. More numerous than the Mallard, being distributed in some numbers throughout those jheels of the Vale which have plenty of reedy cover and water-plants. Can be considered fairly common on the State *rukhs* such as, for instance, Hokra, the northern end of the Anchar Lake, and Haigam. Their numbers are considerably increased in the cold weather by fresh arrivals from the north-west. In winter this duck spreads itself over the whole of the northern half of India and Bengal, some birds reaching the Madras province and, in Burma, as far south as the Arakan.

The White-eyed Pochard nests in South and Eastern Europe, and in Asia from the Kirghiz steppes and South-West Siberia to Tibet. Also in Morocco and Algeria, Syria to Northern Persia and Turkestan and down to Kashmir as its southern limit. Another race extends the species' range to Manchuria.

HABITS AND NESTING. This little duck used to share with the Mallard the reputation of having boat-loads of its eggs marketed in the bazaars of Srinagar. We have already gone into that question when dealing with the Mallard, and, as Whistler says, the practice has now been stopped. Although the open sale of these ducks' eggs may have been prohibited, there is no doubt that outside the protected marshes the birds are much persecuted and the eggs filched from all nests which are detected sufficiently early. Fortunately for them a good proportion of nests are not easy to get at, being well concealed amongst thick matted reeds through which it is difficult to force even the lightest of shikaras. Sometimes it is possible to get to a nest by wading across a quaking morass of bent-down reeds and sodden water-plants as the nests are at times built into such thickets at a small height above the water. They may also be found resting on reed-surrounded stumps and willows and at times on solid ground.

The nest is a circle of coarse grass and rush leaves with a somewhat finer lining, and as incubation proceeds the bird adds considerable quantities of its own dark-brown down. The delicate *café-au-lait* shells, thus warmly embedded in the protective flock, are a most pleasing sight. We once found a nest, from which the bird rose when we were some 20 yards off, in which all the eggs had been carefully covered over with down. Wilson observed much the same thing except that in his case the eggs were covered with reed as if the bird were copying the Dabchick's ideas on egg-protection. Up to 11 eggs are laid. Eight to 10 are more usual, while as few as 6 have been recorded. In size they average 50·5 by 36·9 mm. This duck nests rather later than the Mallard, throughout May, though eggs may be found up to the end of June.

INDIAN LITTLE GREBE, OR DABCHICK

Podiceps ruficollis capensis Salvadori

(See Plate 74 facing p. 293.)

KASHMIRI NAME. *Pind*.
BOOK REFERENCES. *F.B.I.*, VI, 481; *Nidification*, IV, 521; *Handbook*, 539.
FIELD IDENTIFICATION. (3) Not unlike a small tailless duck. Appears a uniform dark brown, swims low in the water and often dives, disappearing with a little plop and remaining submerged for a considerable period. Usually found where there is cover of reeds or aquatic plants into which it can retreat. Has a querulous bubbling call.
DISTRIBUTION. Very common around all the lakes and in the marshes of the Vale wherever there is water deep enough for it to dive in and cover nearby into which it can retreat. It is resident in Kashmir in winter in diminished numbers, when it appears on more open waters in severe frosts, even being seen on the Jhelum River.

Podiceps ruficollis is a widespread species with various races which are found from the British Isles throughout Europe, Asia, and Australia to New Zealand. Our Indian form is also found over a great part of Africa and eastwards it extends as far as Yunnan and Siam. Although mainly a resident and local migrant, its short wings can and do carry it considerable distances.

HABITS AND NESTING. Although a shy cautious little bird, the Dabchick does not by any means spend all its time in or close to the reed beds, but may often be seen swimming about open water far from cover. Every now and then it dives beneath the surface with a little spring so that all that remains to remind one of its presence is a widening circle of ripples. For it is a powerful swimmer, its lobed feet propelling it swiftly below the surface, and as it can submerge for a considerable period, 10 or 15 seconds, it may come up again some way from where it disappeared from view. It is this ability to swim submerged which accounts for this shy bird so often being seen well out in the open, for at the first sign of danger it makes for the cover of the reeds, or any aquatic plants capable of hiding it, in a series of dives. Its approach to the nest is habitually carried out in the same manner ; a cautious glance around, a dive, up to the surface for another peep, and so on to the cover where its floating pad is anchored. From the cover of the hiding-tent we once made out the form of a suspicious Dabchick circling its nest some inches below the surface. It came up for but the fraction of a second to slap the water with its little wings with a resounding smack which nearly frightened the life out of an over-curious reed-warbler which had alighted on the nest's rim.

When not actually breeding, Dabchicks collect in some numbers, for they are as numerous in Kashmir as elsewhere, but when the nesting season is in full swing they will generally be seen in pairs only, each in its own territory, though several pairs may occupy quite a limited area. During the greater part of the daytime the eggs are left to take care of themselves, being covered over in the rotting mass of weedy vegetable matter which constitutes their nesting place. Thus it is that both birds will so often be espied swimming about together, at times giving call and answer in that treble querulous rippling voice which finishes up on a pitch decidedly lower than that on which it starts. Sometimes the call of one bird seems to set off its neighbours for some distance around it. They also emit a sharp high-pitched squeak.

Little Grebes seldom leave the water but once in the air they are strong fliers, for with their abbreviated wings they find a certain amount of difficulty in taking off—they cannot rise from land at all and indeed can scarcely walk. First they paddle furiously, raising quite a bow wave ; then they spatter along the surface until sufficient speed is gained for swiftly beating wings to maintain them in flight.

Although in a measure shy and retiring, provided good cover exists not only will they be found close to the most public and frequented places, but also their nests. Certainly there is no reed bed in the Dal lakes or bordering the canals approaching them which they eschew, not even those nearest to the Dal Gate where houseboats and shikaras are moored gunwale to gunwale and where pandemonium reigns by day and by night. Here their nests will be found in all but the very closest-growing beds, but they certainly prefer those sufficiently open for them to be able to swim easily between the stems. Elsewhere nests will be found out in the lakes and marshes, even in isolated little patches of rushes or aquatic plants, however small and thinly growing they may be, so that the nests can often be seen afar off. At times they even appear to be floating in open water. There is, however, always anchorage of some kind to prevent drifting, but such as will also allow the nests to rise and fall with the water-level.

The nests are soaking pads of water-weeds about 8 to 10 inches across. Thick stiff stems are never used ; they consist of the leaves of the plants and thin stems of various kinds of water-weeds all in such a state of decay that they form a slimy coagulated mass, the central portion of which protrudes an inch or so above the water. The eggs are embedded in the middle, and although pure white when laid, they rapidly become tarnished from a light to a dark chocolate-brown from their contact with the wet vegetable matter. They are oval in shape, stout-shelled and without gloss, and often inclined to be pointed at both ends. They average $36 \cdot 6$ by $25 \cdot 1$ mm. Three to 6 are laid, 5 being a number which is frequently found. Incubation, in which both birds take part, commences with the laying of the first egg, but as the heat generated by the fermentation of the decaying nest-material assists in incubation, they do not sit during long periods of the day. A loose portion of material is usually pulled over the eggs by the departing bird no matter how hurriedly it may leave. This habit has, of course, a double purpose : the eggs are hidden from prying eyes in what then appears but a half-submerged lump of floating rubbish, while their incubation is further assisted by complete enclosure in the rotting matter. The downy young, which are patterned in stripes of blackish-brown and fulvous, leave the nest almost as soon as they are hatched. They are energetic mites, following their parents wherever they go and sometimes climbing upon their backs for a rest, all the time chirruping like a family of little chickens.

GREBE

Few Dabchicks commence building before the third week in May, but the breeding season is prolonged, and our earliest date for eggs is 6 May. Great numbers are still laying in August and many do so well into September. They are greatly persecuted by the Kashmiri lake-dwellers, who collect quantities of their eggs for food, and these depredations are naturally a considerable factor in this prolongation of the nesting period.

In a short work of this kind generalization is unavoidable, so in this case we find it necessary apparently to contradict ourselves since firstly we have said that the Little Grebe usually nests singly, each pair in its own territory, and secondly comes the matter of covering the eggs on departure. Near Shadipur in 1931 one of us encountered no less than 83 nests on a sheet of water about half a mile square. During a stay in this area of three days 41 cases were noted where the bird made no attempt to cover up its eggs on departure, although on such an open expanse it had plenty of warning and opportunity to do so.

SUPPLEMENT

The inclusion of the Jhelum Valley from Kohala to Baramullah, and also of the Kishenganga Valley from Domel to Gurais, involves the mention of a number of birds which occur and doubtless breed in those areas, but of which we can trace no actual records. Many are species common in the Plains and foothills, but few people tarry on the way to or from the Vale of Kashmir so that the transition which occurs along these routes from birds of the Plains and outer ranges to those of inner Kashmir has received scant attention. There are in addition a number of birds which occur, some freely in the main and side valleys of Kashmir, of which we can trace no breeding records at all or only doubtful ones. Such birds are listed below, being dealt with in brief until such time as further data come to light. This list is not exhaustive and no doubt many of our readers will be able to add to it.

TIBETAN RAVEN, *Corvus corax tibetanus* Hodgson. KASHMIRI NAME: *Botin Kāv*. BOOK REFERENCES : *F.B.I.*, I, 23 ; *Nidification*, I, 4 ; *Handbook*, 2.

We have seen the odd bird during the summer in the Sind Valley between Sonamarg and Baltal and one was shot during a severe winter at Chatragul in the lower part of that valley. Its visits to the south of the Zoji La are in all probability brief, and if it nests at all on the southern face of the main range it will only be on very rare occasions. The chief habitat of this large race of the far-flung raven is across the watershed in Ladakh and the northern provinces of Kashmir, extending thence to Eastern Tibet.

WESTERN HIMALAYAN TREE-PIE, *Dendrocitta formosæ occidentalis* Ticehurst. BOOK REFERENCES: *F.B.I.*, I, 52; *Nidification*, I, 40; *Handbook*, 13.

Ward states that the Himalayan Tree-Pie is found in the Jhelum Valley and that a solitary specimen was recorded from the Vale in 1905. We saw a single tree-pie, probably of this species, on 10 July 1921 between Bijbehara and Pampur. This form is common in the foothills and outer ranges of the Western Himalayas as far as Western Nepal.

TIT, LAUGHING-THRUSHES, BABBLER

Red-headed Tit, *Ægithaliscus concinnus iredalei* Stuart Baker. BOOK REFERENCES: *F.B.I.*, I, 93; *Nidification*, I, 72; *Handbook*, 26.

Osmaston saw this tit in March at 3,000 ft in the Jhelum Valley, that is, presumably, between Garhi and Chenari. The two races found throughout the length of the Himalayas penetrate in places up to 12,000 ft, but breed mainly along the outer ranges between 4,000 and 9,000 ft. It nests commonly around Murree in April and May. Being a strictly sedentary species with but a small altitudinal movement, and an early nester, Osmaston's birds were probably on or near their breeding grounds.

Western White-throated Laughing-Thrush, *Garrulax albogularis whistleri* Stuart Baker. BOOK REFERENCES: *F.B.I.*, I, 154; *Nidification*, I, 117; *Handbook*, 32.

Ward states, ' Is recorded from Domel, Jhelum Valley Road '. We have no other records but Rattray, and Marshall and Cock, took its eggs near Murree, the latter stating that they found all their nests fairly close together, the birds being gregarious even in the breeding season. The Western form of this species nests commonly along the outer Himalayas from Hazara to about Eastern Nepal, mainly between 5,000 and 9,000 ft though we have seen it much lower than this.

Kumaon Rufous-chinned Laughing-Thrush, *Ianthocincla rufogularis occidentalis* (Hartert). BOOK REFERENCES: *F.B.I.*, I, 159; *Nidification*, I, 121; *Handbook*, 34.

There is a single record from the Lolab Valley, and Unwin writes in *The Valley of Kashmir* : ' Found in hillside thickets and woods in Kashmir up to 8,000 or 9,000 ft. Has a low chattering note, much uttered just before it roosts for the night.' It is probable that he is referring to the outer provinces, e.g. Poonch and Jammu, for it is in the main a bird of rather low elevations in the outer hills. This form of the Rufous-chinned Laughing-Thrush appears to be uncommon in the westerly portions of its range, Stuart Baker stating that it is rare in the Murree Hills.

Sind Jungle Babbler, *Turdoides terricolor sindianus* Ticehurst. BOOK REFERENCES: *F.B.I.*, I, 193; *Nidification*, I, 154; *Handbook*, 41.

According to Ward it is found in the Jhelum Valley up to 4,000 ft. We can trace no other mention of this area in connexion with its distribution but it may well occur here. Races of this familiar babbler are found throughout the whole of the Indian peninsula and, although typically birds of the Plains, they do ascend the hills fairly generally up to about the elevation given by Ward.

HODGSON'S TREE-CREEPER, *Certhia familiaris hodgsoni* Brooks.
BOOK REFERENCES: *F.B.I.*, I, 434 ; *Nidification*, I, 414.

This bird has already been mentioned when dealing with the Himalayan Tree-Creeper. A race of the Northern Tree-Creeper, it is rare in Kashmir. In fact the short note by Brooks in Hume's *Nests and Eggs of Indian Birds* to the effect that it was seen at Gulmarg and also at Sonamarg, where Captain Cock took a few nests, remains the only breeding record. Osmaston also records it from Gulmarg and he shot a specimen in the Lidar Valley in July in birch forest at 12,000 ft. It is said to range from Garhwal to the Murree Hills, where Buchanan and Rattray both took its eggs.

BROWN-BACKED INDIAN ROBIN, *Saxicoloides fulicata cambaiensis* (Latham). BOOK REFERENCES: *F.B.I.*, II, 111; *Nidification*, II, 95 ; *Handbook*, 105.

We have seen this bird on the Jhelum Valley road on more than one occasion and up to within a few miles of Baramullah. As it is a strictly resident species, it will almost certainly nest within the area. Four races of this pert little robin cover the whole of India and Ceylon to the outer fringe of the Himalayas, ascending the hills commonly to about 2,000 ft, occasionally higher.

COMMON VERDITER FLYCATCHER, *Eumyias thalassina thalassina* (Swainson). BOOK REFERENCES: *F.B.I.*, II, 239 ; *Nidification*, II, 206 ; *Handbook*, 124.

Ward's notes read, ' Specimens from the Vale marked April and May, mostly from near Srinagar '. There is also a record from Verinag. In April we saw a pair in the lower Kishenganga. As this bird nests freely in the Murree Hills, and indeed all along the Himalayas from about 3,000 to 10,000 ft commencing in the month of April, it can certainly be expected to breed within our area, albeit in small numbers.

FLYCATCHERS, DRONGO, CHIFFCHAFF

SIMLA GREY-HEADED FLYCATCHER, *Culicicapa ceylonensis pallidior* (Ticehurst). BOOK REFERENCES : *F.B.I.*, II, 254 ; *Nidification*, II, 224 ; *Handbook*, 129.

We can trace no breeding records although it is always to be seen in spring and summer in the Jhelum Valley, especially around Domel and Garhi. In addition, there are records of its occurrence both in the Lolab and the Vale of Kashmir where, however, it is undoubtedly rare and possibly only a casual visitor. Two races of this widespread flycatcher nest commonly in the Indian hills between 3,000 and 8,000 ft, many being found in winter in the adjacent plains and lower valleys. Our pale form breeds in the Himalayas from Hazara to Bhutan.

HIMALAYAN BLACK DRONGO, *Dicrurus macrocercus albirictus* (Hodgson)· BOOK REFERENCES: *F.B.I.*, II, 357; *Nidification*, II, 321 ; *Handbook*, 155.

We have no direct evidence that the Black Drongo, so common in the Plains, occurs within our limits. Ward merely says : ' This Drongo ascends the hills to about 7,000 ft, but generally nests not higher than at an altitude of about 6,000 ft.' He mentions no localities. Drongos noted in the lower reaches of the Jhelum Valley may well belong to this species. A skin or two would settle the point.

SIND CHIFFCHAFF, *Phylloscopus collybita sindianus* Brooks. BOOK REFERENCES: *F.B.I.*, II, 456; *Nidification*, II, 413; *Handbook*, 176.

Osmaston in ' A Tour in Further Kashmir ' (*Journal*, xxxiv, 133) under the date 29 July, wrote as follows : ' At Gurais the valley opens out . . . We were surprised to hear Chiffchaffs calling in the willow beds near the river. We did not know they bred in Kashmir proper.' This chiffchaff nests extensively across the divide in Ladakh, so it is possible the birds were also breeding where Osmaston heard them. Proof of this is, however, required, as they pass through Kashmir in large numbers to and from their winter quarters in the Plains of India.[1]

---

[1] There are other leaf-warblers to be seen in Kashmir in summer, small numbers of which may stay to nest within our area, but we have left out all mention of them for lack of data. Lately, H. G. Alexander (*J.B.N.H.S.*, 49, 11) has put forward a strong case in support of *Phylloscopus neglectus*, which he observed on a number of occasions in June and early July at Pahlgam and Lidarwat. We have also seen a willow-warbler of dull plumage in the lower Sind Valley, and twice, in June, came upon nests there whose owners we failed to identify.

WARBLERS, FINCHES

KASHMIR GREY-HEADED FLYCATCHER-WARBLER, *Seicercus xanthoschistos albosuperciliaris* (Jerdon). BOOK REFERENCES: *F.B.I.*, II, 490; *Nidification*, II, 442; *Handbook*, 180.

In spite of its trivial name, this warbler appears to be rare in Kashmir and none of the nest records apply to our area. Its main habitat seems to be the outer hill forests from Murree to Garhwal. Osmaston noted it in spring in the Sind Valley and in the Dachhgam Nullah, where he remarks that it doubtless breeds.

NEPAL BROWN HILL-WARBLER, *Suya criniger criniger* Hodgson. KASHMIRI NAME: *Phitta* (Kishenganga Valley). BOOK REFERENCES: *F.B.I.*, II, 518; *Nidification*, II, 466; *Handbook*, 181.

(See Plate 75 facing p. 300.)

In April and early May we found this bird common, and obviously about to breed, in the lower Kishenganga Valley, the last one being seen at about 5,400 ft near Keran. It is numerous in the Murree Hills and nests there at about 4,000 ft; we have often seen it between Murree and Kohala. It is sure to be found along the Jhelum Valley road from Domel to some point not far short of the Vale. This race nests from the hills of the North-West Frontier Province through the outer Himalayas to Western Assam.

WESTERN RED-BREASTED ROSEFINCH, *Pyrrhospiza punicea humii* Sharpe. BOOK REFERENCES: *F.B.I.*, III, 121; *Nidification*, III, 42.

According to Osmaston this large rosefinch occurs, but not commonly, at high elevations above the limit of tree growth on both the Himalayan and Pir Panjal ranges. He states they undoubtedly breed in such rocky ground but found no nests. It is said to be numerous in the Khagan Valley where Whitehead found a nest being built. At the other extreme Whymper records it as breeding in Garhwal.

GOLD-FRONTED FINCH, *Metoponia pusilla* (Pallas). KASHMIRI NAMES: *Tiok, Taer* (Juvenile). BOOK REFERENCES: *F.B.I.*, III, 158; *Nidification*, III, 66.

This little finch is common in the Sind Valley on both spring and autumn passage, where we have seen it above Gund and around Gagangair as late as June. Unwin says it may be seen in the Vale in winter, and Brooks saw it in flocks at Shupyion in May. It is possible that the odd pair remains in the Sind Valley to nest on the slopes in rose and other small bushes. Rattray is said to have taken a nest with 4 eggs at 6,500 ft near Gund on 26 May. It nests freely in Ladakh, being rather a late breeder and laying mainly in July and August.

CRAG MARTIN, *Riparia rupestris* (Scopoli). BOOK REFERENCES: *F.B.I.*, III, 236 ; *Nidification*, III, 107 ; *Handbook*, 237.

'A common bird in Ladakh, less common in Kashmir.' So writes Osmaston, but we must confess never to having seen it, much less to have found its nest, within our area. No doubt it occurs on passage to and from the hills of Western India, but as it nests in the Murree Hills and is common enough from the moment of crossing the Zoji La, one would confidently expect to find its nest in suitable ground in our area.

INDIAN WIRE-TAILED SWALLOW, *Hirundo smithii filifera* Stephens. BOOK REFERENCES: *F.B.I.*, III, 245; *Nidification*, III, 115; *Handbook*, 237.

This beautiful swallow is said by Ward to be a summer visitor to Kashmir. We have seen it only in the vicinity of Uri, on two occasions in the same year where it was doubtless nesting under a culvert. L. S. White records seeing it in April at Garhi. If it does occur in the Vale, it is probably intermittently as a casual visitor. The Indian race is found throughout the Plains as far south as Mysore and in the hills up to some 5,000 ft.

PURPLE SUNBIRD, *Cinnyris asiatica* (Latham). BOOK REFERENCES: *F.B.I.*, III, 396 et seq. ; *Nidification*, III, 215 et seq. ; *Handbook*, 268.

During a three days' stay in April at Domel this sunbird was seen on two or three occasions and we came to the conclusion that it is probably not uncommon in the lower Jhelum Valley. A pair was also seen at Pateka in the lower Kishenganga Valley. They were probably of the same race as birds from the northern Punjab which Whistler points out are mainly intermediate between *brevirostris* and the typical form. Races of the Purple Sunbird, those in the extreme north and west migratory, nest throughout India both in the Plains and in the hills up to about 5,000 ft.

INDIAN ROLLER, *Coracias benghalensis benghalensis* (Linnæus).
BOOK REFERENCES: *F.B.I.*, IV, 224; *Nidification*, III, 388; *Handbook*, 294.

Ward says a few specimens have been obtained in the Jhelum Valley. Birds noted by us from the car between Domel and Garhi all appeared to be Kashmir Rollers.

BLYTH'S WHITE-RUMPED SWIFT, *Micropus pacificus leuconyx* (Blyth).
BOOK REFERENCES: *F.B.I.*, IV, 331; *Nidification*, III, 455; *Handbook*, 312.

Osmaston includes this bird in his *Notes on the Birds of Kashmir* in the *Journal*, saying that it is occasionally seen on the wing at high elevations on the great Himalayan range. Stuart Baker gives its full range in his *Nidification* as the greater part of the outer Himalayas from Murree to Assam, probably extending to the Baluchistan frontier. Rattray, and later others, took its eggs in the vicinity of Murree from cracks in the faces of precipices.

COMMON INDIAN HOUSE SWIFT, *Micropus affinis affinis* (Gray).
BOOK REFERENCES: *F.B.I.*, IV, 332; *Nidification*, III, 456; *Handbook*, 311.

Although Osmaston says this swift is fairly common in and around Srinagar, we have quite failed to note it in spite of keeping a sharp lookout for it in the last few years. Near Ghori in the lower Kishenganga Valley we saw numbers of a white-rumped swift far below us in a deep gorge which were probably of this species. The typical race breeds practically throughout India and Ceylon except for a limited area in the extreme north-west, and ascends the Himalayas in places to about 6,000 ft.

INDIAN WHITE-THROATED SPINETAIL, *Chaetura caudacuta nudipes* Hodgson. BOOK REFERENCES: *F.B.I.*, IV, 340; *Nidification*, III, 463; *Handbook*, 312.

Reported only from the Kishenganga Valley where we too saw this magnificent swift on two or three occasions over Keran and Bagtor. Unwin also states that he saw it at from 8,000 to 10,000 ft generally in scattered flocks, but he mentions no localities. Found throughout the outer and lower Himalayas from Hazara to Assam, the nesting of this fine swift has not yet been solved. It may possibly nest within large hollow trees, but being one of the fastest birds in the world which may well be seen a couple of hundred miles or more from its roosting and breeding places, it is impossible to say whether it even roosts within our area.

OWLS, OSPREY, VULTURES

INDIAN BARN OWL, *Tyto alba stertens* Hartert. BOOK REFERENCES: *F.B.I.*, IV, 386; *Nidification*, III, 496; *Handbook*, 347.

Osmaston says it is not uncommon in the vicinity of Srinagar where it may be heard calling at night. We have been unable to confirm this. The Indian Barn Owl is widespread in India and also in Burma, where it occurs as far south as Northern Tenasserim.

WESTERN HIMALAYAN BARRED OWLET, *Glaucidium cuculoides cuculoides* (Vigors). BOOK REFERENCES: *F.B.I.*, IV, 444; *Nidification*, III, 528; *Handbook*, 349.

Osmaston saw one pair of these small owls in an olive grove near Garhi, and in early June we observed another pair in the Travellers' Bungalow compound at Domel. In all probability this bird will turn out to be not uncommon along the Jhelum Valley road for it is numerous in the foothills and outer ranges of the Western Himalayas, extending as far as Eastern Nepal whence other forms carry the species eastwards through much of the Oriental region.

OSPREY, *Pandion haliaëtus haliaëtus* (Linnæus). BOOK REFERENCES: *F.B.I.*, V, 3; *Nidification*, IV, 1; *Handbook*, 369.

In summer Ospreys are regularly seen in the Vale; only in small numbers it is true, but one can reckon on noting their presence every year over one or other of the lakes and marshes. We have see nthem on the Dal lakes, near Manasbal, over the Wular, and on one occasion at the lower end of the Ferozpur Nullah. They are not wholly confined to the Vale; Unwin saw this bird on the Sind River and near Aru. The most we can say about the possibility of their nesting in the area is to repeat the vague assertion that they are said to have bred in the vicinity of the Wular Lake. The Osprey, which practically encircles the northern hemisphere, is an infrequent breeder within Indian limits, although it occurs over much of North India in winter. Stuart Baker seems to have taken the one authentic Indian clutch in Cachar. Ludlow found it nesting in some numbers in Turkestan.

CINEREOUS VULTURE, *Ægypius monachus* (Linnæus). BOOK REFERENCES: *F.B.I.*, V, 7; *Nidification*, IV, 3.

This fine vulture has an immense range from South Europe and North Africa across Asia to China, and includes the northern frontiers of India and the Himalayas within its breeding range. Ward mentions that a specimen was obtained in December at

VULTURES, EAGLE, KITE, HARRIER

Manasbal. On the 1st of May we saw one near Keran in the Kishenganga Valley, which made off in the direction of some high cliffs in the river's gorge.

WHITE-BACKED VULTURE, *Pseudogyps bengalensis* (Gmelin). BOOK REFERENCES: *F.B.I.*, V, 19; *Nidification*, IV, 15; *Handbook*, 353.
This common vulture of India ascends the Western Himalayas in places to about 8,000 ft, but is uncommon in Kashmir. Collections of vultures are usually to be seen on the refuse-heaps below Srinagar containing this species as well as Long-billed Vultures (*Gyps indicus nudiceps*). We also saw one on a couple of occasions wheeling over the confluence of the Kishenganga and Jhelum rivers at Domel.

HIMALAYAN GOLDEN EAGLE, *Aquila chrysaëtos daphanea* (Gray). BOOK REFERENCES: *F.B.I.*, V, 68; *Nidification*, IV, 45; *Handbook*, 360.
Although the Golden Eagle is to be seen from the level of the Vale up to the highest upland meadows and ridges, the nearest points to our area at which its nests have been recorded are in Hazara, where Unwin took a nest near Thandiani, and in Bhadarwar, whence C. H. Donald secured a fledgling. The latter informs us that he met with couples in many of the larger Bhadarwar nullahs and in fact actually saw the eyries in more than one of them. The Golden Eagle is not a migrant, and we have seen it in our area in places as far apart as the Lolab and the Wardwan Valley. Although not particularly common in Kashmir, this wide distribution surely warrants the assumption that sooner or later its nest will be recorded from more than one of our precipitous valleys. Elsewhere in Indian limits its nests have been recorded in the Simla Hills and Tehri Garhwal, and also in the Quetta District.

COMMON PARIAH KITE, *Milvus migrans govinda* Sykes. BOOK REFERENCES: *F.B.I.*, V, 122; *Nidification*, IV, 89; *Handbook*, 371.
The identification and status of the kites to be seen in the lower Kishenganga and Jhelum Valleys require careful investigation. Birds from these areas may well turn out to be identical with the Common Kite of the Plains.

MARSH HARRIER, *Circus æruginosus æruginosus* (Linnæus). BOOK REFERENCES: *F.B.I.*, V, 134; *Handbook*, 374.
Although the Marsh Harriers leave India for more northerly breeding grounds about March or April, in the Vale of Kashmir

the odd bird is to be seen throughout the summer. There is no evidence of their nesting in Kashmir, but on one occasion we saw a bird drop two or three times to the same spot in thick reeds on the outskirts of Hokra Jheel. Owing to the nature of the swamp we were unable to reach the place.

NORTHERN BESRA SPARROW-HAWK, *Accipiter virgatus affinis* Hodgson. BOOK REFERENCES: *F.B.I.*, V, 161 ; *Nidification*, IV, 113.

Osmaston secured a Besra Sparrow-Hawk on 29 April in the Lolab Valley. It proved to be laying eggs, but he was unable to find the nest. This race is said to breed throughout the length of the Himalayas, east into China and south-eastwards into the Burmese hills.

SPECKLED WOOD-PIGEON, *Dendrotreron hodgsonii* (Vigors). BOOK REFERENCES: *F.B.I.*, V, 234 ; *Nidification*, IV, 158.

Within our area this pigeon is recorded mainly from Sonamarg and the Wardwan Valley, while Unwin said he met with it in various parts of Kashmir between 7,000 and 9,000 ft and even up to 11,000 ft, but most numerously in the Wangat Nullah where he shot a number of these wary birds as they came down to a salt lick in the forest. We can find no mention of its actual nest in our area, but it stands to reason that it must breed with us. The *Nidification* gives its full range as Kashmir to Eastern Assam and thence through the hills of North Burma to the Shan States.

INDIAN SPOTTED DOVE, *Streptopelia chinensis suratensis* (Gmelin). BOOK REFERENCES: *F.B.I.*, V, 242; *Nidification*, IV, 162; *Handbook*, 396.

The Spotted Dove cannot be particularly uncommon in the Jhelum Valley for we have nearly always encountered it along the road up to little short of Baramullah. Races of this dove cover a great part of the East from China to India, in which it is absent only from the extreme north and north-western parts of the country. It ascends and breeds in the foothills of the Himalayas commonly to about 5,000 ft, and in places higher.

INDIAN LITTLE BROWN DOVE, *Streptopelia senegalensis cambayensis* (Gmelin). BOOK REFERENCES: *F.B.I.*, V, 246; *Nidification*, IV, 165 ; *Handbook*, 397.

This small dove has a more westerly range than the last, various forms being found in Africa and Western Asia in addition to our very common Indian race. It is scarce in the outer Himalayas, and indeed in our area is only occasionally to be met with along the Jhelum Valley road.

CHEER PHEASANT, *Catreus wallichii* (Hardwicke). BOOK REFERENCES: *F.B.I.*, V, 307 ; *Nidification*, IV, 203 ; *Handbook*, 419.

The Cheer Pheasant, which is confined to the Himalayas from Nepal westwards, is a bird of the outer ranges and occurs in no great numbers in the Jhelum and lower Kishenganga valleys. In these valleys Ludlow says it is found between Chenari and Uri, and near Salkalla and Keran, respectively. Between them it also occupies the markhor nullahs of the Kazinag. As regards the Wardwan, Ludlow says it is found below Inshan so it may just come into our area there.

WHITE-CRESTED KALEEJ, *Gennæus leucomelanos hamiltonii* (Griffith & Pidgeon). BOOK REFERENCES: *F.B.I.*, V, 320; *Nidification*, IV, 208 ; *Handbook*, 417.

We have no information regarding this kaleej other than Ward's remarks that it is not found in the Vale, but is abundant on the Murree road. As this Himalayan pheasant and its races are numerous throughout the outer Himalayas from Hazara to Bhutan up to some 10,000 ft, it is probably also found in the lower Kishenganga Valley.

WESTERN TRAGOPAN, *Tragopan melanocephalus* (Griffith & Pidgeon). BOOK REFERENCES: *F.B.I.*, V, 345; *Nidification*, IV, 222; *Handbook*, 419.

The Tragopan is uncommon in Kashmir. A single bird was shot many years ago in the Lolab but, according to Ludlow, the only locality where it now occurs on the rim of the Vale is at the head of the Sandran River east of the Banihal. In the Jhelum Valley it is found near Rampur and Uri, and in the Kishenganga around Keran. Outside our limits this horned pheasant breeds from about 8,000 ft upwards along the Himalayas from Hazara to Garhwal.

COMMON or GREY QUAIL, *Coturnix coturnix coturnix* (Linnæus). BLACK-BREASTED or RAIN QUAIL, *Coturnix coromandelicus* (Gmelin). BOOK REFERENCES : *F.B.I.*, V, 372 and 375 ; *Nidification*, IV, 236 and 238 ; *Handbook*, 422 and 424.

Quail are little more than casual visitors to the Vale. In very dry winters in the Punjab their numbers increase, and while most disappear again in the spring, a few remain in the Vale to nest. Ward states that the eggs of the Common Quail have been taken near Manasbal and we have seen the bird in that area in the breeding season. Stuart Baker's remark in the *Fauna* that in Kashmir the Grey Quail is very common, breeding around the lakes in long grass and fields of crops at elevations between 4,000 and 6,000 ft, is altogether too optimistic an estimate for the Kashmir Valley. In the outer provinces, such as Jammu and Mirpur, they are no doubt more numerous. Osmaston informs us that his shikari brought him several clutches of quails' eggs, while on 24 August two clutches of 7 and 8 eggs were brought to him at Hokra which he assumed at the time to be those of the Common Quail. Subsequently he heard the characteristic calls of both the Grey and Rain Quails on several occasions, and as he never secured a parent bird with his eggs, feels that their identification as belonging to the Grey Quail is not beyond suspicion.

INDIAN LARGE CORMORANT, *Phalacrocorax carbo sinensis* (Shaw & Nodder). KASHMIRI NAME: *Mong.* BOOK REFERENCES: *F.B.I.*, VI, 277; *Nidification*, IV, 425; *Handbook*, 492.

These birds are to be seen in the Vale usually about the Wular and Dal lakes and the larger swamps. As a rule not more than one or two are seen during an outing, but they are certainly widespread, and we once saw 12 in a flight near Ganderbal. Osmaston considered that they may be resident, but there are no records of nesting. A possible nesting site is near Tithwal in the lower reaches of the Kishenganga Valley where we daily saw single birds entering and leaving the narrow rocky gorge. On the other hand, Unwin states categorically that they do not breed in Kashmir, being seen mainly in spring and autumn, as he puts it, ' in the migratory season '.

LITTLE EGRET, *Egretta garzetta garzetta* (Linnæus). INDIAN CATTLE EGRET, *Bubulcus ibis coromandus* (Boddaert). BOOK REFERENCES: *F.B.I.*, VI, 348 and 349; *Nidification*, IV, 462; *Handbook*, 509 and 511.

These lovely little herons are common enough in winter and we have seen them in June in the fields around Ganderbal. Ward states that they breed in Kashmir. Unfortunately he does not elaborate and we can trace no records. It must be remembered that when he writes of Kashmir, he is often referring to the State as a whole and this includes such provinces as Jammu, Mirpur and Poonch which impinge upon the Northern Indian plains.

BITTERN, *Botaurus stellaris stellaris* (Linnæus). KASHMIRI NAME: *Banapoochin*. BOOK REFERENCES: *F.B.I.*, VI, 370; *Handbook*, 514.

May at times be heard booming—we have heard it in June—and local shikaris maintain that it breeds occasionally in thick reed beds in the *rukh* portion of the Anchar Lake and in other secluded State reserves. The Bittern appears in winter in the northern half of India, penetrating in lessening numbers to about Bombay, the North Deccan and Cuttack. It breeds in temperate Europe and Asia from Great Britain to Japan.

COMMON TEAL, *Anas crecca crecca* Linnæus. KASHMIRI NAME: *Keusput*. BOOK REFERENCES: *F.B.I.*, VI, 431; *Handbook*, 530.

A not inconsiderable number of teal are to be seen in summer on secluded jheels in the Vale such as Hokra. There is, however, no evidence that they remain with the intention of nesting. They are probably pricked birds unable to make the effort which migration entails, for they nest in temperate latitudes, mostly many hundreds of miles north of Kashmir, a fact which militates against any remaining from choice to summer in our area.

INDEX

Accentor, 70, 98, 214
—, Garhwal, **96**
—, Jerdon's, **97**
Accipiter nisus melanoschistos, 269
— *virgatus affinis*, 352
Achabal, 6, 13, 39, 65, 66, 138, 139, 219
Acridotheres tristis tristis, 149
Acrocephalus concinens haringtoni, 120
— — *hokrae*, 121
— *stentoreus brunnescens*, 118
Actitis hypoleucos, 317
Adelura coeruleocephala, 79
Ægithaliscus concinnus iredalei, 344
— *niveogularis*, 29
Ægypius monachus, 350
agricola (Acrocephalus), 121
Ahateng, 66, 93
Ahlan, 21, 250
— Nullah, 38, 150, 211, 325
Aishmakam, 117, 146, 185, 231
Alauda gulgula lhamarum, 201
Alcedo atthis pallasii, 234
Alectoris graeca chukar, 282
Alexander, H. G., 346
Alseonax ruficaudus, 107
Amarnath, 181
Amaurornis fuscus bakeri, 292
Anas crecca crecca, 355
— *platyrhyncha platyrhyncha*, 335
Anchar Lake, 6, 38, 118, 120, 215, 275, 289, 296, 298, 299, 323, 328, 333, 335, 336, 338, 355
Anderson, A., 321
Anthus roseatus, 197
— *similis jerdoni*, 196
— *trivialis haringtoni*, 194
Aphawat, 34, 77, 84, 97, 99, 173, 200
Apus melba melba, 241
Aquila chrysaëtos daphanea, 351
Arceuthornis viscivorus bonapartei, 89
Ardea cinerea rectirostris, 325
Ardeola grayii, 327
Argya, 222
Aru, 11, 67, 125, 166, 170, 261, 325, 350
Asio otus otus, 245
Astanmarg, 54, 70, 84, 89, 115, 133, 134, 142, 154, 155, 171, 173, 181, 191, 192, 195, 213, 220, 262, 268
Atawat, 224, 257
aurantiaca (Pyrrhula), 154
Aythya rufa rufa, 338

Babbler, Sind Jungle, **344**
Bagtor, 28, 132, 140, 183, 269, 271, 276, 349
Bailey, Colonel F. M., 316
Baker, E. C. Stuart, 3, 14, 15, 17, 21, 33, 36, 38, 40, 50, 52, 60, 66, 68, 73, 77-81, 84, 85, 92, 97, 101, 104, 113, 118, 123, 126, 127, 129, 132, 140-2, 144, 146, 147, 172, 180, 184, 213-18, 221, 222, 227-30, 238, 245, 247, 250, 262, 268, 270, 271, 275, 282, 285, 287-9, 299, 307, 316, 321, 331, 332, 336, 344, 349, 350, 354
bakeri (Apus melba), 242, 243
Baltal, 4, 25, 57, 61, 123, 154, 155, 162, 163, 178, 192, 248, 268, 343
Bānapoochin, 355
Bandipur, 219, 326, 331
— Nullah, 10, 39, 40, 66, 186, 191, 227, 231, 320
Banihal Pass, 34, 181, 186, 222, 353
Baramullah, 6, 10, 13, 21, 36, 39, 41, 52, 59-61, 82, 111, 113, 144, 169, 171, 174-6, 179, 185, 209, 223-5, 228, 230, 239, 256, 293, 307, 318, 343, 345, 352
Basil-Edwards, S., 45
Basman, 61, 159-61, 211
Batsapur, 329
Bee-Eater, 228
—, European, **229**
Betham, Brig.-General R. M., 147
Betterton, F. A., 300
Bhotkol Glacier, 157
Bijbehara, 6, 343
Bil-bi-chūr, 41
Bittern, 92, **355**
—, Little, 7, 9, 119, 327, 329, 330, **332**
Blackbird, Central Asian, **83**
Bluechat, 214
—, Indian, **54**, 223
Bōher, 260
Bōr, 329
Botaurus stellaris stellaris, 355
Botin Kāv, 343
Boyle, Colonel C. L., 301
Brag, 325
brevirostris (Cinnyris asiatica), 348
Briggs, F. S., 146
Bringh, 9, 211, 307, 325
Broku, 327

357

INDEX

Brooks, W. E., 3, 80, 121, 139, 218, 331, 345, 348
Bubo bubo turcomanus, 248
Bubulcus ibis coromandus, 354
Buchanan, Colonel K., 16, 77, 81, 84, 94, 99, 123, 124, 140, 158, 159, 172, 230, 319, 345
Buck, Sir E. C., 141
Bularu, 317
Bulbul, Himalayan Black, **39**
—, White-cheeked, 34, **41**
Bullfinch, Orange, **154**
Bungus Marg, 195
Bunting, Crested, **179**
—, Grey-headed, 176
—, Indian Grey-headed, **173**
—, Stewart's, 173, 174, 175, 180
—, White-capped, **175**
Burzil Chauki, 50, 195
Burzilkot, 70, 274
Bushchat, Dark Grey, 61, 214
—, Western Dark Grey, **63**
—, Northern Indian Pied, **58**
Bush-Robin, Kashmir Red-flanked, **77**, 81, 218
Bush-Warbler, Large-billed, **123**
—, Pale, **138**, 221
Butcher-Bird, 113
Buteo rufinus rufinus, 267
Butterfly-bird, 46
Buzzard, Long-legged, **267**

Cacomantis merulinus passerinus, 216
Callacanthis burtoni, 163
Calliope pectoralis pectoralis, 76
canorus (*Cuculus*), 215, 216, 219, 220
Capella gallinago gallinago, 323
Carduelis caniceps caniceps, 161
Carpodacus erythrinus roseatus, 159
castanea (*Ceriornis macrolophus*), 279
Catreus wallichii, 353
Cephalopyrus flammiceps flammiceps, 141
Ceriornis macrolophus biddulphi, 278
Certhia familiaris hodgsoni, 44, 345
— *himalayana limes*, 43
Ceryle lugubris guttulata, 233
— *rudis leucomelanura*, 231
Chaetura caudacuta nudipes, 349
Chāh, 323
Chaimarrhornis leucocephalus, 71
Charadrius dubius curonicus, 306
Chatragul Nullah, 277, 343
Chenar Bagh, 210, 259
Chenari, 59, 65, 91, 179, 180, 186, 237, 344, 353
Chengher, 86
Chet Hyot, 124
Chets Tāl, 71

Chiffchaff, Sind, **346**
Chlidonias hybrida indica, 304
Chotta Tōnt, 234
Chough, Alpine, 18
—, Eastern Red-billed, **17**, 20, 21
—, Yellow-billed, 18, **19**
Chukor, **282**
Chut Kotūr, 273
cia par (*Emberiza*), 177
— *stracheyi* (*Emberiza*), 176
Cinclus asiaticus [*tenuirostris*], 51
— *cinclus cashmeriensis*, 50
— *pallasii tenuirostris*, 52
Cinnyris asiatica, 348
Circus aeruginosus aeruginosus, 351
Clamator jacobinus pica, 221
Cock, Major C. R., 130, 131, 153, 271, 336, 344, 345
Columba leuconota leuconota, 273
— *livia neglecta*, 271
concinens (*Acrocephalus*), 121
confusa (*Calliope pectoralis*), 76
Coot, 336, 337
—, Common, 289, 295, 296, **298**
—, Indian Purple, **296**
Copsychus saularis saularis, 81
Coracias benghalensis benghalensis, 349
— *garrula semenowi*, 226
Cormorant, 326
—, Indian Large, **354**
Corvus corax tibetanus, 343
— *corone orientalis*, 3
— *macrorhynchos intermedius*, 4
— *monedula soemmeringii*, 8
— *splendens zugmayeri*, 6
Coturnix coturnix coturnix, 353
— *coromandelicus*, 353
Crake, Baillon's, 289, 292
—, Eastern Baillon's, **290**
—, Ruddy, 289
—, Northern Ruddy, **292**
Crow, 246, 306, 327
—, Eastern Carrion, **3**, 4
—, House, 8, 13, 15, 150, 245, 329, 334
—, Sind House, **6**
—, Jungle, 3, 6, 259, 261, 270, 281
—, Himalayan Jungle, **4**
Cuckoo, 62, 269
—, Asiatic, 58, 64, 69, 113, 120, 178, 194, **212**, 217, 219
—, Himalayan, 64, 137, **217**, 218, 219, 221, 240
—, Pied Crested, 38
—, Northern Pied Crested, **221**
—, Plaintive, 216
—, Small, 70, 137, 215, **218**
Cuculus canorus bakeri, 213
— — *telephonus*, 212
— *optatus optatus*, 217

INDEX

Cuculus poliocephalus poliocephalus, 70, 137, 218
— saturatus, 220
Culicicapa ceylonensis pallidior, 346
curonicus (Charadrius dubius), 307

Dabchick, 119, 293, **339**
Dachhgam, 307
— Nullah, 135, 347
Dal Gate, 341
— lakes, 6, 42, 43, 118, 119, 121, 192, 230, 298, 300, 305, 318, 331, 332, 336, 341, 350, 354
Dangarpur, 313
Dantiwu, 21
Danudra, 325
Davidson, J., 11, 24, 40, 46, 49, 58, 60, 79, 82, 102, 106, 116, 123-5, 130, 131, 137, 142, 162, 174, 177, 179, 212, 224, 228, 262, 268, 273, 279
debilis (Passer rutilans), 170
Delichon urbica cashmeriensis, 181
Dendrocitta formosae occidentalis, 343
Dendrotreron hodgsonii, 352
Dhani, 144, 170, 260, 261
Dicrurus leucophaeus longicaudatus, 116
— macrocercus albirictus, 346
Dider, 201
Didru, 201
Dipper, Brown, 50, 51
—, Indian Brown, **52**
—, Kashmir White-breasted, **50**
Diva Kāv, 3, 4
Doarian, 224
Dobbai, 188
Dodsworth, P. T. L., 94, 167, 257
Dofa Pich, 138
— Tiriv, 60, 63
Domel, 6, 21, 36, 38, 39, 63, 82, 87, 149, 180, 196, 203, 237, 250, 251, 254, 256, 271, 285, 294, 309, 328, 343, 344, 346-51
Donald, C. H., 351
Dove, Indian Little Brown, **352**
—, Indian Spotted, **352**
Drongo, Black, 117
—, Himalayan Black, **346**
—, Indian Grey, **116**
Dryobates brunifrons, 209
— himalayensis albescens, 207
Duddru, 231
Dudnial, 11
Dulai, 38
Duncan, A. B., 307
Dungal, 50, 52

Eagle, Booted, **262**, 266, 306
—, Golden, 287

Eagle, Himalayan Golden, **351**
Egret, 326
—, Indian Cattle, **354**
—, Little, **354**
Egretta garzetta garzetta, 354
Emberiza cia stracheyi, 177
— fucata arcuata, 173
— stewarti, 175
Enicurus maculatus maculatus, 64
Eremophila alpestris longirostris, 199
Erin Nullah, 10, 40, 41, 227, 284
erythropygia (Hirundo rufula), 187
Eumyias thalassina thalassina, 345

Falco severus, 258
— subbuteo centralasiae, 258
— tinnunculus, 260
Ferozpur Nullah, 181, 250, 251, 261, 268-70, 321, 350
Fhāmbasir, 109
Finch, Gold-fronted, **347**
—, Red-browed, 162, **163**
Finn, Frank, 274
Fire-Cap, 142
Firecrest, 141
Fishing-Eagle, Pallas', **263**
—, White-tailed, 263
Flycatcher, Common Verditer, **345**
—, Kashmir Red-breasted, **101**, 103
— —, Sooty, **99**, 109
—, Paradise, 9, 106, 332
—, Himalayan Paradise, **109**
—, Rufous-tailed, 103, 104, **107**, 221
—, Simla Grey-headed, **346**
—, Western Slaty-blue, **103**, 106, 107
—, White-browed Blue, **105**
Flycatcher-Warbler, Kashmir Grey-headed, **347**
Forktail, Little, **67**
—, Western Spotted, **64**, 67, 214
Francolinus francolinus asiae, 284
Fringilauda nemoricola altaica, 171
Fulica atra atra, 298
Fuligula nyroca, 336

Gad Sar, 200
Gāda Grad, 263
gaddi (Columba livia), 272
Gagangair, 39, 40, 50, 166, 347
Gair Kov, 300
Gallinula chloropus indicus, 294
Gallinule, Purple, 296, 297
Gammie, J., 40
Ganderbal, 11, 37, 38, 145, 174, 177, 209, 221, 222, 233, 237, 238, 262, 275, 277, 314, 328, 354
Gangabal Lake, 97, 192, 193
Gānt, 265

INDEX

Garhi, 6, 7, 13, 82, 180, 203, 209, 216, 237, 254, 294, 328, 344, 346, 348-50
Garrulax albogularis whistleri, 344
Garrulus lanceolatus, 13
Gennaeus leucomelanos hamiltonii, 353
Ghori, 233, 349
Gillu, 292
Glacier Valley. *See* Valley of Glaciers
Glaucidium brodiei brodiei, 250
— *cuculoides cuculoides*, 350
Gogaldara, 90, 259, 261
Goi, 332
Goldcrest, Himalayan, **139**, 141, 142, 143
Goldfinch, Himalayan, **161**
Gourkāku, 286
Grad, 252
Gralu, 290
Grebe, Little, 337
—, Indian Little, **339**
Greenfinch, Himalayan, **166**
Grettu, 290
Grosbeak, Black and Yellow, **152**
—, Scarlet, 159
Gug, 302
Gūgi, 275
Gūgū, 302
Gulāb Tsar, 159
Gulmarg, 9, 24, 29, 31, 33, 34, 38, 57, 61, 78, 90, 133, 137, 140, 143, 162-4, 166, 168, 170, 181, 189, 195, 209, 224, 231, 235, 239, 248, 251, 259, 261, 268, 286, 318, 319, 345
Gulol Gali, 47
Gund, 8, 17, 24, 29, 102, 106, 117, 126, 139, 142, 146, 154, 265, 268, 279, 347, 348
Gūnd Kāv, 300
Gunkots, 116
Gurais, 31, 36, 38, 52, 57, 61, 98, 112, 131, 132, 135, 140, 142, 154, 160, 162, 163, 168, 170, 183, 198, 201, 211, 217, 235, 239, 244-6, 260, 269, 272, 276, 300, 318, 343, 346
gutturalis (*Hirundo rustica*), 184
Gyps himalayensis, 252
— *indicus nudiceps*, 351
Gypaëtus barbatus hemachalanus, 255

Haigam, 296, 338
Halcyon smyrnensis smyrnensis, 237
Haliaëtus leucoryphus, 263
Hām, 280
Handowar, 108, 146, 217, 219
Haramukh, 97, 110, 192, 193, 200

Harawātij, 112
Harawōt, 338
haringtoni (*Acrocephalus concinens*), 121
Harrier, Marsh, 306, **351**
Hatatertu, 309
Hatatūt, 309
Hazār Dastān, 94
Hazratbal, 331
Hemichelidon, 108
— *sibirica gulmergi*, 99
Herbert, E. G., 303
Heron, Grey, 329, 330
—, Eastern Grey, **325**, 331
—, Night, 7, 326, **329**
—, Pond, 7, 332
—, Indian Pond, 327
Hieraëtus pennatus, 262
Hill-Warbler, Nepal Brown, **347**
Himantopus himantopus himantopus, 311
Hingston, Major R. W. G., 315
Hirundo rufula scullii, 186
— *rustica rustica*, 184
— *smithii filifera*, 348
Hobby, Central Asian, **258**
Hodgsonius phoenicuroides, 56
Hokra Jheel, 121, 122, 193, 214, 289, 291, 298, 323, 335-8, 352, 354, 355
hokrae (*Acrocephalus concinens*), 121
Homochlamys pallidus pallidus, 138, 221
Hoopoe, 217
—, European, **239**
—, Indian, 239
Hōr, 149
Hōr Koel, 207
— *Kola Tōnch*, 231
— *Kuk*, 221
Hud-hud, 239
Hume, A. O., 27, 130, 146, 171, 279, 283
Hutton, Captain T., 225, 244
Hyan, 43
Hydrophasianus chirurgus, 300
Hypacanthis spinoides spinoides, 166

Ianthia cyanura pallidiora, 77
Ianthocincla rufogularis occidentalis, 344
Ibidorhyncha struthersii, 314
Ibis-bill, **314**
indicus (*Upupa epops*), 239
Inshan, 61, 98, 121, 122, 166, 191, 201, 211, 212, 235, 268, 273, 307, 314, 317, 353
Islamabad, 6, 168, 318
Ixobrychus minutus minutus, 332

Jaçana, Pheasant-tailed, **300**
Jackdaw, 6, 19, 337
—, Eastern, **8**

INDEX

Jangli Konkli, 275
Janwai, 21, 244
Jay, 34
—, Black-throated, **13**, 245
—, Himalayan, 13
jerdoni (Anthus similis), 196
— *(Charadrius dubius)*, 307
Jhelum River, 6, 7, 10, 52, 73, 147, 169, 193, 224, 237, 255, 263, 307, 313, 329, 339, 351
— Valley, 9, 13, 25, 39, 44, 59-61, 63, 87, 91, 102, 110, 112, 115, 117, 127, 136, 138, 144, 148, 170, 179, 180, 184-7, 201, 203, 205, 213, 216, 217, 224, 233, 237, 244, 249, 254, 256, 265, 269, 275, 309, 328, 329, 343-53
— — road, 21, 59, 82, 150, 187, 196, 204, 227, 284, 345, 347, 350, 353
Jones, A. E., 25, 28, 80, 85, 89, 104, 142, 146, 210, 257
Jynx torquilla japonica, 210

Kabuk, 286
Kāk, 282
Kākov, 282
Kāku, 282
Kaleej, White-crested, **353**
Kamīr, 275
Kamru, 275
Kangan, 40, 42, 82, 155, 209, 231, 262, 263
Kanital, 157
Kantūr, 167
Kao Kumr, 87
Karabudurun, 250, 251
Karkat, 118
Kastūr, 87, 94
Katīj, 184
Kāv, 6
Kāva Kunūr, 94
Kavīn, 8
Kāvput, 298
Kazinag mountains, 9, 16, 24, 31, 98, 106-8, 139, 170, 195, 201, 217, 224, 225, 320, 353
Keran, 15, 25, 39, 41, 65, 87, 88, 91, 106, 110, 112, 125, 138, 203, 207, 209, 210, 224, 225, 233, 239, 244, 273, 284, 347, 349, 351, 353
Kestrel, 258-**60**, 270
Keusput, 355
Khak Dobbai, 190
Khel, 198, 244, 273
Killenmarg, 29, 84, 128, 226
Kingfisher, 9, 22, 110, 184, 228, 332

Kingfisher, Central Asian, **234**
—, Pied, 235, 237
—, Himalayan Pied, **233**
—, Indian Pied, **231**
—, White-breasted, **237**
Kishenganga River, 13, 52, 63, 234, 237, 272, 318, 351
— Valley, 6, 9, 11, 14, 15, 21, 25, 28, 36, 39, 41, 44, 59, 61, 63, 65, 82, 85, 87, 89, 91, 96, 102, 106-8, 110, 112, 115, 117, 125, 127, 136, 138, 144, 150, 167, 168, 170, 175, 179, 181, 184, 186, 187, 190, 191, 196, 198, 203-5, 207, 209, 213, 217, 224, 233, 239, 242, 244, 249, 250, 254, 260, 265, 269, 273, 275, 279, 280, 282, 285, 320, 325, 343, 345, 347-9, 351, 353, 354
Kishtwar, 34, 72, 122
Kitardarji, 102, 106
Kite, 150, 306
—, Black-eared, 117, 169, 262, 263, **265**, 268
—, Pariah, 265
—, Common Pariah, **351**
Koel Dider, 43
— *Ku-kīr*, 207
— *Makōts*, 205
Kohala, 21, 41, 52, 60, 61, 73, 82, 95, 149, 168, 174, 175, 185-7, 209, 227, 271, 343, 347
Koklas, Kashmir, **278**
Kola Katīj, 306
— *Kavīn*, 317
— *Tiriv*, 73
— *Tōnch*, 234
Kolahoi, 50, 51, 71, 110, 154, 165, 192
Kolru, 298
Kolūr, 298
Koolan, 8, 20, 39, 126, 139, 146, 174, 273
Koragbal, 98, 135
korejewi (Rallus aquaticus), 288
Korkuch, 118
Kotwal peak, 51
Krahom swamp, 122, 302, 323
Krew, 304
Krind, 304
Kruhun Bulbul, 39
Kukil, 277
Kukū, 212
Kumīdi, 71
Kurgoiu, 332
Kyunus, 93, 125, 230

La Personne, V. S., 154, 158
Laiscopus collaris whymperi, 96

INDEX

Lamba Dider, 46
Lämmergeier, 252, 254, **255**
Lanius schach erythronotus, 112
Lapwing, Red-wattled, 113, 313, 314
—, Mekran Red-wattled, **309**
Lark, Long-billed Horned, **199**
Laughing-Thrush, Black-gorgeted, 222
—, Kumaon Rufous-chinned, **344**
—, Necklaced, 222, 223
—, Streaked, 35, 214, 223
—, Simla Streaked, **36**, 222
—, Western Variegated, **34**
—, Western White-throated, **344**
Lawrence, Sir Walter, 7, 42, 336
Ledor Dobbai, 192
leucopsis (Sitta leucopsis), 31
Lidar River, 47, 156, 157, 234, 309, 312, 314, 325
— —, East, 170, 274, 314
— Valley, 8, 11, 18-20, 22, 39, 51, 85, 90, 117, 125, 139, 140, 142, 146, 154-8, 160, 166, 170, 184, 185, 198, 211, 224, 227, 231, 233, 257, 265, 292, 310, 345
— —, East, 38, 47, 50, 70, 84, 98, 128, 158, 192, 205
— —, West, 154
Lidarwat, 20, 49, 54, 57, 72, 106, 107, 128, 154, 163, 219, 346
Lily Trotter, 300
Literāz, 11, 109
Littledale, H., 242
Livesey, Captain T. R., 29, 30, 233, 257, 291, 321
Lobivanellus indicus aigneri, 309
Loke, W. T., 200, 312
Lolab Pass, 29
— Valley, 9, 13, 24, 30, 31, 38, 63, 107, 109, 140, 142, 167, 203, 217, 223, 224, 245, 250, 344, 346, 351-3
longicaudatus (Dicrurus leucophaeus), 117
Longzeyet, 311
Lophophanes melanolophus, 25
— *rufonuchalis rufonuchalis*, 27
Lophophorus impejanus, 280
Lōt Rāza, 11
Lōz, 263
Ludlow, F., 13, 39, 85, 86, 127, 195, 200, 211, 272, 302, 307, 314, 316, 350, 353
Luscinia brunnea brunnea, 54

macrocercus (Dicrurus), 118
macrolophus (Ceriornis macrolophus), 279
Madmatti River, 228, 307, 309
Magpie, Western Yellow-billed Blue, **11**

Magpie-Robin, Indian, **81**
Magrath, Colonel H. A. F., 64, 67, 69, 76, 80, 88, 90, 91, 115, 126, 137, 143, 152, 156, 164, 179, 182, 183, 192, 197, 202, 206, 219, 248, 270, 315
Mahadeo, 250, 279
Makōts, 207
Mallard, **335**, 338, 339
Manasbal, 65, 66, 230, 351, 354
— Lake, 93, 186, 298, 304, 350
Marbal Glen, 38, 57, 150, 162, 163, 181, 250
— Pass, 36
Margan Pass, 34, 80
Marshall, Colonel C. H. T., 40, 156, 225, 245, 344
Martin, Crag, 182, **348**
—, Kashmir House, **181**
Meadow-Bunting, 175, 176, 214, 216, 218
—, Eastern, **177**
Meinertzhagen, Colonel R., 25, 29, 30, 33, 44, 49, 50, 52, 62, 71, 78, 87, 93, 97, 98, 112, 121, 144, 146, 147, 155, 161, 171, 185, 186, 192, 195, 196, 198, 227, 239, 265, 273, 274, 287, 288, 304, 318, 326, 327
melanolophus (Lophophanes), 28
Melophus lathami subcristatus, 179
Merops apiaster, 229
Metoponia pusilla, 347
Microcichla scouleri scouleri, 67
Micropus affinis affinis, 349
— *apus pekinensis*, 243
— *pacificus leuconyx*, 349
Microscelis psaroides psaroides, 39
Milvus migrans govinda, 265, 266, 351
— — *lineatus*, 265
Minivet, 26
—, Indian Short-billed, **114**
Mirpur, 233
Mistle-Thrush, 84
—, Himalayan, **89**
Mojipal, 181
Molpastes leucogenys leucogenys, 41
Monal, 278, 279, **280**
Mong, 354
Monticola cinclorhyncha, 91
— *solitaria pandoo*, 92
Moorhen, 289, 292, 319, 337
—, Indian, **294**
Motacilla alba alboides, 188
— *cinerea melanope*, 190
— *citreola calcarata*, 192
Mountain-Finch, Stoliczka's, 70, **171**
Muscicapula superciliaris superciliaris, 105
— *tricolor, tricolor*, 103

362

Muzaffarabad, 6, 82
Myiophoneus coeruleus temminckii, 94
Myna, 148
—, Black-headed, **148**
—, Common, 148, **149**

Nagmarg, 29, 30
Nanga Parbat, 110
Nasim Bagh, 237
Naubug, 9, 34
— Glen, 211, 325
neglecta (Columba livia), 272
Neophron, **254**
Neophron percnopterus percnopterus, 254
Nichnai Nullah, 122
Nīlakrāsh, 226
Nilam Plateau, 203
Nildori, 16, 98
Nilij, 335
nipalensis (Hirundo rufula), 186
Nishat Bagh, 110, 305, 326, 330
— Gardens, 42
Nucifraga multipunctata, 15
Nun Kun, 110
Nutcracker, Larger Spotted, **15**
Nuthatch, Brooks', 22, **31**, 32, 33
—, White-cheeked, 31, **32**, 157
Nycticorax nycticorax nycticorax, 329

Oates, E. W., 29
Oriole, 9, 110, 144, 332
—, Indian, **143**
Oriolus oriolus kundoo, 143
Osmaston, A. E., 19, 85, 93, 129, 130, 207, 210, 274, 287
—, B. B., 7, 8, 10, 13, 16, 17, 21, 23, 24, 25, 29, 31, 33, 34, 38, 47, 48, 51, 54, 57, 61, 62, 68, 79-81, 84, 90, 97, 108, 116, 122-4, 126, 130, 137-40, 143, 155, 157, 158, 160-4, 166, 168, 171, 173-5, 177, 181, 189, 191-3, 200, 202-4, 209, 211, 212, 221, 224, 226, 232, 233, 239, 240, 244, 245, 247, 249, 251, 253, 255, 259, 261-3, 266, 268, 277, 286, 287, 289, 291, 299, 308, 312, 314, 316, 321, 327, 328, 344-50, 352, 354
Osprey, **350**
Owl, Indian Barn, **350**
—, Long-eared, 13, **245**
—, Scully's Wood, **247**
—, Turkestan Great Horned, **248**
Owlet, Western Collared Pigmy, **250**
—, Western Himalayan Barred, **350**

Paddy Bird, 328
Pahlgam, 8, 9, 12, 21, 24, 25, 34, 38, 39, 54, 57, 85-7, 92, 106, 125, 139, 142, 150, 162, 163, 166, 170, 208, 235, 239, 257, 259, 265, 269, 281, 282, 315, 319, 346
Pāla Tiriv, 91, 92
Pampur, 6, 209, 239, 296, 343
Pandion haliaëtus haliaëtus, 350
Pandon Pathar, 181
Pantsol Kāv, 3, 4
Panzal, 225
Parakeet, Himalayan Slaty-headed, **223**
Partridge, Indian Black, **284**
Parus major caschmirensis, 21
— *monticolus monticolus*, 23
Passer domesticus griseigularis, 167
— *rutilans debilis*, 169
Patan, 230
Pateka, 44, 59, 110, 175, 209, 244, 348
Patyāl, 254, 255
Pericrocotus brevirostris brevirostris, 114
Perissospiza icteroides, 152
Phalacrocorax carbo sinensis, 354
Pheasant, Cheer, **353**
—, Impeyan, **280**
—, Monal, 4
Phillips, Lieut.-Col. B. T., 113, 174, 200, 209, 210, 230, 232, 237, 248, 296, 297, 305, 315, 316, 318, 319
Phitta, 347
Phoenicurus frontalis, 68
— *ochruros phoenicuroides*, 70
Phylloscopus affinis, 127
— *collybita sindianus*, 346
— *inornatus humei*, 133
— *magnirostris*, 134
— *neglectus*, 346
— *occipitalis occipitalis*, 136
— *proregulus simlaensis*, 131
— *tytleri*, 129
Picus squamatus squamatus, 205
Pind, 339
Pinskanni, 188
Pintsakon, 25, 27
Pipit, 70
—, Hodgson's, 194-6, **197**
Pir Panjal Mountains, 3, 11, 13, 18, 23, 31, 34, 50, 57, 63, 65, 69, 70, 83, 89, 98, 129, 131, 133, 142, 154, 158, 159, 164, 171, 194, 198, 199, 200, 213, 219, 239, 242, 244, 259, 261, 280, 286, 318, 320, 347
Pisu, 128
Pitt, Frances, 213
Plās, 278

INDEX

Plover, 315
—, Little Ringed, 317
—, European Little Ringed, **306**
Pochard, White-eyed, 335-7, **338**
Podiceps ruficollis capensis, 339
Pohru River, 9, 23, 146, 224
poliocephalus (Cuculus poliocephalus), 215
Porphyrio poliocephalus poliocephalus, 296
porphyronotus (Sturnus vulgaris), 146
Porzana pusilla pusilla, 290
Poshinūl, 143
Poshnūl, 143
Praslun, 12, 34, 38, 125, 314
Propasser rhodochrous, 157
— *thura blythi*, 156
proregulus (Phylloscopus), 130
Prunella strophiata jerdoni, 97
Pseudogyps bengalensis, 351
Psittacula himalayana himalayana, 223
Pyrrhocorax graculus forsythi, 19
— *pyrrhocorax himalayanus*, 17
Pyrrhospiza punicea humii, 347
Pyrrhula aurantiaca, 154

Quail, Black-breasted or Rain, **353**
—, Common or Grey, **353**

Rallus aquaticus korejewi, 288
Rām Chukār, 286
Rampur, 11, 12, 25, 39, 65, 82, 170, 209, 224, 233, 353
Rampur-Rajpur, 30, 102, 110, 125, 214, 226, 247, 248, 283, 284
Ramsay, Colonel R. G. Wardlaw, 33, 106
Ranga Bulbul, 109, 110
— *Tsar*, 21
Rangmarg, 19, 47, 166
Rāta Mogul, 247, 248
Rāt Monglu, 247
Ratan Pir, 218
Rattray, Colonel R. H., 11, 36, 109, 137, 163-5, 174, 183, 206, 263, 275, 344, 345, 348, 349
Raven, Tibetan, 4, **343**
Redstart, Blue-fronted, **68**, 213
—, Kashmir, 68, 69, **70**
—, Plumbeous, 53, **73**, 214
—, White-capped, **71**
Reed-Warbler, 113, 215, 216, 332, 340
—, Indian Great, **118**, 120-3, 214, 215
Regulus regulus himalayensis, 139
Reshna, 39, 272
Rewil stream, 51
— Nullah, 154, 155

Rhodophila ferrea ferrea, 63
Rhyacornis fuliginosa, 73
Ribbon-bird, 111
Rikinwas, 160, 161
Ring-Dove, 43, 247
—, Indian, **277**
Riparia rupestris, 348
Robin, Blue-headed, **79**
—, Brown-backed Indian, **345**
Rock-Pigeon, Blue, 273, 275
—', Hume's Blue, **271**
Rock-Pipit, Brown, **196**
Rock-Thrush, Blue-headed, **91**, 108
—, Indian Blue, **92**
Rodabubru, 229
Roller, Indian, **349**
—, Kashmir, **226**, 349
roseatus (Carpodacus erythrinus), 159
Rosefinch, Common, 157, 158, **159**
—, Hodgson's, **159**
—, Kashmir White-browed, **156**
—, Pink-browed, **157**
—, Western Red-breasted, **347**
Rostratula benghalensis benghalensis, 302
Rubythroat, 70, 214
—, Western Himalayan, **76**
ruficaudus (Alseonax), 105
rufilatus (Laiscopus collaris), 97
rustica (Hirundo rustica), 184

Sabōz Tsar, 166
Sain Nullah, 192
Sálim Ali, 60, 227, 335
Salkalla, 11, 13, 41, 175, 224, 233, 234, 353
Sāma Sonatsar, 154
Sandpiper, 314-16
—, Common, 307, **317**
Sandran River, 353
Sanzipur, 108, 170, 225
Sarah, 177
Satūt, 239
Saxicola caprata bicolor, 58
— *torquata indica*, 60
Saxicoloides fulicata cambaiensis, 345
Scolopax rusticola, 319
Scott, J. E., 120, 215
Scully, J., 16, 81, 262
Sehāra, 161
Seicercus xanthoschistos albosuperciliaris, 347
Sekwas, 18, 20, 52, 128
Sēra, 161
Shadipur, 10, 120, 230, 263, 312, 313, 342
Shāh Kuk, 212
Shahabad, 261
Shakhel Lōt, 64

INDEX

Shalabug, 37, 120, 145, 193, 209, 222, 233, 328, 333, 337
Shalimar Gardens, 42, 110, 121, 326, 331
Shalput Nullah, 183
Shaltin, 293
Sharda, 21, 85, 89, 138, 239, 272, 273, 279
Sheen-a-pī-pin, 36, 37
Shelley, G. E., 245, 246
Shishram Nag, 47, 70
Shoga, 223
Shortwing, Hodgson's, **56**
Shrike, Rufous-backed, 9, **112**, 214, 261
Shupyion, 121, 348
Sind River, 37, 66, 122, 126, 156, 157, 191, 222, 275, 309, 314, 328, 350
— Valley, 4, 5, 9, 17, 18, 20, 39-41, 49, 78, 80, 82, 88, 90, 106, 108, 117, 122, 123, 139, 140, 142, 146, 154, 157, 160, 166, 170, 171, 173, 184, 185, 198, 204, 211, 224, 227, 231, 244, 248, 250, 257, 262, 263, 265, 268, 272, 273, 279, 300, 310, 318, 343, 346-8
Sinthan Pass, 34
Siphia hyperythra, 101
Sitta caesia cashmirensis, 31
— *leucopsis leucopsis*, 32
Skinner, R. A., 164, 165
Skylark, Kashmir, **201**
Snipe, 111
—, Fantail, 302, 319, **323**
—, Jack, 323
—, Painted, **302**, 337
Snow-Cock, Himalayan, **286**
Snow-Pigeon, Nepalese, **273**
Sona Sar, 50, 84
Sonamarg, 3, 9, 11, 16, 18, 46, 50, 57, 58, 68, 90, 106, 123, 124, 130, 133, 134, 136, 144, 150, 154, 158, 162-4, 170, 172, 173, 192, 201, 212, 239, 244, 248, 268, 269, 271, 274, 343, 345, 352
Sonamuss, 157
Sonarwain, 40, 66, 144, 228, 231, 307
Sopor, 296, 328, 331
Sparrow, Cinnamon, 22, 168
—, Kashmir Cinnamon, **169**
—, House, 159, 169-71, 267
—, Kashmir House, **167**
Sparrow-Hawk, 212, 258, 259, 261, 276
—, Indian, **269**
—, Northern Besra, **352**
'Spectacle Bird', 204
Spinetail, Indian White-throated, 243, **349**

Srinagar, 4, 6, 7, 10, 11, 18, 37, 39, 41, 47, 49, 87, 94, 95, 97, 113, 125, 126, 144, 163, 174, 177, 184, 185, 189, 200, 203, 205, 207, 209, 227, 230, 239, 247, 249, 252, 259, 262, 265, 279, 293, 330, 336, 338, 345, 349-51
Starling, Himalayan, 113, **145**, 151
Stevens, H., 80
Stilt, Black-winged, **311**
Stonechat, Indian, 59, **60**, 63, 64, 77, 214
Streptopelia chinensis suratensis, 352
— *decaocto decaocto*, 277
— *orientalis meena*, 275
— *senegalensis cambayensis*, 352
Strix aluco biddulphi, 247
Sturnus vulgaris humii, 145
Sufaid Tōnt, 231
Suknes, 57, 58, 61, 80, 81, 85, 99, 121, 122, 160, 161, 166, 201
Sumbal, 122, 202, 262, 267, 291, 296, 312, 323, 336
Suna Mūrg, 280
Sunāl, 280
Sunbird, 230
—, Purple, **348**
Surphrar, 5, 86, 110
— Nullah, 80, 143, 157, 250, 320
Suya criniger criniger, 347
Swallow, 181, 183, 185
—, Common, 182, **184**, 186, 187
—, Indian Wire-tailed, **348**
—, Red-rumped, 187
—, Scully's Red-rumped, **186**
—, Striated, 182, 186, 187
Swift, Alpine, **241**
—, Blyth's White-rumped, **349**
—, Eastern, **243**
—, House, 243
—, Common Indian House, **349**
Swinhoe, C., 244
Sylvia althoea, 124

Taer, 347
Tājdār Tsar, 25, 27
Takht (Takht-i-Suleman), 25, 47, 93, 95, 97, 126, 163, 174, 176, 230, 249, 254, 262
Tangmarg, 9, 205, 261
Tanin, 38, 49, 79, 98, 128, 130, 153-5, 158, 163, 170, 181, 205, 274
Taobat, 217, 318
Teal, Common, **355**
Tech, 294
Techal Kastūr, 89
Telakots, 116

INDEX

Telbal, 209
telephonus (Cuculus canorus), 215
Temenuchus pagodarum, 148
Tern, 337
—, Indian Whiskered, **304**
Terpsiphone paradisi leucogaster, 109
Tetraogallus himalayensis himalayensis, 286
Thajwas, 122, 171
— Marg, 164, 172
Theobald, W., 261
Thiun, 5, 248, 272
Thrush, Grey-headed, **85**, 88, 90
—, Tickell's, 43, **87**, 93, 113
Thuj, 335
Ticehurst, C. B., 33, 184, 200
Tichodroma muraria, 46
Tiok, 347
Tit, 155, 204
—, Crested Black, **25**, 28
—, Simla Black, 25, 26, **27**
—, Green-backed, 21
—, Simla Green-backed, **23**
—, Grey, 23-5
—, Kashmir Grey, **21**
—, Red-headed, **344**
—, White-throated, **29**
Tit-Warbler, Fire-capped, **141**
Tithwal, 110, 150, 168, 170, 175, 242, 272, 354
— Gorge, 186, 242
Tōnt Kon, 317
Tosha Maidan, 191, 244, 286
Tota, 223
Tragbal, 29, 125, 195
Tragopan melanocephalus, 353
Tragopan, Western, **353**
Tree-Creeper, 22, 26, 32, 155, 204
—, Himalayan, **43**
—, Hodgson's, 44, **345**
Tree-Pie, Western Himalayan, **343**
Tree-Pipit, Witherby's, **194**, 196
Tribura major, 123
Trochalopteron lineatum griseicentior, 36
— *variegatum simile*, 34
Troglodytes troglodytes neglectus, 48
Tsar, 167
Tsari Suh, 269
Tsinihangŭr, 145
Tsuidraman, 235, 276, 314
Tulamal Jheel, 307, 309
Tuleri Kāv, 229
Turdoides terricolor sindianus, 344
Turdus merula maximus, 83
— *rubrocanus rubrocanus*, 85
— *unicolor*, 87
Turtle-Dove, 4, 269
—, Himalayan Rufous, **275**
Tyto alba stertens, **350**

Unwin, Colonel W. H., 7, 11, 86, 146, 155, 164, 224, 231, 282, 336, 344, 347, 349-52, 354
Upupa epops epops, 239
urbica (Delichon urbica), 181
Uri, 9, 65, 82, 91, 148, 149, 162, 174, 175, 179, 186, 209, 254, 255, 284, 285, 328, 348, 353
Uriwan, 85
Urocissa flavirostris cucullata, 11

Vale, 4, 6-9, 11, 13, 20-2, 24, 28, 34, 36, 37, 39, 41-4, 47, 49, 50, 52, 59-61, 63, 65, 72, 73, 82, 87, 88, 89, 92, 95, 98, 102, 104, 108, 110, 112, 113, 115, 117-21, 123, 125, 127, 129, 131, 133, 136, 138, 140, 142-7, 150, 152, 154, 161, 162, 168, 169-71, 173-6, 184-6, 188-90, 192-4, 196, 198, 200-4, 209-11, 213-16, 219, 221-3, 225-7, 229-31, 233-5, 237, 239, 244, 247, 249, 250, 252, 254, 256, 258, 260, 261, 263, 265, 266, 268, 269, 272, 273, 275, 277, 282, 283, 288-90, 292, 294, 296, 300, 302, 304, 307, 309, 312, 317, 318, 323, 325, 328, 330, 332, 335, 336, 338, 339, 343, 345, 346, 348, 350, 351, 353-5
Valley of Glaciers, 171, 172
Verinag, 138, 345
Vihom, 170
Viri Mōt, 210
— *Tiriv*, 127, 129, 131, 133, 134, 361
Vishan Sar, 200, 300, 312
Vulture, Cinereous, **350**
—, Egyptian, 254
—, Griffon, 256
—, Himalayan Bearded, **255**
—, Himalayan Griffon, **252**
—, Large White Scavenger, **254**
—, Long-billed, 252, 351
—, White-backed, 252, **351**

Wagtail, Grey, 193, 194
—, Eastern Grey, **190**
—, Pied, 193, 198
—, Hodgson's Pied, **188**, 214
—, Yellow-headed, 191, 214
—, Hodgson's Yellow-headed, **192**
Waite, H. W., 89, 98
Wall-Creeper, **46**
Wān Bulbul, 39
— *Kastūr*, 85

INDEX

Wān Kavīn, 17, 19
— Kotūr, 271
— Kukil, 275
— Tech, 296
— Tsar, 177
Wangat Nullah, 185, 204, 325, 352
Wanhoi, 159, 211, 212
Warbler, 9, 204
—, Witherby's Paddy-field, **120**
Ward, Colonel A. E., 5, 16, 19, 20, 29, 36, 68, 69, 77, 80, 92, 97, 135, 137, 140, 141, 143, 146, 154, 155-7, 159, 164, 165, 169, 204, 205, 224, 230, 248, 253, 261, 262, 279, 287-9, 320, 343-6, 348-50, 353, 354
Wardwan River, 212, 248, 307, 314
— Valley, 37, 47, 54, 57, 58, 61, 80, 85, 86, 98, 112, 121-3, 131, 140, 154, 156, 157, 159, 160, 162-4, 154-66 pass., 190-2, 201, 202, 211, 212, 217, 235, 250, 254, 268, 273, 276, 279-81, 287-9, 318, 351-3
Waterhen, 119, 292, 294, 295, 297
Waters, Major H. P. E., 164, 165
Water-Rail, 319
—, Turkestan, **288**
Water-Robin, 74
Watlab, 125
Whistler, Hugh, 13, 36, 59, 80, 89, 112, 121, 125, 133, 146, 152, 159, 168, 171, 172, 180, 184, 188, 200, 227, 236, 248, 254, 268, 338, 348
Whistling-Thrush, Himalayan, **94**
White, L. S., 279, 326, 330, 348
Whitehead, Captain C. H. T., 4, 15, 33, 47, 58, 76, 80, 85, 106, 129, 133, 156, 205, 224, 247, 347
Whitethroat, Hume's Lesser, **124**
White-Eye, 143
—, Western, **203**

Whymper, S. L., 29, 30, 51, 69, 75, 77, 97, 109, 158, 173, 316, 347
Wild Duck, **335**
Willow-Warbler, 139
—, Crowned, 130
—, Large Crowned, 133, **136**, 218
—, Large-billed, **134**
—, Hume's Yellow-browed, **133**, 218
—, Ticehurst's, 130, **131**, 218
—, Tickell's, **127**, 220
—, Tytler's, **129**
Wilson, F., 287
—, N. F. T., 122, 174, 192, 212, 339
Witherby, H. F., 246
Woodcock, 317, **319**
Woodpecker, 14, 31, 32, 45
—, Pied, 205, 206, 226, 251
—, Brown-fronted Pied, **209**
—, Kashmir Pied, **207**, 209, 210
—, Scaly-bellied Green, **205**
Wood-Pigeon, Speckled, **352**
Woyil, 166
Wozij Tōnti Kavīn, 17
Wozul Mini, 114
Wren, Kashmir, **48**, 98
Wryneck, Japanese, **210**
Wular Lake, 29-31, 44, 63, 93, 98, 102, 110, 125, 140, 176, 186, 192, 193, 214, 217, 224, 230, 256, 263, 268, 269, 283, 298, 307, 312, 313, 320, 328, 330, 331, 350, 354
Wurjwan, 162, 164, 268
Wyet Tōnt, 152

Yamhar Pass, 192
Yāquat Hōt, 76

Zar Batchi, 319
Zoeb Kuk, 212
Zoji La, 163, 343, 348
Zojpal, 50, 128
Zosterops palpebrosa occidentis, 203